INSECT SAMPLING
IN FOREST
ECOSYSTEMS

METHODS IN ECOLOGY

SERIES LIST

Geographic Information Systems in Ecology
1997, Carol A. Johnston, University of Minnesota
Researchers will find this an invaluable guide to applying and getting the most out of Geographical Information Systems, one of the most revolutionary and important tools that have become available to ecological researchers in recent years.

An Introduction to Ecological Modelling: Putting Practice into Theory
1997, M. Gillman & Rosie Hails, Open University & CEH Oxford
"Teachers of courses on ecological modelling will find [this book] *a useful source-book at a competitive price."*
This book aims to open up the exciting area of ecological modeling to a much wider audience.

Stable Isotopes in Ecology and Environmental Science
1994, edited by Kate Lajtha & Robert Michener, Oregon State University & Boston University
This book, written by two of the leading researchers in the field, explains the background to stable isotope methodology and discuss the use of the methods in varying ecological situations.

Geographical Population Analysis: Tools for the Analysis of Biodiversity
1994, Brian A. Maurer, Michigan State University
This book discusses methods and statistical techniques that can be used to analyze spatial patterns in geographic populations. These techniques incorporate ideas from fractal geometry to develop measures of geographic range fragmentation, and can be used to ask questions regarding the conservation of biodiversity.

Molecular Methods in Ecology
2000, edited by Allan J. Baker, Royal Ontario Museum
This book provides both postgraduates and researchers with a guide to choosing and employing appropriate methodologies for successful research in the field of molecular ecology.

Biogenic Trace Gases: Measuring Emissions from Soils and Water
1995, edited by P.A. Matson & R.C. Harriss, University of California & University of New Hampshire
"The present volume . . . will serve as an important tool box for researchers and graduate students in this discipline, and will provide both a range of techniques for field measurements and a conceptual framework for extrapolation strategies."
This how-to guide details the concepts and techniques involved in the detection and measurement of trace gases, and the impact they have on ecological studies.

Ecological Data: Design, Management and Processing
2000, edited by William Michener & James Brunt, University of New Mexico
This book provides a much-needed resource for those involved in designing and implementing ecological research, as well as students who are entering the environmental sciences.

Population Parameters: Estimation for Ecological Models
1999, Hamish McCallum, University of Queensland
This book brings together a diverse and scattered literature, to provide clear guidance on how to estimate parameters for models of animal populations.

Insect Sampling in Forest Ecosystems

EDITED BY

SIMON R. LEATHER

Department of Biological Sciences
Imperial College of Science, Technology and Medicine
Silwood Park
Ascot
UK

SERIES EDITORS

J.H. LAWTON CBE, FRS

Natural Environment Research Council
Swindon, UK

G.E. LIKENS

Institute of Ecosystem Studies
Millbrook, USA

Blackwell
Publishing

© 2005 by Blackwell Science Ltd
a Blackwell Publishing company

BLACKWELL PUBLISHING
350 Main Street, Malden, MA 02148-5020, USA
108 Cowley Road, Oxford OX4 1JF, UK
550 Swanston Street, Carlton, Victoria 3053,
Australia

The right of Simon Leather to be identified as the
Author of the Editorial Material in this Work has
been asserted in accordance with the UK
Copyright, Designs, and Patents Act 1988.

First published 2005 by Blackwell Science Ltd

Library of Congress Cataloging-in-Publication Data

Insect sampling in forest ecosystems / edited by
Simon R. Leather.
    p. cm. – (Methods in ecology)
    Includes bibliographical references (p.).
    ISBN 0-632-05388-7 (pbk. : alk. paper)
    1. Forest insects–Research–Methodology.
2. Forest surveys.   3. Ecological surveys.
4. Sampling (Statistics)   I. Leather, S. R.
(Simon R.)   II. Series.

SB761.I56 2005
634.9'67'072–dc22
                              2004009772

A catalogue record for this title is available from the
British Library.

Set in 9^1/$_2$ on 12pt Meridien
by SNP Best-set Typesetter Ltd., Hong Kong
Printed and bound in the United Kingdom
by MPG Books, Bodmin, Cornwall

The publisher's policy is to use permanent paper
from mills that operate a sustainable forestry
policy, and which has been manufactured from
pulp processed using acid-free and elementary
chlorine-free practices. Furthermore, the publisher
ensures that the text paper and cover board used
have met acceptable environmental accreditation
standards.

For further information on
Blackwell Publishing, visit our website:
www.blackwellpublishing.com

Contents

Contributors, vii

Methods in Ecology series, ix

Preface, xi

1 Sampling theory and practice, 1
Simon R. Leather and Allan D. Watt

2 Sampling insects from roots, 16
Alan C. Gange

3 Pitfall trapping in ecological studies, 37
B.A. Woodcock

4 Sampling methods for forest understory vegetation, 58
Claire M.P. Ozanne

5 Sampling insects from trees: shoots, stems, and trunks, 77
Martin R. Speight

6 Insects in flight, 116
Mark Young

7 Techniques and methods for sampling canopy insects, 146
Claire M.P. Ozanne

8 Sampling methods for water-filled tree holes and their artificial analogues, 168
S.P. Yanoviak and O.M. Fincke

9 Sampling devices and sampling design for aquatic insects, 186
Leon Blaustein and Matthew Spencer

10   Methods for sampling termites, 221
David T. Jones, Robert H.J. Verkerk, and Paul Eggleton

11   Parasitoids and predators, 254
Nick Mills

Index, 279

Contributors

Leon Blaustein
Community Ecology Laboratory, Institute of Evolution, University of Haifa, Haifa 31905, Israel

Paul Eggleton
Termite Research Group, Department of Entomology, The Natural History Museum, Cromwell Road, London, SW7 5BD, UK

O.M. Fincke
Department of Zoology, University of Oklahoma, Norman, Oklahoma 73019, USA

Alan C. Gange
School of Biological Sciences, Royal Holloway, University of London, Egham, Surrey, TW20 0EX, UK

David T. Jones
Termite Research Group, Department of Entomology, The Natural History Museum, Cromwell Road, London, SW7 5BD, UK

Simon R. Leather
Department of Biological Sciences, Imperial College of Science, Technology and Medicine, Silwood Park, Ascot, Berkshire, SL5 7PY, UK

Nick Mills
Insect Biology, Wellman Hall, University of California at Berkeley, Berkeley, California 94720-3112, USA

Claire M.P. Ozanne
Centre for Research in Ecology and the Environment, School of Life Sciences, Roehampton University of Surrey, West Hill, London SW15 3SN, UK

Martin R. Speight
Department of Zoology, University of Oxford, South Parks Road, Oxford, UK

Matthew Spencer
Community Ecology Laboratory, Institute of Evolution, University of Haifa, Haifa 31905, Israel
Current address: Department of Mathematics and Statistics, Dalhousie University, Halifax, Nova Scotia, B3H 3J5, Canada

Robert H.J. Verkerk
Department of Biology, Imperial College of Science, Technology and Medicine, Silwood Park, Ascot, Berkshire, SL5 7PY, UK

Allan D. Watt
Centre for Ecology and Hydrology, Hill of Brathens, Glassel, Banchory, Aberdeenshire AB31 4BW, UK

B.A. Woodcock
Centre for Agri-Environment Research, Department of Agriculture, University of Reading, RG6 6AR, UK

S.P. Yanoviak
Department of Zoology, University of Oklahoma, Norman, Oklahoma 73019, USA
Current address: Florida Medical Entomology Laboratory, 200 9th Street SE, Vero Beach, FL 32962, USA

Mark Young
Culterty Field Station, Department of Zoology, University of Aberdeen, Newburgh Ellon, Aberdeenshire, AB41 OAA, UK

Methods in Ecology series

Series editors

Professor John H. Lawton is Chief Executive of the UK Natural Environment Research Council, and holds honorary professorships at Imperial College London and the University of York. He is a Fellow of the Royal Society, and has received numerous national and international prizes. Professor Lawton is author, co-author, or editor of six books, a former editor of *Ecological Entomology*, and has published over 300 scientific articles.

Dr Gene E. Likens is President and Director of the Institute of Ecosystem Studies in Millbrook, New York, and also holds professorships at Cornell University, Yale University, and Rutgers University. He received the 2001 National Medal of Science and has received eight honorary degrees. Dr Likens is also author, co-author, or editor of 15 books, and of over 450 published scientific articles.

About the series

The *Methods in Ecology* series is a useful and ever-growing collection of books aimed at helping ecologists to choose and apply an appropriate methodology for their research. The series is edited by two internationally renowned ecologists, Professor John H. Lawton and Dr Gene E. Likens, and aims to address the need for a set of concise and authoritative books to guide researchers through the wide range of methods and approaches that are available to ecologists.

Each volume is not simply a recipe book, but takes a critical look at different approaches to the solution of a problem, whether in the laboratory or in the field, and whether involving the collection or the analysis of data.

Rather than reiterate established methods, authors are encouraged to feature new technologies, often borrowed from other disciplines, that ecologists can apply to their work. Innovative techniques, properly used, can offer particularly exciting opportunities for the advancement of ecology.

The series strives to be at the cutting edge of the subject, introducing ecologists to a wide range of techniques that are currently rarely used, but deserve to be better known, or it seeks to provide up-to-date methods in more familiar areas. Its main purpose is not only to provide instruction in basic methods (the "how to"), but also to explain the benefits and limitations of each method (the "why this way?"), as well as showing how to interpret the results, what they mean, and generally to put them in the context of the discipline.

Much is now expected of the science of ecology, as humankind struggles with a growing environmental crisis. Good methodology alone never solved any problem, but bad or inappropriate methodology can only make matters worse. Ecologists now have a powerful and rapidly growing set of methods and tools with which to confront fundamental problems of a theoretical and applied nature. We hope that this series will be a major contribution towards making these techniques known to a much wider audience.

Preface

Insect sampling, although firmly based on standard ecological census techniques, presents special problems that are not faced by other ecologists. With the small size, varied life cycles, rapid rates of increase, and ingenious adaptations to habitats of insects, ecological entomologists face problems that are somewhat different to those faced by vertebrate or plant ecologists. That said, these same features make working with insects more amenable than working, for example, with large mammals.

Within the entomological world there are many different groups of specialists —those that work in agricultural systems, in desert systems, or with particular groups of insects. Many of these overlap in their approach and methodology, but some are unique and require specialist knowledge. One such specialization is forest entomology, another is aquatic entomology.

Forest ecosystems, whether natural or manmade, present special problems to the ecologists working beneath their canopies. In contrast to grassland, arable, and moorland ecosystems, where the scientist can stand above the study area and view the system in large patches, forest ecologists are towered over by their study substrate. Trees are large, dominate the canopy, and are not as amenable to sampling as herbaceous plants. The forest floor, often criss-crossed by surface or near-surface roots, also presents its own particular hazards to the researcher. In plantation forests, ridges, furrows, and drains mean that soil sampling, although superficially a similar exercise to that conducted in an arable ecosystem, is again not quite as simple. Root grafting makes sub-soil sampling onerous in the extreme.

Tropical forests are perhaps even more difficult to work in; the profusion of endophytic vegetation and the multi-layered structure of the canopy in many types of forest can make sampling a nightmare.

Study in forest ecosystems is an important part of ecology. In tropical natural forest ecosystems much work is performed in attempts to quantify the diversity of these unique systems. In temperate and boreal forests equally important work is conducted. Furthermore, with the massive increase in plantation forestry (tropical plantation forestry has increased more than threefold in the last decade), the need to sample for survey and protection purposes has dramatically increased. This book, although covering all aspects of insect sampling within all ecosystems, has a definite bias towards forest ecosystems. There are, however, many common features of insect sampling that can be applied to other ecosystems and every chapter brings these together in an integrated whole. Special cases do of course exist and each of these gets a chapter to itself.

This book brings together the collective expertise obtained over many years of intensive fieldwork in tropical and temperate ecosystems by a number of well-known entomologists. Each chapter, as well as dealing with sampling a particular stratum of the canopy or specialized group of insects, presents a comprehensive guide to running experiments within and beneath the forest canopy. Many potentially useful pieces of work conducted in forest ecosystems have fallen at the final hurdle – the translation of field data to the printed page. Unless surveys and field experiments are realistically designed within a sound but manageable framework they are doomed to failure. In addition, the failure of many ecologists working in agricultural ecosystems or on parkland trees to recognize the constraints imposed on ecologists working within large scale forest ecosystems must be redressed.

This book attempts to highlight the problems faced by entomologists working in different ecosystems and to suggest ways in which their methodology can be modified so as to be understood by ecologists and become accepted within the general fields of ecology and entomology.

Simon Leather graduated from the University of Leeds in 1977 with a first-class honors degree in Agricultural Zoology. He followed that with a PhD in aphid ecology at the University of East Anglia. He is currently Reader in Applied Ecology in the Department of Biological Sciences at Imperial College's Silwood Park campus. He has been researching the population biology of agricultural and forest pests, particularly insects, for over 25 years. Ten of these years were spent with the British Forestry Commission, where he learnt how to canopy-sample the hard way! He has written and edited several books and has, since 1996, been editor (latterly co-editor) of *Ecological Entomology*.

CHAPTER 1

Sampling theory and practice

SIMON R. LEATHER AND ALLAN D. WATT

Introduction

This chapter deals with the need to sample insects, the theory underlying sampling, the need to calibrate samples, and the design of sampling programs, and it evaluates the use of different sampling techniques.

Why sample?

Sampling is a scientist's way of collecting information, and the majority of sampling is undertaken to answer specific questions. This was not always the case. Sampling as we know it was first done in a haphazard manner and bore little relation to what we would call sampling today. The first samples taken were basically a by-product of the desire of natural historians to collect information about the world around them.

A brief history of information collection

The history of information collection can be classified into three main stages. There is a little overlap, but in the main we can recognize three separate phases.

1 The collectors

This can be classified as the pin, stuff, and draw era. As travel became relatively safer and people became more interested in what lay beyond their horizons there was a rapid expansion both in the number of naturalists traveling to other continents and in the number of people employed by naturalists to collect and return specimens to Europe. Drawing was also a popular activity and to a certain extent filled the niche now occupied by photography. Many ships' officers were accomplished amateur artists and many had an interest in the flora and fauna of the countries they visited. This phase resulted in the acquisition of many thousands of specimens of plants and animals, either stuffed, pickled,

1

pressed, or pinned, accompanied by many sketches of the organisms in their native settings, although as the majority of the artists had no scientific training these drawings and paintings sometimes bear only a passing resemblance to reality.

Although this resulted in the garnering of many examples of plants and animals there was little knowledge of the biology or ecology of the organisms. This led to a great deal of confusion, particularly in the field of entomology where the sometimes complicated life cycles such as those occurring in dimorphic and polymorphic species such as aphids led to the misclassification of many species. For example, several aphid species were classified as being more than one species, depending on which host plant they were removed from or depending on which stage of the life cycle they were in at the time of their collection. Other similar mistakes occurred in the Lepidoptera where confusion over the identity of members of several mimic species lasted for some time until the larval stages were recognized. Ladybird beetles such as *Adalia bipunctata* and *Adalia decempunctata*, now well known as being extremely variable in their different color forms were also once misidentified as separate species until their life histories were fully elucidated.

2 The observers

There were of course some collectors who also collected observations. Many natural historians, as well as having a keen eye for the chase and for sketching, also felt the need to observe the behavior of the animals that they were collecting. These are exemplified by Darwin and Fabre who, as well as making detailed collections of specimens, also spent many hours observing and recording patterns of behavior. These observations provided plenty of information on the biology of the species, but as much of it was centered on individuals and their interactions with other individuals of the same species did not provide a great deal of information on their place in ecosystems, did not always provide accurate information about mortality factors and was confounded by a great deal of unrecognized environmental "noise."

3 Experimental/controlled sampling

The next great step forward in the field of information collecting was the use of experimental studies in controlled conditions. For example, by studying the biology of an insect in the laboratory, it is possible to obtain detailed knowledge of life history parameters such as fecundity, longevity, etc., and it is also possible to assign specific values to mortality factors, albeit in a far from natural environment. The main drawback of this type of study is that environmental variability is lost and the natural impact of mortality and natality factors is compromised.

The best option is to combine laboratory methods and natural conditions, and to do experimental and manipulative work in the field. The need to obtain accurate estimates of animal numbers in the field led to the development of the theory of sampling and, incidentally, to the use of statistics in the biological sciences.

Estimating abundance and predicting population dynamics

The major use of sampling in entomology is to determine the number of insects in a given area or location, usually for pest control or conservation purposes. The other main reason for wanting information on insect numbers is to increase our understanding of the population dynamics of the insect(s) in question and to make predictions of their future abundance.

Before one can make a prediction, one needs to know how many insects there are in the first place. This is equally true, whether one is going to control a pest or to conserve an endangered species. It is not a sound practice (although some modelers do it) to conjure a number out of thin air. There is also a need to know what factors affect those numbers. There are basically six facts about the population of an insect that are required before sensible predictions of the population dynamics can be made:

1 density—an expression of the species' abundance in an area;
2 dispersion (distribution)—the spatial distribution of individuals of a species;
3 natality—birth rate;
4 mortality—death rate;
5 age structure—the relative proportions of individuals in different age classes;
6 population trend—the trend in the abundance of the study species.

It is only from this sort of information that one can start to make some sort of inferences about the population dynamics of the insect. The only reliable way to obtain this type of information is to sample.

Sampling methods

To sample an insect requires both a sampling technique and a sampling program. These are different things, although it is noticeable that even in the scientific literature the two terms are quite often used interchangeably.

Sampling techniques

A sampling technique is the method used to collect information from a single sampling unit. Therefore the focus of a sampling technique is on the equipment and/or the way the count is accomplished.

Sampling programs

A sampling program, on the other hand, is the procedure for employing the sampling technique to obtain a sample and make an estimate. Sampling programs direct how a sample is to be taken, including sampling unit size, number of sample units, spatial pattern of obtaining sampling units, and timing of samples.

Before, however, one starts to think about either the sampling unit or the sampling program it is necessary to know something about the insect that is going to be sampled.

An important starting point is to find out about the life cycle and biology of the insect and especially about where it is likely to be found. There is no point in sampling terrestrial habitats for something that lives in water. Some insects have marked changes in distribution during the course of a year, so it is important that this is taken into account before any sampling program is undertaken. For example, the bird cherry aphid *Rhopalosiphum padi* lives on grasses in summer and on the bird cherry tree *Prunus padus* in autumn, winter, and spring (Dixon & Glen 1971). For those insects that show seasonal changes in habitat use, it is essential to know when the changeover from habitat to habitat takes place, at least approximately. Sampling will of course have to be conducted in both habitats for some period of time to pinpoint this changeover. Thus, a good knowledge of the biology and ecology of the insect is very important. Another important consideration is the likely cost of the sampling in terms of both time and money.

Deciding on the approach

Sampling tools/techniques

There are a number of tools that can be used to sample insect populations. One can sample aerially, for example using suction traps. These are used throughout Britain by Rothamsted Insect Survey (Knight et al. 1992) and in many other parts of the world. They are primarily used to trap aphids, and sample at two standard heights, 1.2 m and 12.2 m. Sticky traps, either with or without attractants, can be used for almost anything that flies and is too weak to get off the sticky board. Light traps are also commonly used to sample aerial populations, although the insects mainly caught are night-flying Lepidoptera. There are various intercept traps that are used to catch beetles, flies, aphids, and other insects, such as yellow water traps, Malaise traps and window traps. These are discussed further in other chapters. It is useful to note that the range of technology is quite vast. A great deal of effort can go into the design and evaluation of traps, and this is often an essential part of the design of a sampling program. For example, some insects are more readily caught by certain types of trap (Heathcote 1957,

Niemelä et al. 1986). The behavior of the insect will largely determine the type of trap or sampling tool used (see later chapters).

Passive traps versus active traps

Sampling and trapping techniques can broadly be classified into two types—passive or active.

A passive trap is one that should be neutral and depends entirely on chance. An active trap depends on the behavior of the insect but takes advantage of the behavior and attracts the insect to the trap by chemical lures, baits, or even color—all of which can be varied to give different trapping efficiencies and targets (Finch 1990).

What is the advantage of a passive trap over an active trap?

Passive traps allow unbiased estimates of insect populations because the insects are neither attracted nor repelled by the traps. For example, although aerial suction traps are powered by a motor and draw air into the collecting tube they are essentially passive in action as they depend on the insect flying into the ambit of the trap and do not depend on it being attracted to the area. A big drawback to the use of passive traps is that they are not very useful at low densities. This is a particular problem when programs have been designed to monitor the abundance of occurrence of pests—for example insects on quarantine lists. In those cases an attractant trap is a much better alternative as they are better detection tools. They do, however, give a biased estimate of the density per unit area and conversion factors then have to be applied. Thus, when using attractant traps, particularly if they are being used to obtain population estimates, it is vitally important to know over what range the trap is effective and whether there are directional as well as distance effects.

Direct habitat sampling

Sometimes, particularly if one is working with a pest species, the most useful method of sampling is one that estimates the population size in the habitat—e.g. a crop or nature reserve. Indirect methods of sampling—e.g. aerial sampling with a suction trap or pan trapping in a field—only indicate what is present in the area, and do not tell you what is actually on the plant or in the soil. It will tell you what is there and gives some idea of whether there are many or few, but unless it has been backed up by calibration studies it does not tell you how many insects there are per plant or per unit area of habitat, or whether they are actually present on the area that you are concerned with; they may just have been en route somewhere when they were caught. This is particularly true of migratory insects.

What methods are available and what determines their use?

Particular methods are dealt with in the following chapters. Here, we consider the rationale behind the selection of available techniques. When sampling on the ground there are a number of methods available. Quadrats may be used for some insects—e.g. predatory surface-active beetles, aphids on plants, etc. However, searching a surface quadrat is no good for cryptic, soil-dwelling nocturnal insects such as the large pine weevil *Hylobius abietis*. Whole-plant searches or part-plant searches are also useful, and if the insects are relatively sedentary can give good population estimates. Pitfall trapping is useful for surface-active insects, particularly those active during the night, (see Chapter 3). Soil extraction methods can be used for soil or root-dwelling species (see Chapter 2).

Destructive versus non-destructive sampling

If whole plants or small areas of habitat are to be sampled, two approaches can be used—destructive and non-destructive. Destructive sampling involves the removal of the sample unit for later assessment (see below), whereas with non-destructive sampling the sample unit is searched or sampled *in situ*.

Both these approaches have their merits and disadvantages. For example, suppose you are counting aphids on a plant. You could sample destructively—i.e. remove the plant or part of the plant from the ground or from the main stem, and bring it back into the laboratory. Alternatively, you could sample non-destructively—for example, examine 100 leaves and record what is found.

Destructive sampling is more accurate as the insects are less likely to escape during the counting process. One cannot, however, go back and sample the same plant or area again. This is a particular problem if there are only a limited number of plants to begin with, or if the habitat type is rare and easily disturbed. If one is sampling from a large number of uniform plants such as a field of leeks or a forest plantation, destructive sampling may be a useful technique. A disadvantage of destructive sampling is that it is more time consuming, and is thus not useful in situations where a quick estimate of insect numbers is required, say for a control operation. It is possible with destructive sampling, however, to postpone sampling by storage, be it in the freezer or in some sort of preservative. This is particularly useful in those situations where a large number of samples have to be taken in a limited time period and where there is no need for a swift result. It means that the actual counting of the insects can be saved for a less busy time of year—e.g. the winter.

Non-destructive sampling, on the other hand, does allow re-sampling of the plants and habitats on a frequent or regular basis. This is very useful in sensitive areas and when local population dynamics are being studied. Non-destructive sampling tends to be quicker than destructive sampling and causes less disturbance to the habitat. It does however depend on the insects being relatively sedentary or slow to respond to disturbance. Thus the counts will tend to be underestimates. To counter this, as non-destructive sampling is fairly quick, more samples can be taken, although this does not entirely solve the problems of underestimation.

How many samples?

There are a number of factors that determine the number of samples that are taken. The first requirement is to be sure that the sample taken is representative of the population that is being sampled. To ascertain this it may be necessary to perform stratified sampling. It is not always safe to assume that insects are systematically distributed. A number of different distributions are possible. The population could be randomly distributed, uniformly distributed, or even in an aggregated (clumped or contagious) distribution (Fig. 1.1). These factors all need considering. It is possible to determine what distribution the population has by using the following approach.

Variance—mean ratios

The dispersion of a population determines the relationships between the variance s^2 and the arithmetic mean μ thus:

1 random distribution—the variance is equal to the mean—$s^2 = \mu$;

2 regular (uniform) distribution—the variance is less than the mean—$s^2 < \mu$;

3 aggregated (clumped or contagious) distribution—the variance is greater than the mean—$s^2 > \mu$.

The distribution of the organism can have a marked influence on the way in which you might sample. Take, for example, a site in which the organism you are going to sample has a soil-dwelling pupae. The easiest approach is to do a simple line transect from one corner of the field to another, or if you are

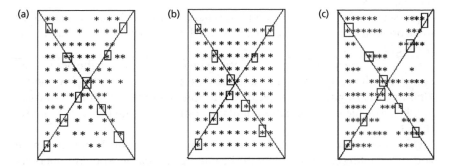

Fig. 1.1 Sampling different distributions with a common sample plan: (a) random, (b) uniform or regular, and (c) aggregated distribution. Note that the values returned, even in this simple example, are very different for the aggregated distribution.

concerned about the slope or topography you might do another transect across from the other two corners in the form of an X. Depending on the distribution of the organism you may get totally different answers (Fig. 1.1).

Stratified sampling

Suppose we know that the population we are going to sample varies systematically across the area we are going to sample.

We may know, for example, that the insect does not occur in high densities in particular areas—e.g. where there are lots of stones—but does occur in high numbers in areas where there is a lot of sand. It may even be something more specific: for example, if we were sampling trees for insects, we might know that the distribution of the insect within the crown of the tree is not uniform. The pine beauty moth, for example, lays most of its eggs in the upper third of the tree, so one can get a good estimate of the population by just counting eggs found on the first five whorls and then either multiplying up or just taking the figure obtained to be representative of the population (Watt & Leather 1988a). It really depends on what one is sampling for. If one is sampling for predictive purposes than the first five whorls is good enough; on the other hand, if the sampling is part of a detailed population study, then the sampling needs to be more thorough and to take more account of the distribution of the organism. Thus, for the pine beauty moth, a branch is taken from every other whorl, the number of branches per whorl counted, and the counts are then multiplied up accordingly. If one had a very large scale study, one might just take a third-whorl branch at random and multiply up from there (Leather 1993). Of course one would have to have done some whole-tree sampling first to determine what all the various multiplication factors were going to be. For example, with winter moth eggs on Sitka spruce there is a marked difference in egg distribution, not just in relation to tree height, but also within the branches (Watt et al. 1992) (Fig. 1.2). One could therefore work out various sampling schemes to use.

In essence, though, before a sampling scheme can be devised, one needs to do some preliminary sampling to get a feel for what number or size of sample one will require.

In general, the more samples that are taken the more precise the population estimates will be. However, time and expense are always constraining factors. Thus the usual approach is to decide on the lowest number of samples that can be taken to achieve a reasonable population estimate within the error limits set (Box 1.1).

One should make such calculations throughout the season. So for example if you are sampling cereal aphids at the beginning of a season when numbers are low you would start with a thousand tillers per field, and make adjustments as the population rises—but never below 100 tillers per field. There is usually a minimum value that the sampler never falls below and a maximum that

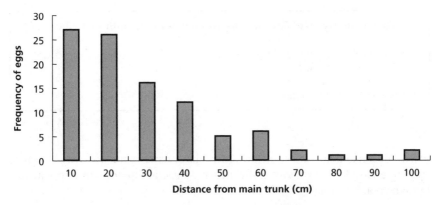

Fig. 1.2 The distribution of winter moth eggs along Sitka spruce branches. Data from Watt et al. (1992).

Box 1.1 To calculate the number of samples required

$n = (\sigma/\mu) \times cv)^2$

where n is the number of samples required, σ (sigma) is the variance, μ (mu) is the unknown mean of the population, and cv is the coefficient of variation of the mean which in turn is defined as

$cv(X) = \sigma\sqrt{n}/\mu$

For practical purposes one needs to collect a series of samples and make preliminary estimates of the mean (Xe) and the variance (Se) and then use this formula

$n = (Se/Xe \times cv)^2$

---

is never exceeded: these are determined by time and the requirement for accuracy. (For a number of case studies see Chapter 9.)

Sampling concepts

Choosing a sample unit

What is a sampling unit?

A sampling unit is a proportion of the habitable space from which insect counts are taken. The units must be distinct and not overlap. A sampling unit can be very variable in form. For example, it could be direct counts of all the caterpillars

---

Box 1.2 Possible types of information required from a sampling scheme

1 Estimates of population density per unit area
2 Assessments of percentage infestation or parasitism
3 Estimation of damage per unit area
4 Absolute population counts

---

in 1 m² of cereal field, or it could be 20 sweeps of a sweep net down a row. More usually, a sample unit is more easily measurable, e.g. a quadrat or template of a known area.

What determines the choice of a sampling unit?

The choice of sampling unit is dependent on what information is required. In an agricultural system the grower most frequently wants to know whether a particular insect pest has reached a threshold level which requires action (e.g. spraying), or what proportion of the crop is infested. In other circumstances (e.g. a population study) the observer may be more interested in the number of insects per given unit area (Box 1.2).

Criteria for sampling units

Sample units must meet a number of criteria if they are to be useful.

1 Each sample unit should have an equal chance of selection

This is where it is important to know what type of distribution individuals within the population display. Unfortunately using a totally random sampling scheme in some situations, even agricultural, can be too expensive in terms of time. Certainly in some situations—e.g. in a dense forest—it may not be logistically possible to apply a totally random sampling pattern. Therefore most fields and plots are sampled on a prearranged pattern—e.g. two X's, a V, a W, or whatever, with the samples collected along the transects. A degree of randomization can then be introduced, for example by varying the distance between sampling stations or by taking samples from either side of the transect on a random basis. It is important to avoid bias when sampling. This is particularly easy to introduce when sampling in crops. It is difficult to avoid selecting the leaves that look infested, e.g. discolored or curling. In cases like that, an element of chance should be built into the process when arriving at the sample station—e.g. take the first plant on the left, or throw a quadrat to standardize the sampling unit.

2 The proportion of an insect population using the sample unit as habitat should remain constant throughout the sampling event

If for example the insect moves around depending on time of day, then your population estimates will vary accordingly — e.g. the beet armyworm *Spodoptera exigua* is phototactic so sampling at midday will produce lower counts than during twilight or dawn as the larvae move into the ground or center of plants at high light intensities. The large pine weevil *Hylobius abietis* is another example — it is night active so sampling should be done at the same time each day to keep things constant. Sampling should therefore be planned to take these factors into account.

3 There should be a reasonable balance between the variance produced when data are collected from a given sample unit and the cost (time, labor, or equipment) in assessing that unit

Generally a preferred sample unit would be the minimum size which would allow an adequate number of replications on a given date to produce averages with meaningful variance. Sampling all the leaves on a plant would provide very accurate information on that plant but as one would only be able to sample a few plants then the population estimate for the site would be extremely poor. Incidence counts are also useful (Ward et al. 1985). These rely on intensive sampling over a number of seasons so that one has a robust relationship between the numbers of insects present and the infestation rate of those plants. This is a very useful technique for non-experts such as farmers. It is however, not a feasible option unless the preliminary studies have been completed. Caution should also be exercised with this method as the relationship between incidence and population can change.

4 Whenever possible or practical the sample unit should be as near as possible to the natural habitat unit

In other words the area within which the insect is likely to spend most of its time in a given developmental stage — e.g. a cereal plant for an aphid, a leaf on a tree for an aphid or leaf miner, a branch for a defoliating caterpillar, and so on. Insects without discrete habitats — e.g. soil dwellers, predatory beetles, etc. — are somewhat more problematic and in such cases it is probably wise to rely on random quadrats etc.

5 A sample unit should have stability

Or, if not, then its changes should be easily and continuously measured — e.g. the number of shoots in a cereal crop.

6 The sampling unit must be easily delineated or described

For example, buds on a branch, leaves, or plants, or quadrats of standard size.

7 Ideally a sample unit should be able to be converted to some measure of unit area

Thus it is important to count the number of trees in a compartment or plants in a field, etc., and then to be able to convert the counts obtained to numbers per m^2, for example (Box 1.3). What conversion is used, however, is less important that the fact that a conversion of some type is required in order to compare the density of different stages of the same insect species. This is essential if the mortality occurring between different stages is to be estimated.

8 The number and location of sample units should be selected according to the purpose of the sampling

Thus one could just sample the ears of cereal plants if one was interested in *Sitobion avenae* for prediction purposes (George & Gair 1979), whereas whole-plant counts would be needed for population estimates (Leather et al. 1984).

Box 1.3 Sampling the pine beauty moth

The pine beauty moth *Panolis flammea* has a typical univoltine lifecycle. The adult lays eggs on pine needles which hatch into larvae that pass through five instars whilst feeding in the canopy. The fifth-instar larva stops feeding and passes into a pre-pupal stage that spins to the ground, burrows into the litter layer, and pupates (Watt & Leather 1988b).

Sampling is carried out at all stages of the life cycle. Although each sampling technique gives a different output, they are all easily converted to a common measure, in this case individuals per square meter.

Stage	Method	Output	Conversion
Adult	Pheromone trap	Males per trap	Calibrated to area covered by trap
Eggs	Needle counts	Eggs per whorl	Converted to projected area covered by tree
Larvae	Funnel traps	Head capsules per funnel	Collecting area of funnel known
Pre-pupae	Basin traps	Pre-pupae per basin	Collecting area of basin known
Pupae	Soil sample	Pupae per $15\,cm^2$	Converted to per m^2 measure

Informed sampling and collecting

As one works more and more with insects, one gains a knowledge or feeling of where to find particular groups or species. Although this is not strictly sampling, it does help inform the sampling process, and when one requires insects to start cultures or laboratory and field experiments it is certainly useful to be able to locate relatively large numbers of specimens quickly and easily.

In general, insects are small and relatively fragile, their reproductive and development rates are highly influenced by environmental factors, in particular temperature, and many of them, especially in their larval stages, are likely to feature in the diets of birds and other vertebrates as well as arthropod predators. This tends to mean that insects, except for the brightly colored highly mobile species such as butterflies, are more likely to spend most of their day in sheltered or concealed habitats, and in fact many insect species have taken this to the extreme and spend much of their life cycle living and feeding within plant parts — e.g. gall insects, leaf miners, bark beetles. Therefore, if looking for a ready supply of various insect species, dense clumps of grass, piles of leaves, under rocks and stones, in tree hollows and crevices, under loose bark, under logs, or even in fungi, will prove rewarding sites to search. Very dry habitats are unlikely to yield large numbers of individuals or species, but a moist, sheltered hollow under a broad-leaved tree is a sure source of a myriad of different species, albeit not all insects.

Insects, particularly herbivorous ones, are of course closely associated with their host plants, and certain times of year and sites on the plant are more likely to yield results than others. Certain plant species naturally potentially harbor more insect species than others. Oaks, willows, and birches are natural hot spots for insects of all descriptions from bark-dwelling Pscoptera to gallers, miners, general defoliators, and sap suckers. Many herbivorous insects depend on a ready supply of nitrogen to enable them to develop quickly at the beginning of the year. Check meristems, developing buds, young shoots, and flower buds for caterpillars and sap suckers. Birch aphids *Euceraphis punctipennis* closely follow growing shoots. Curled or distorted leaves are often signs that sap suckers or leaf tiers are in the vicinity, although be warned that these deformations will persist long after the insect has completed its life cycle and departed. Similarly, sooty mould, sticky leaves, and silken threads are often signs that aphids, other sap suckers, and web-spinning Lepidoptera are or have been present. Swellings on stems and sap and resin flows may also indicate the presence of stem borers, gallers, and bark beetles.

In temperate parts of the world insects spend a large proportion of their life cycle overwintering (Leather et al. 1993). Many have behavioral adaptations that cause them to seek out specific overwintering sites — e.g. negative phototaxis that causes them to search for dark crevices or thigmotactic responses that make them aggregate. If looking for ladybirds during the winter, it is often useful to look under loose bark, under window sills, or even on fence posts. Aggregations often form in such situations. If your insect overwinters in the soil, avoid wet places and look for well-drained sites, preferably under trees rather than in the open. Overwintering is a costly business and insects attempt to minimize costs by overwintering in sites where the soil is unlikely to freeze, below about 10 cm depth. During winter, searching under hedges, in the middle of rotting logs, and in dense clumps of grass is also likely to repay one's efforts.

In general, think shelter, food, and protection and you are likely to find some insects in a relatively short space of time.

Conclusions

In this chapter we have tried to give an overview of the philosophy of sampling, the rationale behind the choice of sample unit and technique, and some pointers towards what is the best approach to use in particular situations. We have not provided detailed mathematical and statistical formulae or numerous worked examples. Those wishing to acquire more of the mathematical background should consult two excellent textbooks that provide a wealth of such information, Southwood and Henderson (2000) and Sutherland (1996). Chapters within this book provide more specific mathematical and theoretical approaches for specific cases, but in the main deal with the practicalities of sampling either in specific habitats or with problematic guilds or groups.

References

Dixon, A.F.G. & Glen, D.M. (1971) Morph determination in the bird cherry-oat aphid, *Rhopalosiphum padi* (L). *Annals of Applied Biology*, **68**, 11–21.

Finch, S. (1990) The effectiveness of traps used currently for monitoring populations of the cabbage root fly (*Delia radicum*). *Annals of Applied Biology*, **116**, 447–454.

George, K.S. & Gair, R. (1979) Crop loss assessment on winter wheat attacked by the grain aphid *Sitobion avenae* (F.). *Plant Pathology*, **28**, 143–149.

Heathcote, G.D. (1957) The comparison of yellow cylindrical, flat and water traps, and of Johnson suction traps for sampling aphids. *Annals of Applied Biology*, **45**, 133–139.

Knight, J.D., Tatchell, G.M., Norton, G.A., & Harrington, R. (1992) FLYPAST: an information management system for the Rothamsted aphid database to aid pest control research and advice. *Crop Protection*, **11**, 419–426.

Leather, S.R. (1993) Influence of site factor modification on the population development of the pine beauty moth (*Panolis flammea*) in a Scottish lodgepole pine (*Pinus contorta*) plantation. *Forest Ecology & Management*, **59**, 207–223.

Leather, S.R., Bale, J.S., & Walters, K.F.A. (1993) *The Ecology of Insect Overwintering*. Cambridge University Press, Cambridge.

Leather, S.R., Carter, N., Walters, K.F.A., et al. (1984) Epidemiology of cereal aphids on winter wheat in Norfolk, 1979–1981. *Journal of Applied Ecology*, **21**, 103–114.

Niemelä, J., Halme, E., Pajunen, T., & Haila, Y. (1986) Sampling spiders and carabid beetles with pitfall traps: the effect of increased sampling effort. *Annales Entomologici Fennici*, **52**, 109–111.

Southwood, T.R.E. & Henderson, P.A. (2000) *Ecological Methods*. 3rd edn. Blackwell Science, Oxford.

Sutherland, W.J. (1996) *Ecological Census Techniques*. Cambridge University Press, Cambridge.

Ward, S.A., Rabbinge, R., & Mantel, W.P. (1985) The use of incidence counts for estimation of aphid populations. 1. Minimum sample size for required accuracy. *Netherlands Journal of Plant Pathology*, **91**, 93–99.

Watt, A.D. & Leather, S.R. (1988a) The distribution of eggs laid by the pine beauty moth *Panolis flammea* (Denis & Schiff.) (Lep., Noctuidae) on lodgepole pine. *Journal of Applied Entomology*, **106**, 108–110.

Watt, A.D. & Leather, S.R. (1988b). The pine beauty in Scottish lodgepole pine plantations. In *Dynamics of Forest Insect Populations: Patterns, Causes, Implications* (ed. A.A. Berryman), pp. 243–266. Plenum Press, New York.

Watt, A.D., Evans, R., & Varley, T. (1992) The egg-laying behaviour of a native insect, the winter moth *Operophtera brumata* (L.) (Lep., Geometridae), on an introduced tree species, Sitka spruce, *Picea sitchensis*. *Journal of Applied Entomology*, **114**, 1–4.

CHAPTER 2

Sampling insects from roots

ALAN C. GANGE

Introduction

There are relatively few ecologists who dare to venture below ground, to study the effects of subterranean insects on plants. If one examines the insect–plant interaction literature for the last 20 years, fewer than 2 percent of studies deal with root-feeding insects. From this paucity of information, one is tempted to conclude that subterranean insects are of little consequence in natural systems. However, a quick glance at the agricultural and horticultural literature shows that there is a rich array of studies involving these insects, since many of them are pests of considerable economic importance. Indeed, root-feeding insects can be so destructive that several species have been introduced in biological control programs against weeds (e.g. Blossey 1993, Cordo et al. 1995, Sheppard et al. 1995).

Why is there this apparent lack of interest in ecological studies involving subterranean insects? The answer undoubtedly lies in the difficulty of sampling these animals. Unlike their foliar counterparts, rhizophagous insects are often invisible for part or all of their life cycles. Furthermore, excavation of soil may not always be sufficient to detect them, since some species feed internally in the root system. Experiments involving these insects often end in failure, as non-destructive monitoring of the system is difficult and problems may go undetected. To add to these physical problems, various aspects of the biology of the species may also hinder sampling methods. In some cases, the stage in the soil is long-lived and the time span involved may be greater than that allotted to standard research projects, which are generally of three years' duration. The end result of these problems is that sampling for rhizophagous insects is generally a laborious, time-consuming, and often tedious operation. However, it need not always be so and a number of ingenious methods have been developed.

The most recent comprehensive review of rhizophagous insects and their effects on plants is that of Brown and Gange (1990). This documents that only six of the 26 orders of insects are well represented as below-ground herbivores, and of these the most important order is the Coleoptera. Diptera and Lepidoptera also contain species with rhizophagous larvae, while within the Hemiptera the Aphididae (aphids), Cercopidae (spittle bugs), Cicadidae (cicadas), and Pseudococcidae (mealy bugs) contain economically important root-feeding species.

The Collembola also have representatives which feed on roots, though the majority probably feed on microorganisms or decaying leaf litter (Hopkin 1997). Collembola apart, the majority of insects associated with roots have a stage of their life cycle above ground and these mobile adults can easily be used to identify the presence of subterranean stages in a particular area. A good way to start with rhizophagous insect sampling is to understand the visible signs of their presence, manifest in the terrestrial environment.

External clues

To determine if a species is present in a location, a variety of trapping methods for adults can be used. Suction sampling (e.g. Arnold 1994) can be particularly effective, but a number of species have nocturnal adults. Many of these seem to be attracted to light, and mercury vapor (MV) light traps have been used to monitor adult numbers of chafer grubs (Coleoptera: Scarabaeidae) near pastures (Roberts et al. 1982b). Interestingly, adults of the wingless black vine weevil *Otiorhynchus sulcatus* are also attracted to light, but generally to tungsten bulbs, rather than MV (Labuschagne 1999). Water traps have been used to capture adults of the cabbage root fly *Delia radicum* (Bracken 1988), while pheromone traps have been developed for some species (e.g. the pea and bean weevil *Sitona lineatus* [Smart et al. 1994]). If the biology of the species is well known, then emergence traps (described in Southwood & Henderson 2000) can be very effective (e.g. for *S. discoideus* [Goldson et al. 1988]). Some species on eclosion leave characteristic evidence, and the empty emergence skins of various cicada species have been used to estimate nymphal densities below ground (White & Sedcole 1993). Sticky traps, with the sticky side facing downwards, have been used to estimate numbers of grape phylloxera *Daktulosphaira vitifoliae* emerging from grape rootstocks (Hawthorne & Dennehy 1991).

Adults of many species feed on foliage in a characteristic manner. A good example of this is the leaf-notching produced by *O. sulcatus* and this can be used as an excellent method of detecting the pest (Labuschagne 1999). However, the effects of subterranean larval feeding are also often apparent, most commonly manifest in wilting of foliar tissues, because the main effect of root removal by larvae is the imposition of drought stress in a plant (Masters 1995). In natural plant populations, individuals which show unusual drought stress or which die for reasons not attributable to foliar insects or pathogens (e.g. Strong et al. 1995) should be suspected of having insects attacking the roots. In some cases, internal root borers produce quantities of frass at the exterior end of their tunnels and this can be visible at or just below the soil surface. Maron (1998) gives an example with ghost moth *Hepialus californicus*, where frass can be easily seen at the base of infested bush lupine plants.

Subterranean aphids often live in close proximity to ant colonies and a number of species live entirely within the nest of the ants. In grassland systems, one must first find the ant mounds and then sample within these to find the aphids (Pontin 1978). Although the aphids are "cultured" by the ants, a significant number are eaten too, and a further method of deciding whether subterranean insects are present in any given location is to look for the signs of predation. For example, in pasture grassland and amenity turf, birds such as rooks, crows, and magpies can do significant damage, when searching for large subterranean larvae of chafer grubs (Coleoptera: Scarabaeidae) or "leatherjackets" (Diptera: Tipulidae). Indeed, for turf managers, birds represent the best early warning system that subterranean larvae are present and may need to be controlled (Fermanian et al. 1997).

Field extraction methods

Chemical methods

Extraction of insects from soil without disturbance of the soil profile must involve some form of chemical expulsion or the use of an attractant. Various chemicals have been used over the years to expel insects from soil, with varying degrees of success. These include St Ives fluid (a mixture of disinfectant and other chemicals), potassium permanganate, mustard, formalin, petrol (gasoline), diesel fuel, ammonia, nitric acid, acetic acid, soapy water, and brine. In the early years, the chemical was poured onto the surface of soil and the appearance of larvae awaited. There are of course many problems with this approach, not least toxicity of the chemicals to the operator and to any plant life present. Furthermore, the method is not quantifiable, as the area from which larvae have appeared is unknown.

Of these chemicals, only brine has any merit and is worth consideration. Stewart and Kozicki (1987) developed a successful sampling method for tipulid larvae in grassland, termed the "brine pipe method." This involved hammering 10 cm diameter plastic pipes into soil to a depth of about 5 cm, and filling the pipes with strong brine solution. The brine slowly percolates into the soil, and on contact with the larvae causes these to rise to the surface, where they float in the pipe. The method can produce comparable results with more conventional laboratory-based techniques (below) and can be quantified, by treating the pipe as a soil "core." Figure 2.1 shows the efficacy of the method. Here, 16 different fields, all under permanent ryegrass *Lolium perenne* / clover *Trifolium repens* pasture were sampled in the spring of 1999 (Gange, unpublished). Twenty 10 cm diameter brine pipes were placed randomly in each field. Within 30 cm of each pipe, a 10 cm diameter × 10 cm deep soil core was taken and tipulid larvae were extracted from each in the laboratory by wet-sieving (see below). It can be seen that the brine pipe method provides a good estimator of total abundance when larval numbers are high, but tends to underestimate abundance when total numbers are low. The most likely reason for this is that the pipe method relies on

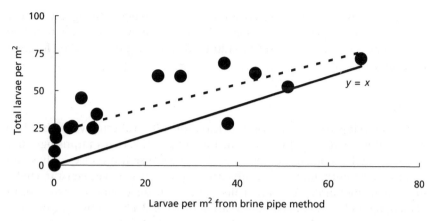

Fig. 2.1 Relation between tipulid larvae extracted by the brine pipe method and by exhaustive hand-sampling (total numbers). Dashed line is the fitted regression ($y = 0.817x + 21.89$), solid line is the line of equality ($y = x$). At low densities, brine pipes underestimate total numbers, but the accuracy of the method improves with increasing density. The regression predicts that brine pipes will record the total population when density is about 120 larvae per m². Data from Gange (unpublished).

the percolation of brine into the soil and if larvae are at low density, it is likely that not all will be affected by the solution. However, if larval numbers are high, then a higher proportion of larvae are likely to come into contact with the solution. Furthermore, the number of pipes required and time to check them means that this is a less efficient method from the labor point of view, if larvae are rare or patchy. Interestingly, no other subterranean insects seem to appear in the pipes, but earthworms can also be sampled by this method.

Behavioral methods

Perhaps because of their economic importance, tipulids (Diptera: Tipulidae) seem to have been the subject of more published sampling methods than any other root-feeding insect. The brine pipe method outlined above is particularly useful because it enables farmers or turf managers to sample for the insects *in situ*, and, as it is quantifiable, indices of infestation have been produced against which field counts can be compared. Farmers can then decide whether it is economically viable to spray a field to control their numbers (Clements 1984). However, if a source of salt, or water, or pipes is not available, it is still possible to determine if tipulid larvae are present in a field, by taking advantage of their nocturnal behavior. An area of grassland is thoroughly soaked with water and a tarpaulin or similar item (polyethylene bin liners are an acceptable substitute) is laid over the soil surface (Gratwick 1992). Inspection beneath the tarpaulin in the early morning should reveal larvae, which have emerged at night to feed on the surface, but which do not return to the soil because it remains dark under the cover. These must be collected quickly, because exposure to light will cause them to burrow rapidly into the soil. If the researcher merely wishes to obtain larvae for experiments or to start a culture, this is a very easy method for their collection.

Baits

Instead of trying to persuade insects to leave the soil, an alternative method is to provide them with an attractant in the form of a bait. Perhaps surprisingly, this is not a widely adopted method, most likely because it produces only semi-quantitative information, as the area from which larvae have been attracted is difficult to measure. However, baits have been developed for wireworms (Coleoptera: Elateridae) (Ward & Keaster 1977) and a bait consisting of a 1 : 1 mixture of wheat and corn was used by Belcher (1989) to estimate the proportion of corn fields infested with wireworms in Missouri. An example of the kind of data one can obtain by this method is given in Fig. 2.2. While the method may be of little use for quantifying insect density on a local scale (e.g. per m^2), it is useful for recording density on a regional scale (e.g. proportion of fields infested, etc.) Belcher (1989) mentions that white grubs (Coleoptera: Scarabaeidae) (otherwise known as chafer grubs) were also attracted to the bait. However, this fact does not appear to have been used in any subsequent sam-

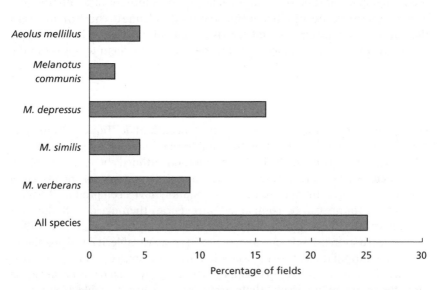

Fig. 2.2 An example of the data that can be obtained by baiting for subterranean larvae. Belcher (1989) sampled cornfields in Missouri and was able to record the percentage of fields infested by different species of wireworm (Coleoptera: Elateridae). Drawn from data in Belcher (1989).

pling program for these often injurious insects. Baits have been used to control one insect pest, the black field cricket *Teleogryllus commodus*, which can cause serious damage to pastures in Australia. Williams et al. (1982) describe the success of cereal baits impregnated with insecticide in the fight against this insect. Recipes for baits for attracting adults of *O. sulcatus* are given by Labuschagne (1999) and can be most successful when scouting for the presence of this pest.

Most baits for larvae and adults (Labuschagne 1999) appear to be based on a cereal/bran mixture, but probably the main criterion for a successful bait is the evolution of CO_2. This is because it is thought that CO_2 is the primary stimulus used by insects to orientate themselves to roots in the soil (Brown & Gange 1990, Bernklau & Bjostad 1998). This probably explains why another excellent bait for wireworm larvae in ex-pasture is a buried potato. Apart from being an acceptable food source, the potato gives off CO_2 and the larvae aggregate towards it. While not being of much use from a quantitative point of view, this method can be used to determine if the insects are present.

Hand-sorting

The most laborious, but also probably the most accurate method of extraction in the field is hand-sorting of extracted soil cores. An excellent example of this is provided by Penev (1992) who hand-sorted soil cores measuring 25 × 25 cm and 30–40 cm deep in the field when sampling for wireworms. For large insects which are abundant, this method is often quite rewarding. Gange et al. (1991) hand-sorted turf when sampling for larvae of *Phyllopertha horticola* (garden chafer) infesting a golf tee. They used 25 × 25 cm × 10 cm deep quadrats and found that the number of larvae varied between 1 and 49 per quadrat, equivalent to a range of 16–784 per m². The distribution of larvae was highly aggregated, conforming to a negative binomial distribution (Fig. 2.3).

Highly aggregated distributions are observed commonly with subterranean insects and result from clumped ovipositional patterns, feeding preferences, and the heterogeneous nature of the soil environment (Brown & Gange 1990). This means that in any situation a large number of quadrats may contain zero or very few insects, and the overall process of accurately measuring the population and its spatial distribution may be an extremely time-consuming business. The time taken largely depends on the ease of visibility of larvae and their size. For example, Harcourt and Binns (1989) hand-sorted soil cores measuring 3600 cm³ when searching for larvae of the alfalfa snout beetle *Otiorhynchus ligustici* and it took them nine minutes for each core. The distribution of larvae was also highly aggregated, again conforming to a negative binomial distribution (Fig. 2.4). Meanwhile, Seastedt (1984) sorted soil cores from prairie grassland measuring 2000 cm³ and it took 40 minutes per core. The best option is to organize a team of people to perform the sampling together. Thus, in the study of Gange et al. (1991), seven people managed to sort 100 cores, each measuring 6250 cm³, in five hours (equivalent to 21 person minutes per core).

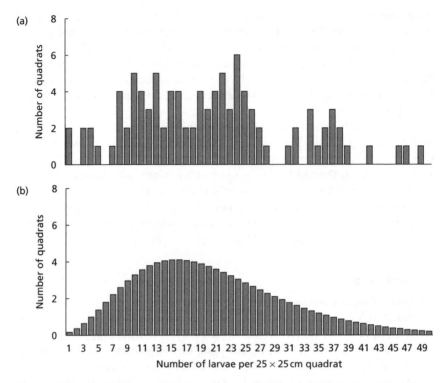

Fig. 2.3 The spatial distribution of chafer grub larvae (Coleoptera: Scarabaeidae) in a golf tee, as revealed by hand-sorting of quadrats. (a) recorded distribution; (b) fitted negative binomial. Redrawn from Gange et al. (1991).

The number and size of cores are generally determined by the identity of the species being sampled. The depth of cores needs to be such that virtually all of the root system is sampled, but must also take into account the biology of the insect. Unless published information on a species is available, it is best to perform a preliminary experiment to develop a sampling strategy (e.g. De Barro 1991) which minimizes variance, but with a replicate number which is feasible in the time available. Good examples of the use of binomial sequential sampling methods are provided by Allsopp (1991) for a sucking insect and Badenhausser and Lerin (1999) for a chewer. In any sampling program, it must be remembered that insect vertical distribution in soil can vary in time and space within a season (e.g. Hanula 1993), and over the course of several seasons (Brown & Gange 1990).

To speed up the extraction process, sieving of soil may be used, but this of course depends upon the soil texture. Sieving has been used successfully to record insects as disparate in size as white grubs *Phyllophaga* spp (Coleoptera: Scarabaeidae) in pine plantations (Fowler & Wilson 1971) and sugar beet root

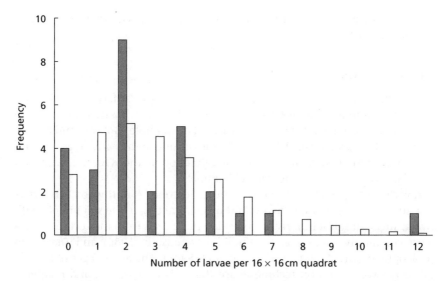

Fig. 2.4 The spatial distribution of larvae of the alfalfa snout beetle *Otiorhynchus ligustici* in field soil. Solid bars represent the recorded larval abundance, open bars represent a fitted negative binomial distribution. Drawn from data in Harcourt and Binns (1989).

maggot *Tetanops myopaeformis* (Diptera: Otitidae) in cultivated fields (Whitfield & Grace 1985).

For large-scale sampling of insects across whole fields, plough transects have been used. This is simply where a tractor plough cuts a furrow and the insect larvae exposed are counted and expressed as numbers per unit length of furrow. The technique has generally been used in grassland where the destructive nature of the method is not too much of a problem (Roberts et al. 1982a, East & Willoughby 1983).

Laboratory extraction methods

Dissection of roots

In the majority of studies with rhizophagous insects, sampling involves removing soil cores from the field and extracting these in the laboratory. In this situation it is possible to make detailed examinations and dissections of roots to determine larval numbers. In cases where the insect lives internally in the root, this may be the only way in which accurate records of numbers can be obtained. Dissection has been used to record insect attack in a range of plant species, including grape (Dutcher & All 1979), sunflower (Rogers 1985), purple viper's bugloss (Forrester 1993), and bush lupine (Strong et al. 1995, Maron 1998). In one case, careful dissection has enabled the entire entomofauna associated with the roots of *Centaurea* species to be determined (Müller et al. 1989).

Flotation methods

As with field studies, hand-sorting has been commonly used. However, it is noticeable in a number of long-term studies that this process has then given way to other, more automated forms of extraction. For example, Goldson and Proffitt (1988) and Goldson et al. (1988) used hand-sorting in the early stages of their work, but subsequently changed to using flotation methods for the extraction of *S. discoideus* larvae from lucerne field samples.

A variety of flotation methods have been described (Southwood & Henderson 2000); these generally involve a thorough mixing of the soil sample with water, sugar, or salt solution and collecting the insects from the surface. Salt is most commonly used—e.g. for tipulids (Lauenstein 1986) and *Sitona* spp (Coleoptera: Curculionidae) (Goldson et al. 1988). The advantages of this so-called passive extraction technique are that it is inexpensive and relatively quick. De Barro (1991) compared hand-sorting of sugarcane roots with flotation in water for the wax-covered mealybug *Saccharicoccus sacchari*. Flotation took half the time of direct counting and produced identical counts of insects. Furthermore, flotation is particularly useful for the extraction of inactive stages such as pupae and eggs. Indeed, this is the standard method for obtaining egg counts of a range of subterranean Coleoptera (Elvin & Yeargan 1985, Blank et al. 1986) and Diptera (Dosdall et al. 1994).

However, if one is trying to obtain an accurate estimate of the spatial distribution of eggs in soil, then flotation is not ideal, particularly for friable soils, in which cores easily break up. An ingenious egg sampling method for onion fly *Delia antiqua* eggs was therefore developed by Havukkala et al. (1992). In this, Petri dishes 15 cm diameter × 2 cm deep with wire gauze bottoms were filled with soil and then exposed to ovipositing flies in various situations. After oviposition, the dishes were filled with molten agar, from below. After cooling, the resulting solid was cut into sections and mixed with hot water, and the position of eggs accurately determined following extraction with flotation. In this way, it was possible to show how eggs of this insect were distributed within the soil profile (Fig. 2.5). Most eggs were deposited within the top 8 mm of soil, a fact which can be used to improve the targeting of insecticides against this pest (Havukkala et al. 1992).

The other disadvantages of flotation are that it is often difficult to get "clean" samples of insects and that dead animals in the soil will also be extracted. It may therefore be misleading in terms of producing estimates of active population sizes for some species (McSorley & Walter 1991).

To overcome the problem of obtaining clean samples, chemicals such as magnesium sulfate may be added to the water to ease separation of insects from the soil material. However, one extraction method that is unique for arthropods is

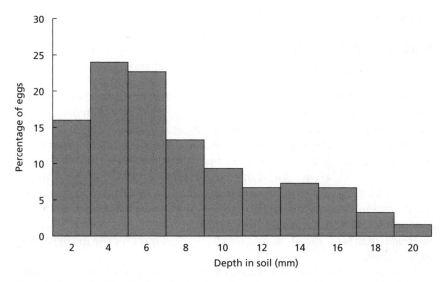

Fig. 2.5 The vertical distribution of onion fly *Delia antiqua* eggs in soil, as revealed by the molten agar technique. Redrawn from Havukkala et al. (1992).

hydrocarbon adhesion. As the cuticle of most species is lipophilic, it adheres to petroleum derivatives and makes for a very efficient extraction process. The soil is mixed with a solution of water and a hydrocarbon (usually heptane) and allowed to settle. The insects will be found in the heptane layer. The procedure was first described by Walter et al. (1987) and has since been improved by Geurs et al. (1991) and Kethley (1991).

A variation of the "wet" method involves the sieving of insects from the soil/water solution. With the aid of a continuous stream of water, this method can be an improvement on the simple act of flotation and has been used successfully when the root-feeding community is sought (e.g. Clements et al. 1987). Sieving can also be combined with subsequent flotation in magnesium sulfate (Murray & Clements 1995) to separate small larvae from the debris remaining on the sieve. Another refinement to the flotation method is elutriation, in which air is bubbled through the soil/water mixture in an effort to improve separation of the insects from vegetative and soil material (e.g. House & Alzugaray 1989).

Behavioral methods

In contrast to passive extraction methods, a variety of active techniques are also available, which rely on behavioral mechanisms of the insects. As all subterranean insects shun light and avoid high temperatures, these methods rely on the production of a temperature gradient to drive them out of soil samples.

Possibly the most common method is use of the Berlese–Tullgren funnel, the history and development of which is described by Southwood and Henderson (2000). Briefly, this technique involves the use of heat and light to drive insects out of a soil sample into a collecting container. The collecting receptacle is usually filled with a 70% alcohol solution, to preserve specimens. This method can be used to extract the active stages of any subterranean insect, but it is especially useful for microarthropods, such as Collembola, which cannot easily be obtained by any of the preceding methods.

Various modifications to the basic design have been made, usually for particular root-feeding insects. For example, the soil may be contained within a canister, which allows for regulation of temperature gradients through the core (Lussenhop 1971). This method can easily be adapted to collect live insects and in this way it is particularly useful for obtaining microarthropod "communities." Indeed, Klironomos and Kendrick (1995) used the canister method to extract Collembola and mites from leaf litter for use in subsequent experiments.

One of the most widely used and efficient variations of the funnel is the Blasdale version, for tipulid larval extraction (Blasdale 1974). In this, turf cores are held in metal cylinders and positioned turf surface downwards in a dish of cold water. Heat is applied to the soil end of the core and this drives the larvae through the core and into the water. It is likely that this method would be of little use for most rhizophagous insects, as many species will only move downwards through a soil profile. Tipulids are an exception, as they often leave the soil at night to feed on the surface (Gratwick 1992).

The use of active extraction methods for root-feeding insect density estimates is widespread and in general they are inexpensive and produce clean samples. However, their efficiency is often questioned (e.g. McSorley & Walter 1991), as the number of insects extracted can be affected by soil moisture content, whether the soil core is inverted or in its original position, and whether it is intact or broken up. Hammer (1944) found that to extract maximum numbers of Collembola it was necessary to maintain the core intact and to invert it. Inversion appears to allow animals to leave the soil by natural passages, such as earthworm burrows, which open to the surface. Another problem is that condensation can form on the inside of the soil container and small animals can become trapped in this and so not be counted (Haarløv 1947). Furthermore, a particular problem, especially with Tullgren-type extractors with high temperature gradients (e.g. Crossley & Blair 1991) is that the temperature generated inside the core may be detrimental to the insect being sampled. It is a fact that big funnels extract relatively more large invertebrates, which Ausden (1996) attributed to the desiccation of microarthropods in large funnels. It is best to run the extractor with a low temperature gradient and to prevent the soil from drying out. A simple alteration to the standard Tullgren funnel is to use a very low wattage light bulb (e.g. 10 W), and to place polyethylene film over the sample container. Indeed, a very simple demonstration of the importance of these

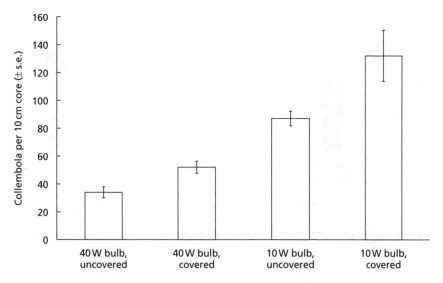

Fig. 2.6 Collembola abundance in a rye grass *Lolium perenne* pasture soil, as measured by Tullgren extraction. Each sample was left for one week. Use of too powerful a bulb (40 W) significantly reduces the numbers of animals recorded, while the use of a polyethylene protective covering over the sample increases the numbers obtained. Data from Gange (unpublished).

modifications can be seen in Fig. 2.6. Here, the use of too powerful a bulb and no protective covering dramatically reduced the number of Collembola obtained from soil samples.

With any behavioral extraction method, there must be a trade-off between extraction time and the accuracy of the method. Thus, the use of a high temperature gradient will speed up the extraction process, but may underestimate numbers if many microarthropods die without leaving the soil. Use of a lower temperature gradient means a longer extraction period, but higher estimates of abundance. However, a further problem with these systems is that in the time it takes for the insects to be persuaded to leave the soil, considerable reproduction can have taken place. The slower the process, the worse this situation is likely to become. This problem was noted by Pontin (1978), who suggested that subterranean aphids could produce a large number of offspring while still in the sampling containers, leading to overestimates of population size.

An excellent comparison of a behavioral and a passive extraction method for root aphids is provided by Salt et al. (1996). Here, Tullgren funnels were compared with flotation to extract the subterranean aphids *Pachypappa* spp and *Pachypappella* spp from Sitka spruce *Picea sitchensis* plantation soil (Fig. 2.7). In Tullgren funnels, the majority of the aphids extracted were first-instar nymphs, while in water flotation the majority of the aphids were adults and late-instar nymphs. Tullgren estimates of total abundance were significantly higher than

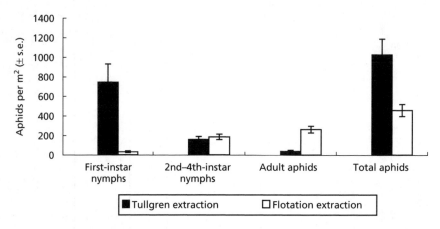

Fig. 2.7 A comparison of Tullgren extraction with flotation, for measuring the abundance of spruce root aphids *Pachypappa* spp and *Pachypappella* spp. Tullgren extraction produces higher overall estimates, but these are almost entirely first-instar nymphs. Flotation reveals very few small, first-instar aphids, but records more adults. Drawn from data in Salt et al. (1996).

flotation estimates (Fig. 2.7), caused by the large numbers of first instars obtained by this method. Salt et al. (1996) suggest that reproduction had occurred in the funnels, leading to the high proportion of first instars, but also acknowledge that flotation underestimated first-instar numbers, because it is virtually impossible to separate such small animals from the organic debris floating on the surface. This paper emphasizes why neither active nor passive extraction methods are ideal for small root-feeding insects. In general, the method used should be commensurate with the biology and size of the insect being sampled. Thus, for larger larvae, which are not close to pupation, Tullgren funnels or their equivalent are very efficient. However, for smaller larvae or insects, inactive stages, or actively reproducing adult insects, flotation is a better choice, with the proviso that great care must be taken to ensure that all individuals, no matter how small, are found. For the latter scenario, wet-sieving is likely to represent the best way of ensuring that (for example) first-instar aphid nymphs are sampled efficiently.

Laboratory visualization methods

While not strictly sampling methods, a number of techniques have been used for the examination of insect distribution and behavior in soils. These methods have not been widely used, but offer a lot of promise for the understanding of insect responses to soil parameters such as moisture content and temperature. Improved knowledge of rhizophagous insect response to biotic and abiotic factors in the soil will enable improved targeting and efficacy of pesticides against injurious species and a clearer understanding of the interactions between these insects and their host plants in natural situations.

To understand which insects were feeding on clover roots, Baylis et al. (1986) labeled roots with ^{32}P and then used autoradiography to see which members of the soil fauna were radioactive. This method would be useful if one were simply trying to determine the structure of the rhizophagous community associated with one plant species, but it is of little use for the determination of host plant preferences in a given community. More recently, Briones et al. (1999) have used carbon stable isotope analysis to determine the feeding of two collembolan species associated with leaf litter. This method assumes that the isotopic composition of the body tissue of microarthropods gives an accurate estimate of the δ^{13}C value of their diet. With this approach, Briones et al. (1999) were able to show feeding preferences for organic matter derived from maize, a C_4 plant (or microorganisms growing on it), compared with matter derived from C_3 plants. This technique represents an important advance in soil biological research and should be applicable to rhizophagous insects, as has already been achieved with earthworms (e.g. Schmidt et al. 1997).

A technique for insect behavioral observation was presented by Lussenhop et al. (1991), who advocated the use of video technology for the observation *in situ* of subterranean insects. The method does allow for a considerable amount of visual observation time to be achieved. However, the manner in which the experimental units (biotrons) are set up may be open to criticism in that they generally involve some form of glass observation plate, against which insects and roots may show unnatural behavior. Nevertheless, for the observation of small organisms and the detection of their feeding behavior, this method does offer a number of opportunities. Direct observation of southern corn rootworm larvae *Diabrotica undecimpunctata* was successfully used by Brust (1991) to monitor predation of larvae in the soil and enabled a species of *Lasius* (Hymenoptera: Formicidae) to be identified as the main predator.

To overcome the problem of soil disturbance or insertion of observation chambers into soil, Villani and Gould (1986) and Villani and Wright (1988) developed the use of radiography for direct observation. Intact blocks of soil were subjected to X-ray analysis and, as the pictures in Villani and Wright (1988) demonstrate, individual scarab larvae could be clearly seen. The larvae used in these experiments were all large; smaller individuals or species may be impossible to detect by this method, so a recourse to hand-sorting must be made (Villani & Nyrop 1991). Nevertheless, Harrison et al. (1993) applied X-ray computed tomography to the study of the smaller pecan weevil *Curculio caryae* and were able to record the burrowing activity of this insect. The X-ray technique is very useful for documenting the responses of larvae to changes in abiotic parameters, such as soil moisture, and is considerably less time-consuming than handsorting the soil to determine larval positions. Such observation methods are particularly important for documenting the behavior of larvae within a soil profile. Results such as those of Villani and Nyrop (1991) clearly show differences in the behavioral patterns of two species of chafer grub (Coleoptera: Scarabaeidae) and could be used to target insecticides more efficiently in time and space in the field.

Conclusions

The difficulty of sampling subterranean insects has undoubtedly led to a lack of study by ecologists. However, a number of methods are available for their study, and a summary of the decisions needed to be taken is given in Fig. 2.8. Before starting any sampling program, it is wise to understand as much about the biology of the species involved as possible. Many species have adult stages which are free-flying above ground.

Developing a sampling program for these is a good start, and considerably easier than searching for the larval forms in the soil. In situations such as grassland, predators such as birds and mammals provide an excellent indicator of larval presence in the soil. Other visible signs in the host plant are drought stress (though not necessarily in hot dry conditions) and poor plant growth not attributable to foliar feeders or pathogenic fungi.

There are few *in situ* extraction methods in the field. Brine pipes work well for tipulid larvae and baits are an under-used method of determining larval presence. All other sampling methods are destructive. Hand-sorting of soil, whether in field or laboratory, is the most accurate method, but is time-consuming and tedious. For internal root-feeders, there is no other way than excavating the root system and dissecting it. Passive extraction methods, generally involving some form of flotation, are useful for inactive stages, very small insects, and actively reproducing adults. Great care must be taken to separate things such as Collembola or first-instar aphids from soil debris; the use of wet-sieving may help in the capture of these individuals. Hydrocarbon adhesion is excellent, though surprisingly under-used.

Active extraction methods rely on heat and light to drive insects out of the soil. They are good for large, active insects but do not sample inactive stages. They have been widely used for the extraction of small insects, but there are several problems with this approach. Adult insects such as aphids can produce considerable numbers of offspring within the apparatus, leading to erroneous estimates of population size and structure. Too high a temperature gradient in the soil core can kill small insects such as Collembola, leading to underestimates of abundance.

Several methods of subterranean insect observation have been developed, the most promising of which is radiographic imaging. This is very good for larger insects, but needs to be refined to detect small individuals. The use of carbon stable isotopes offers great promise for the future.

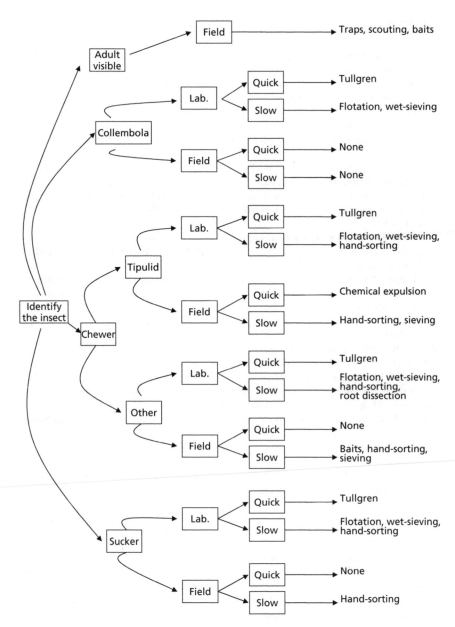

Fig. 2.8 Schematic diagram showing the decisions that need to be taken to decide on a particular sampling method appropriate for any subterranean insect. First, identify the insect; having done so, the nature of its life history (sucker, chewer, etc.) needs to determined. Then, one must ask if the extraction procedure will take place in the laboratory (lab.) or field. The final step is to decide whether answers are required with relative ease (quick), perhaps at the expense of complete accuracy, or whether some time can be devoted to the procedure, to ensure that is as accurate as possible (slow). Each extraction procedure, with respective advantages and disadvantages, is described in the text. For some groups in some situations (e.g. Collembola in the field) there is no realistic method available.

References

Allsopp, P.G. (1991) Binomial sequential sampling of adult *Saccharicoccus sacchari* on sugarcane. *Entomologia Experimentalis et Applicata*, **60**, 213–218.

Arnold, A.J. (1994) Insect sampling without nets, bags or filters. *Crop Protection*, **13**, 73–76.

Ausden, M. (1996) Invertebrates. In *Ecological Census Techniques: a Handbook* (ed. W.J. Sutherland), pp. 139–177. Cambridge University Press, Cambridge.

Badenhausser, I. & Lerin, J. (1999) Binomial and numerical sampling for estimating density of *Baris coerulescens* (Coleoptera: Curculionidae) on oilseed rape. *Journal of Economic Entomology*, **92**, 875–885.

Baylis, J.P., Cherrett, J.M., & Ford, J.B. (1986) A survey of the invertebrates feeding on living clover roots (*Trifolium repens* L) using ^{32}P as a radiotracer. *Pedobiologia*, **29**, 201–208.

Belcher, D.W. (1989) Influence of cropping systems on the number of wireworms (Coleoptera: Elateridae) collected in baits in Missouri cornfields. *Journal of the Kansas Entomological Society*, **62**, 590–592.

Bernklau, E.J. & Bjostad, L.B. (1998) Reinvestigation of host location by western corn rootworm larvae (Coleoptera: Chrysomelidae): CO_2 is the only volatile attractant. *Journal of Economic Entomology*, **91**, 1331–1340.

Blank, R.H., Bell, D.S., & Olson, M.H. (1986) Differentiating between black field cricket and black beetle damage in Northland pastures under drought conditions. *New Zealand Journal of Experimental Agriculture*, **14**, 361–367.

Blasdale, P. (1974) A method of turf sampling and extraction of leatherjackets. *Plant Pathology*, **23**, 14–16.

Blossey, B. (1993) Herbivory below ground and biological weed control: life history of a root-boring weevil on purple loosestrife. *Oecologia*, **94**, 380–387.

Bracken, G.K. (1988) Seasonal occurrence and infestation potential of cabbage maggot, *Delia radicum* (L.) (Diptera: Anthomyiidae), attacking Rutabaga in Manitoba as determined by captures of females in water traps. *Canadian Entomologist*, **120**, 609–614.

Briones, M.J.I., Ineson, P., & Sleep, D. (1999) Use of δ^{13}C to determine food selection in collembolan species. *Soil Biology and Biochemistry*, **31**, 937–940.

Brown, V.K. & Gange, A.C. (1990) Insect herbivory below ground. *Advances in Ecological Research*, **20**, 1–58.

Brust, G.E. (1991) A method for observing belowground pest–predator interactions in corn agroecosystems. *Journal of Entomological Science*, **26**, 1–8.

Clements, R.O. (1984) Control of insect pests in grassland. *Span*, **27**, 77–80.

Clements, R.O., Bentley, B.R., & Nuttall, R.M. (1987) The invertebrate population and response to pesticide treatment of two permanent and two temporary pastures. *Annals of Applied Biology*, **111**, 399–407.

Cordo, H.A., DeLoach, C.J., & Ferrer, R. (1995) Host range of the Argentine root borer *Carmenta haematica* (Ureta) (Lepidoptera: Sesiidae), a potential biocontrol agent for snakeweeds (*Gutierrezia*) in the United States. *Biological Control*, **5**, 1–10.

Crossley, D.A. & Blair, J.M. (1991) A high-efficiency, "low-technology" Tullgren-type extractor for soil microarthropods. *Agriculture, Ecosystems and Environment*, **34**, 187–192.

De Barro, P.J. (1991) Sampling strategies for above and below ground populations of *Saccharicoccus sacchari* (Cockerell) (Hemiptera: Pseudococcidae) on sugarcane. *Journal of the Australian Entomological Society*, **30**, 19–20.

Dosdall, L.M., Herbut, M.J., & Cowle, N.T. (1994) Susceptibilities of species and cultivars of canola and mustard to infestation by root maggots (*Delia* spp.) (Diptera: Anthomyiidae). *Canadian Entomologist*, **126**, 251–260.

Dutcher, J.D. & All, J.N. (1979) Damage impact of larval feeding by the grape root borer in a commercial Concord grape vineyard. *Journal of Economic Entomology*, **72**, 159–161.

East, R. & Willoughby, B.E. (1983) Grass grub (*Costelytra zealandica*) population collapse in the northern North Island. *New Zealand Journal of Agricultural Research*, **26**, 381–390.

Elvin, M.K. & Yeargan, K.V. (1985) Spatial distribution of clover root curculio, *Sitona hispidulus* (Fabricius) (Coleoptera: Curculionidae) eggs in relation to alfalfa crowns. *Journal of the Kansas Entomological Society*, **58**, 346–348.

Fermanian, T.W., Shurtleff, M.C., Randell, R., Wilkinson, H.T., & Nixon, P.L. (1997) *Controlling Turfgrass Pests*. Prentice Hall, Upper Saddle River, NJ.

Forrester, G.J. (1993) Resource partitioning between two species of *Ceutorhynchus* (Coleoptera: Curculionidae) on *Echium plantagineum* in a Mediterranean habitat. *Bulletin of Entomological Research*, **83**, 345–351.

Fowler, R.F. & Wilson, L.F. (1971) White grub populations, *Phyllophaga* spp. in relation to damaged red pine seedlings in Michigan and Wisconsin plantations (Coleoptera: Scarabaeidae). *Michigan Entomologist*, **4**, 23–28.

Gange, A.C., Brown, V.K., Barlow, G.S., Whitehouse, D.M., & Moreton, R.J. (1991) Spatial distribution of garden chafer larvae in a golf tee. *Journal of the Sports Turf Research Institute*, **67**, 8–13.

Geurs, M., Bongers, J., & Brussard, L. (1991) Variations of the heptane flotation method for improved efficiency of collecting microarthropods from a sandy loam soil. *Agriculture, Ecosystems and Environment*, **34**, 213–221.

Goldson, S.L. & Proffitt, J.R. (1988) The effect of lucerne age on its sensitivity to damage by *Sitona discoideus* Gyllenhall (Coleoptera: Curculionidae) in Canterbury. In *Proceedings of the 5th Australasian Conference on Grassland Invertebrate Ecology* (ed. P.P. Stahle), pp. 323–331. University of Melbourne, Victoria.

Goldson, S.L., Frampton, E.R. & Proffitt, J.R. (1988) Population dynamics and larval establishment of *Sitona discoideus* (Coleoptera: Curculionidae) in New Zealand lucerne. *Journal of Applied Ecology*, **25**, 177–195.

Gratwick, M. (1992) *Crop Pests in the UK*. Chapman & Hall, London.

Haarløv, N. (1947) A new modification of the Tullgren apparatus. *Journal of Animal Ecology*, **16**, 115–121.

Hammer, M. (1944) Studies on the Oribatids and Collemboles of Greenland. *Meddelelser om Grønland*, **141**, 1–210.

Hanula, J.L. (1993) Vertical distribution of black vine weevil (Coleoptera, Curculionidae) immatures and infection by entomogeneous nematodes in soil columns and field soil. *Journal of Economic Entomology*, **86**, 340–347.

Harcourt, D.G. & Binns, M.R. (1989) Sampling technique for larvae of the alfalfa snout beetle, *Otiorhynchus ligustici* (Coleoptera: Curculionidae). *Great Lakes Entomologist*, **22**, 121–126.

Harrison, R.D., Gardner, W.A., Tollner, W.E., & Kinard, D.J. (1993) X-ray computed tomography studies of the burrowing behavior of fourth instar pecan weevil (Coleoptera, Curculionidae). *Journal of Economic Entomology*, **86**, 1714–1719.

Havukkala, I., Harris, M.O., & Miller, J.R. (1992) Onion fly egg distribution in soil: a new sampling method and the effect of two granular insecticides. *Entomologia Experimentalis et Applicata*, **63**, 283–289.

Hawthorne, D.J. & Dennehy, T.J. (1991) Reciprocal movement of grape phylloxera (Homoptera: Phylloxeridae) alates and crawlers between two differentially phylloxera-resistant grape cultivars. *Journal of Economic Entomology*, **84**, 230–236.

Hopkin, S.P. (1997) *Biology of the Springtails (Insecta: Collembola)*. Oxford University Press, Oxford.

House, G.J. & Alzugaray, M.D.R. (1989) Influence of cover cropping and no-tillage practices on community composition of soil arthropods in a North Carolina agroecosystem. *Environmental Entomology*, **18**, 302–307.

Kethley, J. (1991) A procedure for extraction of microarthropods from bulk samples of sandy soils. *Agriculture, Ecosystems and Environment*, **34**, 193–200.

Klironomos, J.N. & Kendrick, W.B. (1995) Stimulative effects of arthropods on endomycorrhizas of sugar maple in the presence of decaying litter. *Functional Ecology*, **9**, 528–536.

Labuschagne, L. (1999) Black vine weevil—the Millennium bug? *Antenna*, **23**, 213–220.

Lauenstein, G. (1986) Leatherjackets as pests of grasslands: their biology and control. *Zeitschrift für Angewandte Entomologie*, **73**, 385–432.

Lussenhop, J. (1971) A simplified canister-type soil arthropod extractor. *Pedobiologia*, **11**, 40–45.

Lussenhop, J., Fogel, R., & Pregitzer, K. (1991) A new dawn for soil biology: video analysis of root–soil–microbial–faunal interactions. *Agriculture, Ecosystems and Environment*, **34**, 235–249.

Maron, J.L. (1998) Insect herbivory above- and belowground: individual and joint effects on plant fitness. *Ecology*, **79**, 1281–1293.

Masters, G.J. (1995) The impact of root herbivory on plant and aphid performance: field and laboratory evidence. *Acta Oecologia*, **16**, 135–142.

McSorley, R. & Walter, D.E. (1991) Comparison of soil extraction methods for nematodes and microarthropods. *Agriculture, Ecosystems and Environment*, **34**, 201–207.

Müller, H., Stinson, C.S.A., Marquardt, K., & Schroeder, D. (1989) The entomofaunas of roots of *Centaurea maculosa* Lam., *C. diffusa* Lam., and *C. vallesiaca* Jordan in Europe. *Journal of Applied Entomology*, **107**, 83–95.

Murray, P.J. & Clements, R.O. (1995) Distribution and abundance of three species of *Sitona* (Coleoptera: Curculionidae) in grassland in England. *Annals of Applied Biology*, **127**, 229–237.

Penev, L.D. (1992) Qualitative and quantitative spatial variation in soil wire-worm assemblages in relation to climatic and habitat factors. *Oikos*, **63**, 180–192.

Pontin, A.J. (1978) The numbers and distribution of subterranean aphids and their exploitation by the ant *Lasius flavus* (Fabr.). *Ecological Entomology*, **3**, 203–207.

Roberts, R.J., Ridsdill Smith, T.J., Porter, M.R., & Sawtell, N.L. (1982a) Fluctuations in the abundance of pasture scarabs over an 18-year period of light trapping. In *Proceedings of the 3rd Australasian Conference on Grassland Invertebrate Ecology* (ed. K.E. Lee), pp. 75–79. SA Government Printer, Adelaide.

Roberts, R.J., Campbell, A.J., Porter, M.R., & Sawtell, N.L. (1982b) The distribution and abundance of pasture scarabs in relation to *Eucalyptus* trees. In *Proceedings of the 3rd Australasian Conference on Grassland Invertebrate Ecology* (ed. K.E. Lee), pp. 207–214. SA Government Printer, Adelaide.

Rogers, C.E. (1985) Bionomics of *Eucosma womonana* Kearfott (Lepidoptera: Tortricidae), a root borer in sunflowers. *Environmental Entomology*, **14**, 42–44.

Salt, D.T., Major, E., & Whittaker, J.B. (1996) Population dynamics of root aphids feeding on Sitka spruce in two commercial plantations. *Pedobiologia*, **40**, 1–11.

Schmidt, O., Scrimgeour, C.H., & Handley, L.L. (1997) Natural abundance of ^{15}N and ^{13}C in earthworms from a wheat and wheat–clover field. *Soil Biology and Biochemistry*, **29**, 1301–1308.

Seastedt, T.R. (1984) Belowground macroarthropods of annually burned and unburned tallgrass prairie. *American Midland Naturalist*, **111**, 405–408.

Sheppard, A.W., Aeschlimann, J.P., Sagliocco, J.L., & Vitou, J. (1995) Below-ground herbivory in *Carduus nutans* (Asteraceae) and the potential for biological control. *Biological Control Science and Technology*, **5**, 261–270.

Smart, L.E., Blight, M.M., Pickett, J.A., & Pye, B.J. (1994) Development of field strategies incorporating semiochemicals for the control of the pea and bean weevil, *Sitona lineatus* L. *Crop Protection*, **13**, 127–135.

Southwood, T.R.E. & Henderson, P.A. (2000) *Ecological Methods*. 3rd edn. Blackwell Science, Oxford.

Stewart, R.M. & Kozicki, K.R. (1987) DIY assessment of leatherjacket numbers in grassland. *Proceedings of Crop Protection in Northern Britain Conference*, pp. 349–353. Scottish Crop Research Institute, Dundee.

Strong, D.R., Maron, J.L., Connors, P.G., Whipple, A., Harrison, S., & Jefferies, R.L. (1995) High mortality, fluctuation in numbers, and heavy subterranean insect herbivory in bush lupine, *Lupinus arboreus*. *Oecologia*, **104**, 85–92.

Villani, M.G. & Gould, F. (1986) Use of radiographs for movement analysis of the corn wireworm, *Melanotus communis* (Coleoptera: Elateridae). *Environmental Entomology*, **15**, 462–464.

Villani, M.G. & Nyrop, J.P. (1991) Age-dependent movement patterns of Japanese beetle and European chafer (Coleoptera: Scarabeidae) grubs in soil-turfgrass microcosms. *Environmental Entomology*, **20**, 241–251.

Villani, M.G. & Wright, R.J. (1988) Use of radiography in behavioral studies of turfgrass-infesting scarab grub species (Coleoptera: Scarabaeidae). *Bulletin of the Entomological Society of America*, **34**, 132–144.

Walter, D.E., Kethley, J., & Moore, J.C. (1987) A heptane flotation method for recovering microarthropods from semiarid soils, with comparison to the Merchant–Crossley high-gradient extraction method and estimates of microarthropod biomass. *Pedobiologia*, **30**, 221–232.

Ward, R.H. & Keaster, A.J. (1977) Wireworm baiting: use of solar energy to enhance early detection of *Melanotus depressus*, *M. verberans* and *M. mellillus* in midwest cornfields. *Journal of Economic Entomology*, **70**, 403–406.

White, E.G. & Sedcole, J.R. (1993) A study of the abundance and patchiness of cicada nymphs (Homoptera: Tibicinidae) in a New Zealand sub-alpine shrub-grassland. *New Zealand Journal of Ecology*, **20**, 38–51.

Whitfield, G.H. & Grace, B. (1985) Cold hardiness and overwintering survival of the sugarbeet root maggot (Diptera: Otitidae) in southern Alberta. *Annals of the Entomological Society of America*, **78**, 501–505.

Williams, P. Stahle, P.P., Gagen, S.J., & Murphy, G.D. (1982) Strategies for control of the black field cricket, *Teleogryllus commodus* (Walker). In *Proceedings of the 3rd Australasian Conference on Grassland Invertebrate Ecology* (ed. K.E. Lee), pp. 365–369. SA Government Printer, Adelaide.

Index of methods and approaches

Methodology	Topics addressed	Comments
Field extraction methods		
Chemical	Use of irritant chemicals to expel insects from soil.	May be toxic to user; hard to produce density estimates; may kill specimens.
Behavioral	Use of dark covers on a damp soil surface.	Only works for tipulid larvae; not quantifiable; produces live specimens for culture.
Baits	Use of food attractants to obtain active larvae or adults.	Not quantifiable; good for obtaining live specimens for culture.
Hand sorting	Systematic sifting through a defined volume of soil.	Quantifiable, but laborious; not really suitable for small insects.
Laboratory extraction methods		
Root dissection	Removal of insects from inside a root system.	Quantifiable, but laborious.
Flotation	Immersion of a defined volume of soil in a liquid (usually water or brine).	Quick; good for inactive stages, but quantification hampered because dead specimens are obtained too.
Behavioral	Use of temperature and light to expel insects from soil.	Quantifiable, but does not sample inactive stages; Soil factors and operating conditions affect results.
Laboratory visualization methods		
	Use of stable isotopes, video or X-ray observation techniques.	Not quantifiable; good for behavioral studies.

CHAPTER 3

Pitfall trapping in ecological studies

B.A. WOODCOCK

Introduction

Pitfall trapping is one of the oldest, most frequently used, and simple of all invertebrate sampling techniques, and yet it is also one of the most frequently misused. This is because pitfall trapping for surface-active invertebrates is prone to producing non-quantitative data, particularly if used without considering the problems associated with this sampling technique. This chapter considers the practical aspects of pitfall trap design and installation, and then discusses the theoretical basis of pitfall trapping that must be incorporated into experimental design and analysis if this method is to be used successfully in ecological studies. This chapter provides a comprehensive review of the methods and theory behind pitfall trapping and provides information on suitable experimental protocols to be applied in pitfall trapping sampling programs.

The choice of sampling technique in any invertebrate sampling program is integral to the success of the project. The decision will not only determine the type of invertebrates that are sampled, but also in what numbers, over what spatial scale, and how quantitative the data produced are. What sampling method will also be influenced by more pragmatic decisions based on the money and time available to each project. Sampling epigeal invertebrates, those species active on the soil surface, is a good example of where these problems must be considered carefully to produce an effective sampling program within the means of the project.

The most commonly sampled epigeal invertebrates are the ground beetles (Coleoptera: Carabidae), rove beetles (Coleoptera: Staphylinidae), wandering spiders (e.g. Aranae: Lycosidae and Clubionidae), and ants (Hymenoptera: Formicidae). These groups are characterized as highly active, mostly polyphagous, invertebrate predators (Greenslade 1973, Uetz & Unzicker 1976, Thiele 1977, Frank 1991). These characteristics can make these groups hard to sample using many techniques. Their active nature means that they while they may show specific habitat associations, a spatially and temporally restricted sampling technique may fail to catch many species. Also, polyphagous predators are not associated with either a particular host plant, or specific prey species. Any sampling strategy that could be used to target such an association would also be ineffective.

The sampling technique used most frequently to collect epigeal invertebrates is pitfall trapping. The technique was first developed by Hertz (1927), and shortly after by Barber (1931), who used open-top containers buried with the rim level to the substrate surface, so that anything falling into the container becomes trapped. While originally conceived as a qualitative technique, the potential of the method for quantitatively sampling epigeal invertebrate populations was soon realized (Fichter 1941). From this inauspicious start pitfall traps have come to dominate epigeal invertebrate sampling (Uetz & Unzicker 1976, Thiele 1977). They have been used in practically every terrestrial habitat, from deserts (Thomas & Sleeper 1977, Faragalla & Adam 1985), to forests (Niemelä et al. 1986, Spence & Niemelä 1994), to caves (Barber 1931). The technique has also been used to obtain information on the structure of invertebrate communities (Hammond 1990, Jarosík 1992), habitat associations (Honêk 1988, Hanski & Niemelä 1990), activity patterns (Ericson 1978, Den Boer 1981), spatial distribution (Niemelä 1990), relative abundances (Desender & Maelfait 1986a, Mommertz et al. 1996), total population estimates (Gist & Crossley 1973, Mommertz et al. 1996), and distribution ranges (Barber 1931, Giblin-Davis et al. 1994). Pitfall trapping also plays a role in some pest monitoring programs (Kharboutli & Mack 1993, Obeng-Ofori 1993, Rieske & Raffa 1993, Simmons et al. 1998). For the last three-quarters of a century pitfall traps have proved to be one of the most versatile, useful, and widely used invertebrate sampling techniques.

The wide-scale adoption of this technique is due to a number of factors. Basic traps are cheap, and normally require no specialized manufacturing process. Traps are also easy to transport (Lemieux & Lindgren 1999), and quick to install. Perhaps one of the greatest advantages of pitfall traps is that they will sample continuously, requiring only periodic emptying. This not only removes biases associated with other techniques that sample at one point in time (Topping & Sunderland 1992), but also allows large numbers of invertebrates to be caught over an entire season with minimal effort. This makes the technique particularly useful for sampling invertebrate occurring at low densities (Melbourne 1999). The low levels of disturbance, both physically and aesthetically, which pitfall trap installation and collection causes has made them useful for sampling environmentally sensitive areas (Melbourne 1999).

Unfortunately, while Fichter (1941) was the first to recognize the values of pitfall trapping as a quantitative tool, he was also the first to acknowledge its failings. As each species has the potential to respond uniquely to pitfall traps, the rates at which they are caught can vary. The proportion of each species in the traps no longer necessarily represents their relative abundance in the sampling habitat. If pitfall traps are to be used in ecological studies it is necessary that field biologists have a comprehensive understanding of both the advantages and disadvantages of this method. This must include an understanding not only of different trap designs, and sampling strategies, but also of what can be done to improve the quantitative nature of the data.

This chapter will first consider the various designs and modifications that have been developed for pitfall traps, and the implication of how design impacts on the capture rates of different species. Secondly the integral role that sampling strategy plays in reducing biases associated with pitfall traps is reviewed. Finally the concept of activity–abundance as a tool for the quantitative interpretation of pitfall traps is described. Pitfall traps are a valuable tool in ecology, and like any tool they must be used carefully with an understanding of their flaws if their use is not to be open to criticism.

Pitfall traps, their designs and application

It would seem that every study uses a novel design of pitfall trap (Table 3.1); different sizes, shapes, and construction material are normal. This is often due to the immediate availability of materials for each study, and has led to a high level of inconsistency between different research projects. However, pitfall trap design can influence the capture and retention of different species. Although there is no right or wrong design, knowledge of the effectiveness of different trap types will allow traps to be tailored to the experimental requirements and

Table 3.1 Examples of the application of various modifications that have been developed for pitfall traps.

Trap type	Use in ecological studies	Examples
Conventional	Habitat associations; spatial patterns; community structure; mark–recapture studies	Ericson (1978, 1979), Niemelä et al. (1990), Niemelä et al. (1992), Dennis et al. (1997)
Baited traps	For aggregated or rare species; pest monitoring	Walsh (1933), Rieske & Raffa (1993)
Time sorting traps	Determination of diel activity patterns	Luff (1978), Kegel (1990), Chapman & Armstrong (1997)
Barrier trapping	Prevents immigration; samples from a defined area; closed mark–recapture experiments	Baars (1979), Desender & Maelfait (1986a), Momertz et al. (1996)
Drift fences	Increasing overall catch; identifying directional movement	Smith (1976), Morrill et al. (1990), Melbourne (1999)
Ramps	Biases catch to larger individuals; reduces flooding	Bostanian et al. (1983)
Gutter traps	Increases overall catch; biasing catch towards larger species	Luff (1975, 1978), Lawrence (1982), Spence & Niemelä (1994)
Subterranean	Useful for soil active species or larval stages	Kuschel (1991), Owen (1995)

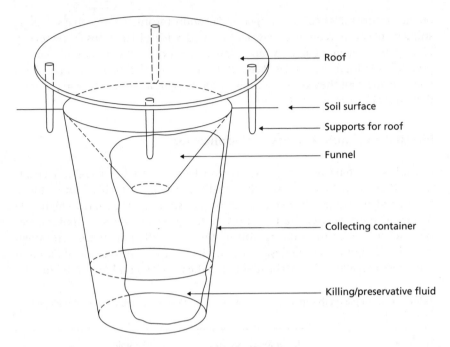

Fig. 3.1 A cut away diagram showing the design of a basic pitfall trap suitable for most epigeal invertebrate surveys. This design is only a guideline and should be modified according to the specific requirements of each sampling program and the materials available for its construction.

practical limitations of each project. This section considers what designs of pitfall trap, and what modifications, are available. Figure 3.1 shows the design of a basic pitfall trap.

Trap material

The material traps have been constructed from has almost always been determined by what is easily available at the time. Plastic is presently the most frequently used material (e.g. Honêk 1988, Niemelä 1990, Dennis et al. 1997), although prior to this both metal (e.g. Hertz 1927, Luff 1975, Smith 1976) and glass (e.g. Briggs 1960, Mitchell 1963a, Greenslade 1964) were frequently used. Species-specific responses to trap material are common (Luff 1975, Obeng-Ofori 1993). Glass is normally found to be the most effective material in terms of numbers of individuals captured, and is almost always superior to metals (Luff 1975, Obeng-Ofori 1993). The superiority of glass over other materials is particularly important if live catches are required. Glass provides few abrasions which insects can use to escape, although if a killing agent or preservative is used other materials like plastic are likely to be as effective (Luff 1975). However, glass is heavy, fragile, and hard to use in the construction of more specialized trap designs. Rieske and Raffa (1993) suggested the use of Teflon to minimize surface grip within traps, and so maximize invertebrate retention where other trap materials are used. Trap color was found by Greenslade (1964) to have no effect on the catch of Carabidae.

Trap shape and size

While default rather than design has dictated that traps are normally circular there is no reason why they should always be this shape. However, there does not appear to be any advantage gained, in terms of overall trap efficiency, by deviating from this standard circular design (Spence & Niemelä 1994). The only exception to this is with some of the more extreme pitfall trap designs like gutter traps, which are described below.

However, different species do respond in different ways to trap shape, even when there is no difference in overall trap perimeter (Baars 1979, Spence & Niemelä 1994). As a rule where communities are being compared in a given study trap shape should therefore be kept constant. If the shape must be altered, then the perimeter should be kept as constant as possible (Luff 1975, Baars 1979). Independent of shape there is considerable variation in the size of traps. For circular traps, diameter may vary from as little as 1.8 cm (Greenslade & Greenslade 1971, Greenslade 1973, Abensberg-Traut & Steven 1995) to over 25 cm (Morrill et al. 1990). A modal diameter determined from the literature is found to be around 6–8 cm. Trap depth is variable but tends to be at least 8–10 cm; anything below this is likely to be particularly prone to escape. Larger pitfall traps catch more individuals than smaller traps (Luff 1975, Baars 1979, Abensberg-Traut & Steven 1995), but this increase in catch is not necessarily proportional to trap diameter (Morrill et al. 1990).

Baars (1979) used simulations to show that when comparing sites the number and shape of traps was not important providing that total perimeter area of traps was constant. However, this will vary with the target taxa. Abensberg-Traut and Steven (1995) suggested that for a comparable area many small traps may be more efficient than a few large ones, particularly for species that are highly aggregated like ants. In the case of very small diameter pitfall traps, larger species may be too big to be trapped, and will be excluded from samples (Luff 1975, Abensberg-Traut & Steven 1995). A theoretical basis exists for correcting catch sizes of traps of different diameters and is discussed by Luff (1975); this may even be applied to traps that differ in shape. Although this correction method has been used to compare traps of dissimilar shapes, and sizes (Luff 1975, Spence & Niemelä 1994), Scheller (1984) found that it was not always effective.

Roofs

Roofs covering the mouth of traps have been used in many pitfall trap studies, providing protection from the elements (e.g. Fichter 1941, Honêk 1988, Hammond 1990). The roofs are supported 3–4 cm from the soil surface to allow free access to the traps. Roofs are useful for traps both with and without preservatives, since rain will cause preservative dilution (Hammond 1990) or may drown live insects (Briggs 1960). Roofs will also prevent debris, which provides escape routes for insects, falling into traps (Uetz & Unzicker 1976, Morrill et al. 1990). Access by birds and small mammals stealing the contents of the traps is prevented (Briggs 1960, Mitchell 1963a), as is their consumption of toxic preservatives (Marshall & Doty 1990, Hall 1991). Wire barriers have also been used to prevent the accidental capture of small vertebrates in traps.

Unfortunately roofs cause bias in the catches of pitfall traps (Joose 1965, Morrill 1975, Baars 1979). The use of transparent materials for roof coverings, however, minimizes the influence of roofs on the catches of invertebrates (Joose 1965, Baars 1979).

Funnels

Funnels have been used in many studies (e.g. Gist & Crossley 1973, Faragalla & Adam 1985, Morrill et al. 1990, Clarke & Bloom 1992), normally to reduce escape where no preservative is used (Vlijm et al. 1961, Uetz & Unzicker 1976). Funnels also reduce desiccation of the trap contents and prevent vertebrate interference (Briggs 1960, Mitchell 1963a). Funnels are placed at the opening of the traps, and work on the same principle as lobster pots, making escape difficult. Both capture rate and trap efficiency will probably be influenced by the presence of a funnel, although as of yet there is no evidence for this.

The trap rim

The protrusion of the trap rim can repel invertebrates, although this is dependent on invertebrate size (Morrill et al. 1990, Good & Giller 1991). It has been suggested that the trap rim should be placed 1–7 mm below the level of the substrate surface (Good & Giller 1991, Obeng-Ofori 1993). However, this is normally awkward, and for large numbers of traps may be impractical. As a general rule it is necessary to at least get the rim of the trap level with the substrate surface. After heavy rain erosion of soil around trap rims can occur, requiring that the soil be replaced (Hammond 1990).

Killing agents, preservatives, and detergents

Mark and recapture experiments require that the sampling procedure does not kill the catch (e.g. Ericson 1977, Parmenter & MacMahon 1989, Thomas et al. 1998). However, once confined within the trap predatory species will feed on anything small enough to eat, and this may include target species of the sampling program (Mitchell 1963a, Greenslade & Greenslade 1971). Even with daily collection of the traps this can still be a problem. One solution is to place soil, or another suitable substrate, in the trap to provide a refuge for smaller species (Ericson 1979, Honêk 1988). Wire meshes have also been used to separate large and small species (Lawrence 1982, Niemelä et al. 1992).

When it is not necessary to keep the catch alive a killing solution is normally used, to stop predation and reduce levels of escape (Uetz & Unzicker 1976, Curtis 1980, Waage 1985, Holopainen & Varis 1986, Lemieux & Lindgren 1999). The solution will also normally act as a preservative, reducing the need for regular collections. The choice of solution (Lemieux & Lindgren 1999) is dependent on: its effectiveness in preventing decay and fouling of specimens; the speed with which it kills insects before they can escape; whether it will remain non-volatile when diluted by rain, or concentrated by the sun. Other considerations include legal or health and safety requirements that may prevent the use of potentially harmful chemicals (Hall 1991) (Table 3.2).

Table 3.2 Killing fluids/preservatives that have been used in ecological studies, giving suggested concentrations and listing the advantages and disadvantages of their use in pitfall traps. The concentrations are only suggestions for a collection interval of between two and four weeks under temperate conditions. It is suggested that to all of these an unscented detergent should be added to reduce surface tension.

Preservative	Concentrations	Advantages and disadvantages
Ethylene glycol	25–50% solution	Freely available as car antifreeze. Good preservative. Toxic to birds and mammals. Attractant to some invertebrates.
Propylene glycol	25–50% solution	More expensive than ethylene glycol, but considered less toxic. Possible attractant?
Water	N/A	Freely available. Poor preservative.
Formalin	5–10% solution	Relatively freely available. Good preservative. Possible health and safety problems. Toxic. Attractant.
Saline solution	1% to saturated solution	Freely available. Reasonable preservative, but damages some specimens. Possible attractant?
Alcohol	70% solution	Freely available. Good preservative. Attractant. Volatile and will evaporate.
Acetic acid	5% solution	Freely available. Good preservative. Attractant.
Chloral hydrate	An additive to above solutions	Relatively freely available and can be used to inhibit bacterial/fungal growth. Toxic. Possible attractant?

A wide variety of solutions have been used in pitfall traps, including propylene or ethylene glycol (Digweed et al. 1995, Dennis et al. 1997), water (Briggs 1960, Holopainen 1992), alcohol (Fichter 1941, Greenslade & Greenslade 1971), formalin (Baars 1979, Desender & Maelfait 1986a), kerosene (Faragalla & Adam 1985), brine (van den Berghe 1992, Lemieux & Lindgren 1999), chloral hydrate (Hammond 1990), and benzoic/acetic acid (Scheller 1984). The quality of these preservatives is variable and has meant that few are in common use. Water for example may be freely available, and will kill invertebrates, but sample decomposition is a problem. At present one of the most commonly used preservatives in ecological experiments is ethylene glycol (antifreeze). Its popularity is based on its low cost, free availability, and good preservative and killing qualities. However, ethylene glycol is sweet-tasting and toxic to both birds and mammals, which actively consume it (Beasley 1985, Marshall & Doty 1990, Hall 1991). The less toxic propylene glycol has been proposed as an alternative, as it shares essentially the same beneficial qualities as ethylene glycol, although it is more expensive (Hall 1991).

Preservative concentration depends on the interval between collection dates. A 50-percent ethylene glycol solution is suitable for most purposes (Epstein & Kulman 1984), although if the trap is to be checked very infrequently, e.g. less than once a month, concentrations as high as 100 percent may be required (Clarke & Bloom 1992). This dilution principle can be sensibly applied to most preservatives. It is also normal to add a small quantity of unscented detergent to the killing solution to reduce surface tension. This will increase the efficiency of traps, as insects slip more easily under the surface of the killing agent/preservative.

It should be noted that almost all preservatives will act as attractants for at least some species of invertebrates. For example, in the Carabidae positive species-specific responses have been found for the preservatives formalin (e.g. Luff 1968, Scuhravy 1970, Adis & Kramer 1975, Ericson 1979, Feoktistov 1980, Scheller 1984, Holopainen & Varis 1986), ethylene/propylene glycol (Hammond 1990, Holopainen 1990, Holopainen 1992), and benzoic / acetic acid (Scheller 1984). While this will influence the relative proportions of different species caught in pitfall traps, the use of preservatives is often a necessary evil.

Baits

The use of baits is one of the only techniques in pitfall trapping that intentionally biases the catch size of different species (Walsh 1933, Greenslade 1964, Greenslade & Greenslade 1971). Their use should be strictly for qualitative analyses, such as determination of habitat association (Hanski & Niemelä 1990), or in producing total species inventories (Romero & Jaffe 1989, Hammond 1990). There is an argument for the quantitative use of baits in the analysis of population size of a single species occurring at low densities, or one that is too aggregated to ensure capture with more passive approaches to pitfall trapping. This is particularly so when monitoring pest populations where warning of population increases in a single species is valuable (Rieske & Raffa 1993, Giblin-Davis et al. 1994, Yasuda 1996). In combination with other sampling methods, baits have proved useful for sampling ants, whose aggregated distribution often makes a determination of total species richness difficult (Greenslade & Greenslade 1971, Romero & Jaffe 1989).

Solid baits, including carrion, fruit, or dung (Romero & Jaffe 1989, Hammond 1990, Hanski & Niemelä 1990, Giblin-Davis et al. 1994), are normally positioned on a platform in the middle of the trap, or suspended immediately above it. Liquid baits, such as beer or honey solutions (Greenslade & Greenslade 1971), can be placed directly into the collecting vessel, although they may be poor preservatives and so the catch may require regular collection. Other types of baits may be highly specific and are used more frequently in pest monitoring programs. Such baits have included pheromones (Yasuda 1996) and alpha- and beta-pinenes (Rieske & Raffa 1993).

Specialized designs

The evolution of the pitfall trap from its initial simple concept (Hertz 1927, Barber 1931) has produced designs that have increased catch sizes, or allowed several normally hard-to-investigate aspects of invertebrate ecology to be considered. These traps include: time-sorting traps for determining diel activity patterns (Houston 1971, Barndt 1976, Luff 1978, Chapman & Armstrong 1997); gutter traps, which are essentially highly elongated conventional pitfall traps, that will increase the overall catch size (Luff 1975, Luff 1978, Lawrence 1982, Spence & Niemelä 1994); drift fences, which are strips of metal or plastic placed on the surface to direct insects towards the pitfall trap, so increasing the overall catch or identifying the directional movement of insects (Smith 1976, Desender & Maelfait 1986a, Morrill et al. 1990, Melbourne 1999); barrier trapping, which uses normal pitfall traps in conjunction with an outer barrier preventing the immigration or emigration of invertebrates from a spatially delimited area (Gist & Crossley 1973, Baars 1979, Holopainen & Varis 1986, Desender & Maelfait 1986a, Mommertz et al. 1996); ramp traps, which use a ramp to lead up to the collection chamber, so that the trap does not need to be buried, and can be used on rocky ground (Bostanian et al. 1983, Spence & Niemelä 1994). Recently, some entomologists have experimented with subterranean pitfall traps. These traps are placed so that they are in a hole some distance below the soil surface, e.g. 10–20 cm. A column of coarse wire mesh encircling the rim of the trap and extending to the surface prevents soil falling in, while allowing insects crawling though the soil to be collected (Kuschel 1991, Owen 1995).

Sampling strategy

Independent of the actual design of the trap, the sampling strategy employed can be used to maximize the quantitative potential of a pitfall trapping program.

Sampling strategy refers to the number of traps, their spatial arrangement, and the duration of sampling. Differences in sampling strategy affect not only the proportion of the community sampled, but also the relative abundances of species caught in traps. For these reasons it is important to carefully consider sampling strategy before initiating any experiment.

Trap number

The number of traps used to obtain information from a particular sampling area is highly variable in the literature, ranging from as few as two (Melbourne 1999) to as many as 300 (Niemelä et al. 1990). This number usually depends both on the size of the area to be sampled and on the specific design of the pitfall traps. For example, Melbourne (1999) used drift fences in conjunction with pitfall traps to increase their effective perimeter, and so used few traps. In the case of Niemelä et al. (1990) 300 traps were used to sample a 19-hectare woodland.

Obrtel (1971) showed that the highest incremental increases in overall species richness occurred for the first five traps in beetle communities. However, Stein (1965) considered that fewer than 20 traps would be insufficient to determine the number of Carabidae species in a site, while Bombosch (1962) found that increasing the number of traps above 70 still caught additional species. It is likely that reliable data on the species at a site can be obtained from 12 pitfall traps when considering common temperate Carabidae fauna (Obrtel 1971). Use of preliminary studies, or previous published work, may be the most reliable method for estimating trap number, particularly in long-term studies (Uetz & Unzicker 1976). As a rule, the more pitfall traps the better an individual community will be sampled, although this must be traded off against the extra work required in processing the data.

Spatial arrangement

Traps are rarely positioned randomly within a single plot or site, due to practical problems of finding them again. Frequently used trapping patterns are linear transects (e.g. Mitchell 1963b, Honêk 1988, Good & Giller 1991, Kharboutli & Mack 1993), and grids (e.g. Ericson 1979, Epstein & Kulman 1984, Niemelä et al. 1992) (Table 3.3).

The arrangement of traps and their number can reduce overall trapping efficiency. Luff (1975) demonstrated theoretically that a correction factor should be applied to traps placed in a grid, as outer traps will shield inner traps and reduce their effective diameter. This will have the effect of reducing the sizes of catches. Scheller (1984) experimentally demonstrated this effect, but for only one carabid species. However, this does have implications when comparing sites using the same number and type of trap but with different spatial arrangements.

The separation between each trap will be dependent on the area of the sam-

Table 3.3 Frequently used spatial arrangements of pitfall traps, giving the application of these arrangements in ecological studies and the advantages and disadvantages of each approach. In all cases it is often advantageous to use an obvious marker for each trap, such as a flag, to aid in relocation.

Spatial layout	Appearance	Advantages and disadvantages
Random	○ ○○ ○ ○○ ○ ○ ○○ ○ ○○ ○	From a practical perspective traps can be extremely hard to find, although each trap can be considered as statistically independent. Trap separation is unpredictable.
Grid	○ ○	Commonly used approach as it provides good even coverage of the sampling area, while the individual plots are relatively simple to relocate. By adjusting trap separation, individual traps' statistical independence can be maintained or avoided.
Transect	○ ○ ○ ○ ○ ○	Suitable for the identification of the effects of environmental gradients on invertebrate communities. For example edge effects in fragmented woodlands or altitudinal gradients.

pling site, the number of traps, and their diameter. It may be desirable to increase trap separations, ensuring that there is an even coverage of traps over the whole of the sample plot. Divisions ranging from 0.3 m (Luff 1975) to 30 m (Honêk 1988) have been used, although separations of between 5 and 10 m are more common (Baars 1979, Holopainen 1990, Niemelä 1990, Kharboutli & Mack 1993). As trap size increases so should separation (Uetz & Unzicker 1976). It is common practice to amalgamate trap contents within a given sampling point to reduce small-scale spatial differences in catch sizes between adjacent traps. If it is required that the catches of different traps are to be independent from each other then large separations are required. Digweed et al. (1995) suggest that a minimum separation of 25 m is required in Carabidae communities if traps are to be statistically independent.

Sampling duration and temporal pattern

The duration over which pitfall traps have been used to sample epigeal invertebrates ranges from as little as little as two days (Greenslade 1973) to over three years (Clarke & Bloom 1992) for an individual experiment. Long-term sampling programs may sample essentially indefinitely. However, following the work of Baars (1979) and Den Boer (1979), it is now acknowledged to be necessary for quantitative work to sample over the entire activity period of the community in question. Baars (1979) considers this period to be a year, although this is somewhat conservative. Most temperate studies ignore at least the winter season, as catches during this period are low. At a minimum, a reasonable sampling period should be greater than four months (e.g. Obrtel 1971, Epstein & Kulman 1984, Niemelä 1990). Pitfall trap catches based on these long sampling periods have been shown to have good correlations with the abundance of several species of Carabidae (Baars 1979, Den Boer 1979, Luff 1982). Sampling over such extended periods is necessary as the activity of a species will vary in its seasonal distribution from site to site; however, the total length and intensity of activity is hypothesized to be approximately the same between sites (Baars 1979, Den Boer 1979).

Trap catches should therefore only be used to infer differences in population size for one species between sites and should not be used to provide information on the relative population sizes of each species (Baars 1979, Den Boer 1979). However, Baars (1979) states that comparisons of the population sizes of several species from different sites may be possible with samples taken over the whole year, providing the relationship between the true density and the size of the pitfall trap catch for each species is known. Such relationships are unknown for most species. Hanski and Niemelä(1990) suggest that, although absolute population sizes may not be known, a large difference in the relative abundances of two species is enough to infer that one is more abundant than the other.

Where it has not been possible to sample for long periods of time, data may still have some quantitative value. Niemelä et al. (1990) showed that shorter sampling periods contain important biological information. Temporal sub-samples of between 10 and 28 days retain the approximate rank and relative proportions of the dominant species when compared to data from a much longer sampling period. However, sample similarity increased and variance decreased when the sampling sub-period was increased. Sampling within these short time periods provides extremely limited and potentially unreliable information, and should be avoided where possible.

Depletion

When a killing agent/preservative is used in the trap, depletion of the local population can occur, which may give the impression of a reduction in population size (e.g. Luff 1975, Ericson 1979, Digweed et al. 1995). In sampling programs that occur over a long period, depletion of larval stages early in the season may result in smaller adult populations. This will remain unnoticed unless larval stages have been specifically identified. In the cases of the Carabidae the subterranean lifestyle of most larvae (Kegel 1990) means that their capture in pitfalls is low.

Although the effects of depletion can be reduced by using traps at moderate densities (Greenslade 1973, Digweed et al. 1995), Digweed et al. (1995) suggest that high trap densities could be useful. By using high trap densities the populations present within the sphere of influence of the traps are likely to be sampled in their entirety. The success of this will depend on the levels of immigration into this sphere of influence (Den Boer 1970).

Surrounding vegetation structure

An increase in structural complexity of vegetation around pitfall traps can result in a reduction in the catch (Greenslade 1964, Melbourne 1999). Melbourne (1999) hypothesized this to be due to an increase in the total surface area available for invertebrates to move on in structurally complex habitats. This causes a decrease in the effective number of pitfall traps per unit area, reducing the overall pitfall trap catch. For larger species, or species primarily active on the soil surface, direct impedance by the vegetation will be more likely to reduce catch size (Greenslade 1964). Since vegetation structure will not be static throughout a growth season the effects may also change apparent population sizes as the dilution/impedance effect of vegetation structure becomes more prominent as the season progresses (Greenslade 1964, Melbourne 1999).

For these reasons, pitfall traps should not be used to compare habitats that have field-layer vegetation that is structurally dissimilar (Greenslade 1964, Maelfait & Baert 1975, Melbourne 1999). If such a comparison is integral to the study, vegetation surrounding each trap should be removed to standardize the immediate area surrounding each trap (Greenslade 1964, Penny 1966, Melbourne 1999).

Digging-in effects

Digging in effects are a temporary increase in the capture rate of pitfall traps in response to the physical disturbance caused by trap installation. These do not represent real increases in the density of surface-active invertebrates (Joose 1965, Joose & Kapteijn 1968, Greenslade 1973, Digweed et al. 1995). Digging-in effects have been recorded for species of Collembola (Joose 1965, Joose & Kapteijn 1968), ants (Greenslade 1973), and carabids (Digweed et al. 1995), although wandering spiders have not been found to exhibit this behavior (Greenslade 1973). The extent to which digging-in effects occur is normally species-specific (Joose 1965, Greenslade 1973, Digweed et al. 1995). The duration of digging-in effects is also variable. In Collembola it can be as short as a single day (Joose 1965). Digging-in effects are considered to be minimal for most groups after a week (Majer 1978). For this reason it is advisable to ignore the first week's catch during a sampling program.

Activity–abundance

As the rate of capture of most invertebrates is proportional to their activity (Maelfait & Baert 1975, Curtis 1980), the numbers of each species caught will not reflect their true abundance. Instead their rate of capture will be proportional to the interaction between their abundance and activity, this is the concept of activity–abundance (Tretzel 1954, Heydemann 1957, Thiele 1977). Species that are largely sessile, but occur at high abundances, may be underrepresented in pitfall traps compared to less abundant but more active species. This redefinition of what pitfall catches represent provides a much sounder conceptual framework for the interpretation of pitfall trap data. However, without information on the activity of each species it is almost impossible to relate pitfall trap catches to the true relative abundances of different species. Information on the activity of different species is relatively infrequent in the literature (e.g. Halsall & Wratten 1988). In addition to this there are additional problems with activity–abundance, as while activity may be correlated with capture rate it is likely to be confounded by behavioral peculiarities of each species. These behavioral differences will influence rates of capture independent of the activity and density of different species (e.g. Den Boer 1981, Halsall & Wratten 1988, Morrill et al. 1990, Obeng-Ofori 1993, Topping 1993, Mommertz et al. 1996). Nonetheless, this concept is valuable in the interpretation of pitfall trap catches.

Conclusions

While the quantitative nature of pitfall traps is likely to remain questionable, they are still one of the most frequently used collecting techniques for surface dwelling invertebrates. While this choice may seem irrational in the face of their many problems, their use is probably no more questionable than most other sampling techniques used for invertebrates. Every sampling technique will have inherent biases resulting from individual species behavior. These individual behaviors will influence not only how often a species comes into contact with a trap, but also how it responds when it encounters it. It would seem unlikely that any trapping method has ever provided a perfect representation of the relative abundances of each species in a habitat. While there will always be some species that are highly misrepresented in pitfall traps, the vast majority are likely to be represented at frequencies that at least reflect their true relative abundances. Determining a priori which species will be highly misrepresented in pitfall trap samples cannot be achieved on the basis of general morphological, or even taxonomic, trends. Essentially these highly misrepresented species can be considered as being uncontrolled for random variation, which every collecting technique is prone to.

It is also important to appreciate the limitations of pitfall trapping in terms of what it does actually catch. The method is not an all-purpose technique suitable for catching every species from a predetermined taxonomic group, e.g. the Carabidae. Instead the method is more likely to be guild-specific, targeting only those species that are highly active on the soil surface. For example, while most species of Carabidae are surface-active insects, member of the genus *Dromius* are arboreal (Terell-Nield 1990). Such species are going to be largely absent from pitfall trap catches: while they may be part of a taxonomic group targeted by pit-fall traps and present in an area being sampled, they are not part of the guild of surface-active invertebrates actually caught by pitfall traps. While species not in this surface-active guild may still occur in low numbers, it is possible to try to remove them from the dataset by ignoring those species representing the bottom 1–5 percent of the total abundance of individuals (Dennis et al. 1997). This removal of some lower percentage of the catch also has the advantage of re-moving species that may not be truly associated with a habitat but are instead in transit through it (Den Boer 1977, Desender & Maelfait 1986b, Dennis et al. 1997).

With a good understanding of the flaws associated with pitfall traps, and with proper precautions taken to deal with these problems, it should be possible to use this method at least semi-quantitatively. This should always be done tenta-tively, and highly questionable results should be treated with caution.

References

Abensberg-Traut, M. & Steven, D. (1995) The effects of pitfall trap diameter on ant species richness (Hymenoptera: Formicidae) and species composition of the catch in a semi-arid eucalypt woodland. *Australian Journal of Ecology*, **20**, 282–287.

Adis, J. & Kramer, E. (1975) Formaldehyd-lösung attrahiert *Carabus problematicus* (Coleoptera: Carabidae). *Entomologica Germanica*, **2**, 121–125.

Baars, M.A. (1979) Catches in pitfall traps in relation to mean densities of carabid beetles. *Oecologia*, **41**, 25–46.

Barber, H.S. (1931) Traps for cave inhabiting insects. *Journal of the Elisha Michell Scientific Society*, **46**, 259–266.

Barndt, D. (1976) Untersuchung der diurnalen und saisonalen Aktivität von Käfern mit einer neu entwickelten Electro-bodenfalle. *Verhandlungen des Botanischen Vereins der Provinz Brandenberg*, **112**, 103–122.

Beasley, V.R. (1985) Diagnosis of ethylene glycol (antifreeze) poisoning. *Feline Practice*, **15**, 41–46.

Bombosch, S. (1962) Untersunchungen über die Auswertbarkeit von Fallenfängen. *Zeitschrift für Angewandte, Zoology*, **49**, 149–160.

Bostanian, N.J., Boivin, G., & Goulet, H. (1983) Ramp pitfall trap. *Journal of Economic Entomology*, **76**, 1473–1475.

Briggs, J.B. (1960) A comparison of pitfall trapping and soil sampling in assessing populations of two species of ground beetles (Col.: Carabidae). *East Malling Research Station Annual Report*, **48**, 108–12.

Chapman, P.A. & Armstrong, G. (1997) Design and use of a time-sorting pitfall trap for preda-tory arthropods. *Agriculture, Ecosystem and Environment*, **65**, 15–21.

Clarke, W.H. & Bloom, P.E. (1992) An efficient and inexpensive pitfall trap system. *Entomological News*, **103**, 55–59.

Curtis, D.J. (1980) Pitfalls in spider community studies (Archnida, Aranae). *Journal of Arach-nology*, **8**, 271–280.

Den Boer, P.J. (1970) On the significance of dispersal power for populations of carabid beetles (Coleoptera, Carabidae). *Oecologia*, **4**, 1–28.

Den Boer, P.J. (1977) Dispersal power and survival: carabids in a cultivated countryside. *Landbouwhogeschool Wageningen The Netherlands Miscellaneous Papers*, 14. H. Veenman & Sons, Wageningen.

Den Boer, P.J. (1979) The individual behaviour and population dynamics of some carabid beetles in forests. *Miscellaneous Papers LH Wageningen*, **18**, 157–166.

Den Boer, P.J. (1981) On the survival of populations in a heterogeneous and variable environment. *Oecologia*, **50**, 39–53.

Dennis, P., Young, M.R., Howard, C.L., & Gordon, I.J. (1997) The response of epigeal beetles (Col.: Carabidae, Staphylinidae) to varied grazing regimes on upland *Nardus stricta* grasslands. *Journal of Applied Ecology*, **34**, 433–443.

Desender, K. & Maelfait, J.P. (1986a) Pitfall trapping with enclosures: a method for estimating the relationship between the abundances of coexisting carabid species (Coleoptera: Carabidae). *Holarctic Ecology*, **9**, 245–250.

Desender, K. & Maelfait, J.P. (1986b) The relation between dispersal power, commonness and biological features of carabid beetles (Coleoptera, Carabidae). *Annales de la Societe Royale Zoologique de Belgique*, **116**, 84–94.

Digweed, S.C., Currie, C.R., Cárcamo, H.A., & Spence, J.R. (1995) Digging out the "digging-in effect" of pitfall traps: influences of depletion and disturbance on catches of ground beetles (Coleoptera: Carabidae). *Pedobiologia*, **39**, 561–567.

Epstein, M.E. & Kulman, H.M. (1984) Effects of aprons on pitfall catches of carabid beetles in forests and fields. *The Great Lakes Entomologist*, **17**, 215–221.

Ericson, D. (1977) Estimating population parameters of *Pterostichus cupreus* and *P. melanarius* (Carabidae) in arable fields by means of capture–recapture. *Oikos*, **29**, 407–417.

Ericson, D. (1978) Distribution, activity and density of some Carabidae (Coleoptera) in winter wheat fields. *Pedobiologia*, **18**, 202–217.

Ericson, D. (1979) The interpretation of pitfall catches of *Pterostichus cupreus* and *Pt. melanarius* (Coleoptera, Carabidae) in cereal fields. *Pedobiologia*, **19**, 320–328.

Faragalla, A.A. & Adam, E.E. (1985) Pitfall trapping of tenebrionid and carabid beetles (Coleoptera) in different habitats in the central region of Saudi Arabia. *Zeitschrift für Angewandte Entomologie*, **99**, 466–471.

Feoktistov, B.F. (1980) Effectivost lovushek Barberaraznoga tipa. *Zooliknesky Zhurnal*, **59**, 1554–1558.

Fichter, E. (1941) Apparatus for the comparison of soil surface arthropod populations. *Ecology*, **22**, 338–339.

Frank, J.H. (1991) Staphylinidae. In *An introduction to Immature Insects of North America* (ed. F.W. Stehr), pp. 341–352. Kendall-Hunt, Dubuque, Iowa.

Giblin-Davis, R.M., Peña, J.E., & Duncan, R.E. (1994) Lethal pitfall trap for the evaluation of semiochemical-mediated attraction of *Metamasius hemipterus sericeus* (Coleoptera: Curculionidae). *Florida Entomologist*, **77**, 247–255.

Gist, C.S. & Crossley, J.D.A. (1973) A method for quantifying pitfall traps. *Environmental Entomology*, **2**, 951–952.

Good, J.A. & Giller, P.S. (1991) The effect of cereal and grass management on the Staphylinidae (Coleoptera) assemblages in south-west Ireland. *Journal of Applied Ecology*, **28**, 810–826.

Greenslade, P. & Greenslade, P.J.M. (1971) The use of baits and preservatives in pitfall traps. *Journal of the Australian Entomological Society*, **10**, 253–260.

Greenslade, P.J.M. (1964) Pitfall trapping as a method for studying populations of Carabidae (Coleoptera). *Journal of Animal Ecology*, **33**, 301–310.

Greenslade, P.J.M. (1973) Sampling ants with pitfall traps: digging in effects. *Insectes Sociaux*, **20**, 343–353.

Hall, D.W. (1991) The environmental hazard of ethylene glycol in insect pit-fall traps. *The Coleopterists Bulletin*, **45**, 193–194.

Halsall, N.B. & Wratten, S.D. (1988) The efficiency of pitfall trapping for polyphagous predatory Carabidae. *Ecological Entomology*, **13**, 293–299.

Hammond, P.M. (1990) Insect abundance and diversity in the Dumoga-Bone national park, N.Sulawesi, with special reference to the beetle fauna of lowland rain forest in the Toraut region. In *Insects and Rainforests of South East Asia (Wallacea)* (ed. W.J. Knight & J.D. Holloway), pp. 197–254. The Royal Entomological Society of London, London.

Hanski, I. & Niemelä, J. (1990) Elevation distributions of dung and carrion beetles in Northern Sulawesi. In *Insects and the rain forests of South East Asia (Wallacea)* (ed. W.J. Knight & J.D. Holloway), pp. 145–153. The Royal Entomological Society of London, London.

Hertz, M. (1927) Huomioita petokuoriaisten olinpaikoista. *Luonnon Ystävä*, **31**, 218–222.

Heydemann, B. (1957) Die Biotopstruktur als Raumwiderstand und Raumfulle für die Tierwelt. *Verhandlungen der Deutschen Zoologischen Gesellschaft Saarbrücken*, **56**, 332–347.

Holopainen, J.K. (1990) Influence of ethylene glycol on the numbers of carabids and other soil arthropods caught in pitfall traps. In *The role of Ground Beetles in Environmental and Ecological Studies* (ed. N.E. Stork), pp. 339–341. Intercept, Hampshire, UK.

Holopainen, J.K. (1992) Catch and sex ratio of Carabidae (Coleoptera) in pitfall traps filled with ethylene glycol or water. *Pedobiologia*, **36**, 257–261

Holopainen, J.K. & Varis, A.L. (1986) Effects of mechanical barriers and formalin preservative on pitfall catches of carabid beetles (Coleoptera, Carabidae) in arable fields. *Journal of Applied Entomology*, **102**, 440–445.

Honêk, A. (1988) The effects of crop density and microclimate on pitfall trap catches of Carabidae, Staphylinidae (Coleoptera), and Lycosidae (Aranea) in cereal fields. *Pedobiologia*, **32**, 233–242.

Houston, W.W.K. (1971) A mechanical time sorting pitfall trap. *Entomologist's Monthly Magazine*, **107**, 214–216.

Jarosík, V. (1992) Pitfall trapping and species abundance relationships: a value for carabid beetles (Coleoptera: Carabidae). *Acta Entomologica, Bohemoslovaca*, **89**, 1–12.

Joose, E.N.G. (1965) Pitfall-trapping as a method for studying surface dwelling Collembola. *Zeitschrift für Morphologie und Ökologio der Tiere*, **55**, 587–596.

Joose, E.N.G. & Kapteijn, J.M. (1968) Activity-stimulating phenomena caused by field-disturbance in the use of pitfall traps. *Oecologia*, **1**, 385–392.

Kegel, B. (1990) Diurnal activity of carabid beetles living on arable land. In *The role of Ground Beetles in Environmental and Ecological Studies* (ed. N.E. Stork), pp. 65–76. Intercept, Hampshire, UK.

Kharboutli, M.S. & Mack, T.P. (1993) Comparison of three methods for sampling arthropod pests and their natural enemies in peanut fields. *Journal of Economic Entomology*, **86**, 1802–1810.

Kuschel, G. (1991) A pitfall trap for hypogean fauna. *Curculio*, **31**, 5.

Lawrence, K.O. (1982) A linear pitfall trap for mole crickets and other soil arthropods. *Florida Entomologist*, **65**, 376–377.

Lemieux, J.P. & Lindgren, B.S. (1999) A pitfall trap for large-scale trapping of Carabidae: comparison against conventional design using two different preservatives. *Pedobiologia*, **43**, 245–253.

Luff, M.L. (1968) Some effects of formalin on the numbers of Coleoptera caught in pitfall traps. *Entomologist's Monthly Magazine*, **104**, 115–116.

Luff, M.L. (1975) Some features influencing the efficiency of pitfall traps. *Oecologia*, **19**, 345–357.

Luff, M.L. (1978) Diel activity patterns of some field Carabidae. *Ecological Entomology*, **3**, 53–62.

Luff, M.L. (1982) Population dynamics of Carabidae. *Annales of Applied Biology*, **101**, 164–170.

Maelfait, J.P. & Baert, L. (1975) Contributions to the knowledge of the arachno- and entomo-fauna of different wood habitats. Part I. Sampled habitats, theoretical study of the pitfall method, survey of the captured taxa. *Biol Jb Dodonaea*, **43**, 179–196.

Majer, J.D. (1978) An improved pitfall trap for sampling ants and other epigaeic invertebrates. *Journal of the Australian Entomological Society*, **17**, 261–262.

Marshall, D.A. & Doty, R.L. (1990) Taste responses of dogs to ethylene glycol, propylene glycol, and ethylene glycol-based antifreeze. *Journal of the American Veterinary Medical Association*, **12**, 1599–1602.

Melbourne, B.A. (1999) Bias in the effects of habitat structure on pitfall traps: an experimental evaluation. *Australian Journal of Ecology*, **24**, 228–239.

Mitchell, B. (1963a) Ecology of two carabid beetles, *Bembidion lampros* (Herbst) and *Trechus quadristriatus* (Schrank). I. Life cycles and feeding behaviour. *Journal of Animal Ecology*, **32**, 289–299.

Mitchell, B. (1963b) Ecology of two carabid beetles, *Bembidion lampros* (Herbst) and *Trechus quadristriatus* (Schrank). II. Studies on populations of adults in the field, with special reference to the technique of pitfall trapping. *Journal of Animal Ecology*, **32**, 377–392.

Mommertz, S., Schauer, C., Kösters, N., Lang, A., & Filser, J. (1996) A comparison of D-vac suction, fenced and unfenced pitfall trap sampling of epigeal arthropods in agroecosystems. *Annales Zoolgica Fennici*, **33**, 117–124.

Morrill, W.L. (1975) Plastic pitfall trap. *Environmental Entomology*, **4**, 596.

Morrill, W.L., Lester, D.G., & Wrona, A.E. (1990) Factors affecting efficacy of pitfall traps for beetles (Coleoptera: Carabidae and Tenebrionidae). *Journal of Entomological Science*, **25**, 284–293.

Niemelä, J. (1990) Spatial distribution of carabid beetles in the southern Finnish taiga: the question of scale. In *The Role of Ground Beetles in Ecological and Environmental Studies* (ed. N.E. Stork), pp. 143–155. Intercept, Hampshire, UK.

Niemelä, J., Halme, E., Pajunen, T., & Haila, Y. (1986) Sampling spiders and carabid beetles with pitfall traps: the effects of increased sampling effort. *Annales Entomologici Fennici*, **52**, 109–111.

Niemelä, J., Halme, E., & Haila, Y. (1990) Balancing sampling effort in pitfall trapping of carabid beetles. *Entomologica Fennica*, **1**, 233–238.

Niemelä, J., Spence, J.R., & Spence, D.H. (1992) Habitat associations and seasonal activity of ground-beetles (Coleoptera: Carabidae) in central Alberta. *The Canadian Entomologist*, **124**, 521–540.

Obeng-Ofori, D. (1993) The behaviour of 9 stored product beetles at pitfall trap arenas and their capture in millet. *Entomologica Experimentalis et Applicata*, **6**, 161–169.

Obrtel, R. (1971) Number of pitfall traps in relation to the structure of the catch of soil surface Coleoptera. *Acta Entomologica Bohemoslavaca*, **68**, 300–309.

Owen, J.A. (1995) A pitfall trap for repetitive sampling of hypogean arthropod faunas. *Entomologist's Record*, **107**, 225–228.

Parmenter, R.R. & MacMahon, J.A. (1989) Animal density estimation using a trapping web design: Field validation experiments. *Ecology*, **70**, 169–179.

Penny, M.M. (1966) Studies on certain aspects of the ecology of *Nebria brevicolis* (F.) (Coleoptera, Carabidae). *Journal of Animal Ecology*, **35**, 505–512.

Rieske, L.K. & Raffa, K.F. (1993) Potential use of baited pitfall traps in monitoring pine root weevil, *Hylobius picivorus*, and *Hylobius radicis* (Coleoptera: Curculionidae) populations and infestation levels. *Forest Entomology*, **86**, 475–485.

Romero, H. & Jaffe, K. (1989) A comparison of methods for sampling ants (Hymenoptera, Formicidae) in savannahs. *Biotropica*, **21**, 348–352.

Scheller, H.V. (1984) Pitfall trapping as the basis for studying ground beetle (Carabidae) predation in spring barley. *Tidsskrift for Planteval*, **88**, 317–324.

Scuhravy, V. (1970) Zur Anlockungsfähigkeit von Formalin für Carabiden in Bodenfallen. *Beitrage Entomologie*, **20**, 371–374.

Simmons, C.L., Pedigo, L.P., & Rice, M.E. (1998) Evaluation of seven sampling techniques for wireworms (Coleoptera: Elateridae). *Environmental Entomology*, **27**, 1062–1068.

Smith, B.J. (1976) A new application in the pitfall trapping of insects. *Transactions of the Kentucky Academy of Science*, **37**, 94–97.

Spence, J.R. & Niemelä, J.K. (1994) Sampling carabid assemblages with pitfall traps: The madness and the method. *The Canadian Entomologist*, **126**, 881–894.

Stein, W. (1965) Die Zusammensetzung der Caribidenfauna einer weisen mit stark wechselnden Feuchtigkeitsverhältnissen. *Zeitschrift für Morphologie und Ökologie der Tiere*, **55**, 83–99.

Terell-Nield, C. (1990) Is it possible to age woodlands on the basis of their carabid beetle diversity? *The Entomologist*, **109**, 136–145.

Thiele, H.-U. (1977) *Carabid Beetles in Their Environment: a Study on Habitat Selection by Adaptation in Physiology and Behavior*. Springer, New York.

Thomas, C.F.G., Parkinson, L., & Marshall, E.J.P. (1998) Isolating the components of activity–density for the carabid beetle *Pterostichus melanarius* in farmland. *Oecologia*, **116**, 103–112.

Thomas, J.D.B. & Sleeper, E.L. (1977) The use of pitfall traps for estimating the abundance of arthropods, with special reference to the Tenebrionidae (Coleoptera). *Annals of the Entomological Society of America*, **70**, 242–248.

Topping, C.J. (1993) Behavioural responses of three linyphiid spiders to pitfall traps. *Entomologica Experimentalis et Applicata*, **68**, 287–293.

Topping, C.J. & Sunderland, K.D. (1992) Limitations to the use of pitfall traps in ecological studies exemplified by a study of spiders in a field of winter wheat. *Journal of Applied Ecology*, **29**, 485–491.

Tretzel, E. (1954) Riefe- und Fortpflanzungszeit bei Spinnen. *Zeitschrift für Morphologie und Ökologie der Tiere*, **42**, 643–691.

Uetz, G.W. & Unzicker, J.D. (1976) Pitfall trapping in ecological studies of wandering spiders. *Journal of Arachnology*, **3**, 101–111.

van den Berghe, E. (1992) On pitfall trapping invertebrates. *Entomological News*, **103**, 149–156.

Vlijm, L., Hartsuyker, L., & Richter, C.J.J. (1961) Ecological studies on carabid beetles. I. *Calathus melanocephalus* (Linn.). *Archives Néerlandaises de Zoologie*, **14**, 410–422.

Waage, B.E. (1985) Trapping efficiencies of carabid beetles in glass and plastic pitfall traps containing different solutions. *Fauna Norvegica*, Series B, **32**, 33–36.

Walsh, G.B. (1933) Studies in British necrophagous Coleoptera. II. The attractive powers of various natural baits. *Entomologist's Monthly Magazine*, **69**, 28–32.

Yasuda, K. (1996) Attractiveness of pitfall traps to West-Indian sweet potato weevil, *Euscepes postfasciatus* (Fairmaire) (Coleoptera: Curculionidae). *Japanese Journal of Applied Entomology and Zoology*, **40**, 97–102.

Index of methods and approaches

Methodology	Topics addressed	Comments
Trap design and installation		
Trap material	Materials used in pitfall trap construction and their relative efficiency in catching invertebrates.	Glass is seen to be one of the most effective at preventing escapes, but plastic-based pitfall traps are more practical.
Trap shape and size	The effect of trap shape and size on the capture rates of different species.	Consistency in size and shape within an experiment is recommended.
Roofs	The use of roofs to protect against weather conditions and preventing damage/capture by birds and mammals.	Roofs and covers can reduce damage to the catch, although may effect relative capture rates of target species.
Funnels	The use of funnels to increase capture rates and reduce damage to the catch.	Where the catch is kept alive funnels are useful in reducing escape rates.
Trap rim	Protruding trap rims influence capture rate.	Trap rims must be flush with the substrate surface.
Killing and preservative agents	The use of killing agents to increase capture rate and prevent decomposition.	Killing agents and preservatives can act as attractants for some species and may be toxic to vertebrates.
Baits	Attractants for infrequently occurring or target species.	Useful for highly aggregated species or those targeted by pest monitoring programs.
Specialized designs	Unconventional designs of pitfall traps and their value in asking specific ecological questions.	Traps considered include drift fences, gutter traps, time sort traps, barrier traps, ramp traps, and subterranean pitfall traps.
Sampling strategy		
Trap numbers	Optimal trap numbers and species accumulation rates.	Twelve pitfall traps are suggested as a suitable number in most situations.
Spatial arrangement	Trap arrangement into grids, transects, and random positioning; their relative benefits and uses.	Trap arrangement is chosen primarily for practical reasons, although it may influence capture rates.
Sampling duration	How duration of trapping will influence the quantitative value of the catches.	Whole-season sampling periods are recommended.
Depletion effects	Long or intensive trapping can reduce natural population sizes.	Depletion may affect larval stages, influencing future demographic patterns.

Continued

Methodology	Topics addressed	Comments
Vegetation impediment	Vegetation structure will influence capture rates due to impedance of movement and dilution effects.	Comparisons between structurally different habitats should be avoided.
Digging-in effects	Immediately after pitfall trap installation, capture rates are unusually high.	After one week most digging in effects have dissipated.
Activity–abundance	The interaction between individual species abundance and their relative activity rates.	This concept is of key importance in the interpretation of pitfall trap catches.

CHAPTER 4

Sampling methods for forest understory vegetation

CLAIRE M.P. OZANNE

Introduction

The understory is a varied and complex habitat, forming a key layer in the forest ecosystem. Lawrence (1995) defines the understory as the "vegetation layer between the tree canopy and the ground cover in a forest." Although drawn from a wide taxonomic range, many understory plants share a number of characteristics associated with shade tolerance, including longer foliation and more efficient photosynthesis per unit leaf area (Spurr 1980). Of course, light-demanding understory plants may be also present, but confined to open gaps or glades where light can penetrate to the forest floor.

Since the plant composition varies from forest to forest and biome to biome we would expect many different groups of insect to inhabit the understory. For the purposes of effective sampling, these can be divided into four categories: (a) insects associated with understory plants; (b) insects associated with the understory environment (e.g. shade, constancy of microclimate, low wind speed); (c) gap specialists; and (d) resource specialists (e.g. dead wood, coppice, parasites).

The understory makes a significant contribution to forest resources because it supports a distinctive fauna; however, it cannot be totally isolated from other forest habitats – neither the litter and soil layer below nor the high canopy above. There are several examples of insects and other invertebrates which move between forest layers, interacting with the communities at several levels, e.g. Hymenoptera, Collembola, and Araneae (Oliveira & Campos 1996, Bowden et al. 1976, Simon 1995). This means that techniques and methodologies described in other chapters in this book may be applicable for investigations of the understory.

In this chapter I shall consider the process of collecting insects that are associated with forest understory vegetation (category a) and insects which are actively moving through the understory layer (categories a, b, c, and d).

How should techniques be chosen?

Collecting insects from vegetation usually has one of two major aims. The first is to generate a species list for the habitat, and the second to obtain subsets or samples of the community that are representative of the whole. The understory is taxonomically and structurally diverse and therefore it can be difficult to make decisions about the number of sampling events, their location, and the techniques to use. No one sampling technique will enable the entomologist to meet fully either of the two major aims noted above, and so it is likely that several techniques will be used in any one investigation. The key to choosing techniques is a set of clearly defined research questions or study aims.

The research questions should define the target insect groups or communities, the type of vegetation from which sampling will be needed, and may dictate the location from which samples should be collected. Each of these will signal the most appropriate collection technique. The boundaries of the community under investigation need to be clearly delineated before a sampling protocol can be set up and appropriate techniques chosen. For example, if the research question focuses on the insect community of a specific understory plant the design demanded will be different from a study in which the research question relates to the impact of edge effects or fragmentation.

In this chapter techniques have been arranged according to the structure of the vegetation in which sampling is carried out. Within that, the insect groups and communities that are most successfully collected with each technique are noted, and location sensitivity is mentioned where appropriate. Understory vegetation can range from low grass swards, through herbs of varying structural complexity, to shrubs and small trees and the associated vascular and non-vascular epiphytes, so putting together a well-designed investigation can be both challenging and exciting.

Sampling from low understory vegetation including grasses and herbs

Suction or vacuum sampling

Suction or vacuum sampling can be used in understory vegetation types from short grass through to shrubs and small trees, but is particularly effective for collecting insects in grasses and herbs. The technique is suited to collecting data on insects associated with specific plant communities, resources, and locations, e.g. gaps and edges, and is not directly dependent on insect activity; thus it can be described as a passive sampling method.

Suction samplers may be divided into two major types according to the nozzle or hose diameter and common modes of use: wide-hosed (>20 cm diameter) and narrow-hosed (<15 cm diameter). Wide-hosed models (W-type) include the Dietrick vacuum sampler (D-vac) (Dietrick 1961) and the Thornhill vacuum sampler (Thornhill 1978). There are quite a number of narrow-hosed models (N-type) which are either purpose-built or converted from garden leaf vacuums (Stewart & Wright 1995, Buffington & Redak 1998). All machines

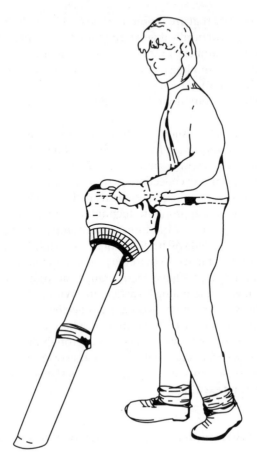

Fig. 4.1 Narrow-hosed vacuum sampler (N-type), constructed from a converted garden leaf vacuum. (Adapted from Sabre® Manual.)

essentially function in the same manner, using a fan to generate a suction current. Air is drawn up a suction tube, one end of which is bisected by a collecting bag constructed of netting, muslin, or cotton (Fig. 4.1). The bag can be detached from the tube and stored or emptied on site.

The effectiveness of vacuum sampling is dependent on two main factors: the speed of airflow and the proficiency with which the operator empties the bag (active insects can easily be lost at this stage!). Narrow-hosed models have been shown to collect samples that are more representative of field communities than wide-hosed models. For example, they are more effective in collecting insect groups such as aphids and associated predatory beetles such as *Tachyporus* sp. (Staphylinidae) (Macleod et al. 1994, Stewart & Wright 1995). This effectiveness has been attributed to the greater airflow rate per unit nozzle area, and to better extraction of insects from the lower parts of the vegetation (Stewart & Wright 1995). Indeed, W-type models have been found to collect aphids poorly from soil level and to favor insects in the top sections of plants (Hand 1986). N-type models are therefore recommended. Additionally they are easier to manage in the field, being lighter to manipulate and much quieter.

Sampling efficiency is also influenced marginally by the method of sample collection and environmental conditions. The vacuum nozzle can be selectively placed over individual plants or directly onto the vegetation sward; the former is clearly easier with wide-hosed models. For N-type models a common mode of use is to lay down quadrats and run the nozzle slowly over and through the vegetation – e.g. 1 minute for a 25 cm^2 quadrat. A more sophisticated method uses a cylinder or tent with known cross-sectional area which can be placed on the ground, the hose inserted through the top, and the contents vacuumed up. Fewer insects will be lost with this method (Stewart & Wright 1995).

Species accumulation curves for vacuum sampling strongly link efficiency to sampling effort (Buffington & Redak 1998), so the number of sampling events needs to be optimized. Rain or dew can adversely affect efficiency. Henderson and Whittaker (1977) demonstrated that when vegetation is wet insects remain attached to the foliage; collection events should therefore be confined to dry conditions. Sampling will only be truly comparative if environmental conditions and vegetation complexity are held constant.

Whatever method is used it will be necessary to clean the samples (i.e. separate insects from debris). This process can be very laborious if done by hand, but some insects will cling to plant material and so large fragments need to be checked carefully. Extraction of samples using Tullgren funnels (Sutherland 1996) or CO_2 and extraction using black fluorescent bulbs (Buffington & Redak 1998) is less time-intensive and very effective. Should samples need to be stored in alcohol prior to cleaning to prevent predation loss, then insects can be extracted subsequently using flotation techniques (Dondale et al. 1971).

Although N-type vacuum samplers are able to collect insects from a wide range of taxonomic groups and produce samples typical of the community, they are particularly suitable for insects with sufficient surface area for suction to act on, for example, medium to large Collembola and Homoptera (although alate aphids are favored over apterous morphs due to larger wing surface area for suction; Hand 1986). Large heavy species that the current cannot carry into the collection bag will be under-sampled; for example, large aculeate Hymenoptera (Buffington & Redak 1998).

How does vacuum sampling compare with other techniques? The other most commonly used technique in low vegetation is the sweep net (see below). Vacuum-sampling has been found more sensitive for detecting community variation than sweep-netting in complex vegetation such as Californian coastal sage scrub (Buffington & Redak 1998) and in cotton (Byerly et al. 1978, Ellington et al. 1984). Differences were particularly notable for animals with a small body size that may not be collected when sweeping because the air draught pushes them away from the net. Buffington and Redak (1998) also found vacuum-sampling to be more effective than sweeping for collecting Diptera and smaller Hymenoptera.

The action of a vacuum sampler is less likely to cause damage to sensitive vegetation, and where several operators are involved in a study it is more consistent than methods such as sweeping, relying less on experience and good technique (Buffington & Redak 1998).

Medium-height vegetation including shrubs

Sweep-net capture

The sweep net can be used in medium-height vegetation for collecting insects associated with specific plants or resources. Nets have been used extensively in crop pest surveys and in some studies for collecting insects in forest understory and even the canopy (Canaday 1987, Noyes 1989, Lowman et al. 1993). The sweep net is a passive sampling method that is suited to collecting insects associated with specific plant communities. (A sweep net could be used for insects associated with specific plant species if the volume of foliage is sufficient to sweep through.)

Sweep nets are constructed from sturdy cotton material that can withstand vigorous movement though vegetation. The mouth of the net is usually circular, although D-shaped nets are more effective in short vegetation (Southwood & Henderson 2000). Net diameter usually approximates 0.5 m. Sweeping is essentially a qualitative method but can be made semi-quantitative by standardizing the number of sweeps in a given area or along a defined transect (25 sweeps being effective; Gray & Treloar 1933), or by standardizing the length of sweeping time (e.g. 5 or 10 minutes). The net should be plied in an energetic figure-of-eight motion, with sufficient forward movement to prevent overlap of sweeps.

The effectiveness of the sweep net is dependent upon the way in which it is manipulated by the operator, the vegetation structure, and the prevailing environmental conditions. For example, speed of net motion has an impact on catch size, with a greater speed collecting more insects (Balogh & Loska 1956), and experienced samplers may capture more insects because they sweep more vigorously and more deeply into the crop (Wise & Lamb 1998). Vegetation density and crop phenology are also significant (Byerley et al. 1978, Wilson & Gutierrez 1980, Ellington et al. 1984). Sweep nets are particularly effective if the insects are known to be present in the top part of the vegetation (e.g. around seed pods), and in less dense vegetation, where insects can easily be knocked into the net. If sweep nets are to be used for population estimation in complex vegetation then calibration is required by means of a pilot study in which netting is compared with some absolute counting method.

As with vacuum sampling, nets do not collect insects from vegetation if the surface is wet or damp, and so the timing of collection (e.g. after the evaporation of dew) is an important factor. However, sweep nets have the distinct advantage that they can be used to collect larger numbers of samples than other methods such as interception traps, and they also allow samples to be collected in a randomized manner meeting some of the essential requirements of parametric statistics. The exact number of samples and their location will depend upon the variability of the community and the study aims.

Sweep nets can collect samples that are completely representative of the population or community of insects (Gadagkar et al. 1990). Wise and Lamb (1998) found that there was stability and repeatability in the relationship between variance and mean, allowing sweep nets to be used to detect whether pest populations were above or below level of control. However, there can be biases, for example towards adults and large nymphs of *Lygus* species (probably due to their location in the upper parts of the plant) (Wise & Lamb 1998). Linders (1995) found the sweep net to be effective for sampling the weevil *Trichosirocalus troglodytes* on *Plantago lanceolata* spikes, but the catches were influenced by diurnal and seasonal activity. In contrast, sweep nets are not good for capturing Lepidoptera (Gadagkar et al. 1990), but sample Hymenoptera, Diptera, small Coleoptera, and arachnids quite well (Canaday 1987).

The major advantage of this technique lies in simplicity and portability, since sweep nets can be easily carried and a large number of samples collected from many locations. When compared to Malaise traps, sweep nets were found to be more effective in collecting representative samples of forest Hymenoptera because they could be used in many parts of the habitat, although the traps were better in wet conditions (Noyes 1989).

Malaise traps

Malaise traps are used to capture insects that are moving about above low vegetation. They have been used in studies of succession (Belshaw 1992) and can be used to detect the direction of movement of insects through a habitat. This type of trap will be particularly effective at collecting insects associated with the understory environment (category b) and if located appropriately could also capture insects associated with specific resources, e.g. gaps and dead wood (categories c and d). Malaise traps will also collect insects associated with specific understory plants, provided these are actively flying through the habitat.

Malaise traps are a form of flight interception trap. They function on the basis that flying insects will not detect the vertical portion of the net and when they collide with it will cease flying and close their wings. Many insects are phototactic (Wigglesworth 1972) and once they have alighted on a surface will move upwards towards the light. The traps make use of this behavior by guiding upwardly moving insects by means of a sloping net roof into a collecting jar.

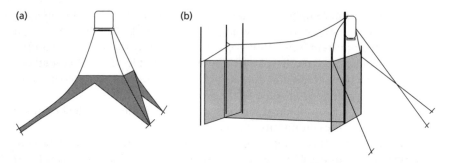

Fig. 4.2 Two Malaise trap designs with taught net walls and collection bottles set at the highest point. (a) Cornell design (4-directional); (b) Townes model (bi-directional). After Matthews & Matthews (1983).

The most common Malaise trap design has two opposing vertical net walls set in a T shape and supported by a pole and guy ropes. The mesh for these walls is usually black to reduce visibility. The walls are topped by an angled net roof which slopes upwards at the junction of the T (Fig. 4.2), the roof may be constructed of white or black mesh. At the highest point a removable collecting pot is set which is filled with preservative fluid. In some designs the trap is baited, for example with dry ice to attract biting flies (Strickler & Walker 1993).

This type of flight interception trap was first designed by Malaise in 1937. There have been a number of subsequent variations in the design with two main types, the four-directional Cornell design (Matthews & Matthews 1983) and the bi-directional Townes model (Townes 1962, 1972) (Fig. 4.2). In a comparative study, Matthews & Matthews (1983) found the Townes model to be more effective partly because it captured more insects from higher up in the air column. Alternative designs have been made to hang in low or high vegetation, constructed again from mesh but with four intersecting sheets and a non-mesh roof and base tray (a composite interception trap; Basset 1985, Winchester & Scudder 1993, Springate & Basset 1996). These can be suspended from branches or ropes within the canopy or across flight paths between trees.

Location and climate can affect the efficiency of Malaise traps. Location is important because insects often follow specific flight paths through vegetation (Hutcheson 1990, Matthews & Matthews 1983). For this reason traps are usually set along vegetation edges and at intersections. Some investigators suggest a north–south orientation with the head of the trap facing the sun's zenith (Noyes 1989). Traps set in exposed and sheltered situations in the same habitat type may have different efficiencies (Noyes 1989), due partly to differences in flight behavior and partly to microclimate variation. However, where several traps are used they should not be located too closely together, or samples will no longer be independent.

Mesh size and color can affect the types of insects captured. For

Hymenoptera, 300 holes per cm^2 are recommended, otherwise very small individuals will be lost (Noyes 1989), but it is more common to use a mesh of 100 holes per cm^2 (Darling & Packer 1988). It has been found that green traps catch fewer tabanids than natural-colored mesh (Roberts 1970), but that the use of a black base with white roof can increase the effectiveness of the trap (Matthews & Matthews 1983). Using a bicolored trap can alter the family balance of the catch (Darling & Packer 1988); for example, coarse-mesh bicolor traps capture greater numbers of aculeates and ichneumonids at the expense of small-bodied Hymenoptera. However, on balance, bicolored traps are recommended for most studies.

Since Malaise traps are activity dependent, captures will be affected by climatic conditions, more specifically by the number of degree days available for flight and on seasonal variations in height of flight (Matthews & Matthews 1983). Scheduling of sample collection therefore needs careful consideration. Malaise traps have the advantage that they are generally inexpensive, and can be easily set up and left in secure locations. The greatest drawback, however, is that the traps are easily subject to windthrow and are therefore difficult to use in exposed sites.

Clearly Malaise traps will capture insects actively flying through the forest understory, including groups such as Hymenoptera, Diptera, Coleoptera, and Lepidoptera. However, they will also capture to a lesser extent insects that are carried passively by air currents and alight on the net, such as Psocoptera and Collembola. They have been used to survey Diptera including Tabanidae (Strickler & Walker 1993), and Disney et al. (1982) found that Calypterates were very effectively collected. Hymenoptera are also successfully surveyed with Malaise traps (Darling & Packer 1988, Noyes 1989). Disney et al. (1982) note that Malaise traps are exceptionally effective at capturing swarms of particular species, and indeed these can overwhelm a sample bottle, making it difficult to sort out rarely occurring species.

Malaise traps only collect a portion of the whole understory community, but they compare favorably with window traps (see below) for collecting insects such as Hymenoptera, probably because they retain more of the insects which alight on them (Noyes 1989). Although climate-dependent, Malaise traps do function when the conditions are damp or wet and therefore have advantages over vacuum-sampling and sweep-netting, where effectiveness is reduced in understory vegetation that is constantly wet (Noyes 1989).

Window traps

Window traps are similar to Malaise traps in principle, i.e. they are flight interception traps. They are usually constructed from a vertical panel of Perspex or mesh with a drop tray below (often plastic piping with drain holes near the top to prevent flooding) (Peck & Davis 1980). They are particularly useful for collecting beetles, which typically close their wings on encountering the wall and therefore fall into the drop tray (Masner & Goulet 1981). Other weakly flying insects are caught, but some will be able to take off again and therefore will not be trapped consistently. Impregnation of the trap with insecticide can increase the capture rate of small insects such as microhymenoptera (Masner & Goulet 1981). In the forest environment these traps are often used to collect insects associated with specific understory resources such as dead wood. They are particularly useful for surveying dead-wood beetles, a tremendously important part of the forest fauna.

Color traps

The response of insects to color, noted for Malaise traps, can be exploited to enhance the efficiency of collection. Color traps, e.g. sticky or water filled, have been used successfully to capture insects in temperate and tropical forest understory as well as in heather and in structurally complex crops such as oilseed rape (Disney et al. 1982, Canaday 1987, Noyes 1989, Usher 1990, Burke & Goulet 1998, Bowie 1999). These traps usually collect insects that are flying above or between plants within the understory, insects that may belong to category a (those associated with particular plants) or category d (resource specialists such as predators and parasitoids). Since color traps are essentially active traps they will be more selective than previously described techniques such as vacuum-sampling.

The principle is that insects are attracted to colors that mimic the spectral reflectance of a habitat resource, e.g. flower, leaf, and stem colors. Predatory and parasitic insects are attracted either to the color of plants on which their prey items feed or to the color of alternative food sources such as pollen-bearing flowers (Bowie 1999). Trap color therefore plays a significant role in the effectiveness with which different insect groups are caught (Kirk 1984).

Although color traps may be sticky or water-filled, here we concentrate on the latter. Water traps are usually constructed from a colored tray, bowl, or bucket that is filled with water (some insect groups may be attracted by the water itself; Noyes 1989). Holes drilled near the top will prevent overflow after rain. The effectiveness of the trap is increased if detergent is added to reduce the surface tension, and it may be necessary to add a preservative such as sodium benzoate or antifreeze, if traps are emptied infrequently. In some habitats and seasons it may be essential to service traps daily to keep up with the capture rates.

Color traps will collect insects that are aerial either because they are actively flying, or because they are part of the air plankton. There will be a higher probability of capture of insects that have control and therefore selectivity over landing location. Typical groups caught are the Diptera, Homoptera, Hymenoptera, and Coleoptera.

Color traps may be set up in lines or grids and are commonly spaced at intervals of a few meters. The choice of trap distribution is dependent on the research question. For example if the aim of the study is to determine species richness for an area then a large grid system may be appropriate, whereas the investigation of a habitat gradient may suggest a line of traps. However, it is clear that trap location within the habitat can significantly affect the catch (e.g. edge effects; Bowie 1999), so this must be "factored out" if it is not of interest.

Height of trap above or within vegetation has been found to influence the catch for insects in rainforest understory, heather and oil seed rape and therefore should be standardized (Usher 1990, Bowie 1999). This could be done in three ways, depending on the target insect groups: (i) standardizing to just above the height of the understory so that visibility is consistent; (ii) standardizing to the mean height of the food resource (e.g. flower or fruit); (iii) standardizing to a defined height from the ground, perhaps determined as optimum in a pilot study.

Selecting trap color will again depend upon the research question because there is some variability in the responses of insect groups (Table 4.1).

Colored water traps have been compared with other collection techniques. In field crops they have been found to be as effective for adult syrphids as sticky traps set in several orientations and more effective for larvae when placed on the ground (Bowie 1999). In woodland, white traps were found to be better than Malaise traps for capturing Syrphidae (Disney et al. 1982). Thus colored water traps are particularly recommended for this group of insects. In comparison with other collection techniques, however, colored traps require much more sampling effort and are therefore recommended for specialist monitoring (e.g. fauna associated with Gramineae or flowering plants) rather than for general surveys. They can be modified to increase their effectiveness for monitoring specific insects either by sheltering (Finch 1992, Coon & Rinicks 1962) or by shape modification. Shape modification allows the traps to mimic plant resources in pattern as well as in color (e.g. yellow spheres on a contrasting background to capture insects in citrus plantations; Cornelius et al. 1999).

Color traps are one example of a bait trap. There is of course a plethora of other baits that will attract insects, including pheromones, CO_2, plant or flower mimics, and there are examples of human bait being used to trap biting flies and mosquitoes (Costantini et al. 1998).

Table 4.1 Preferences of insect groups for specific colored water traps.

Trap color	
White	More effective for Phoridae and non-cereal aphids
Yellow	More effective for Agromyzidae; cereal aphids; Hymenoptera, e.g. Chalcodoidea, Coccinellidae
Orange	Higher densities of Chironomidae

Disney et al. (1982), Noyes (1989), De Barro (1991), Parajulee and Slosser (2003).

Foliage bagging

The techniques described so far for medium-height vegetation have all been relative methods of estimating population parameters, i.e. the captures are recorded relative to trapping effort. These can be converted to absolute estimates by calibration using actual counts of insects on plants or by using techniques that produce absolute estimates such as foliage bagging. Foliage or branch bagging is very straightforward, but usually involves removing part of the vegetation, an activity that may not be desirable at sensitive forest sites.

Bagging involves drawing a net, or a cotton or plastic bag, over the foliage in such a way as to disturb as few insects as possible. In a survey of insects on cotton, Byerly et al. (1978) devised a cotton bag that could be introduced over a branch and drawn back to the branch base. The insects were then allowed to settle for 24 hours and the bag rapidly drawn up over the foliage, trapping a high proportion of the insect community. The branch is then clipped off. In some surveys anaesthetizing chemicals are introduced into the bag to prevent loss of specimens when the bag is opened to extract the catch (Basset et al. 1997).

The technique has the advantage of being absolute, and when carried out efficiently can produce representative samples (Byerly et al. 1978). However, there are also a number of disadvantages. Where branches are long and bags do not stretch the whole length, samples are biased towards herbivores and towards species that feed in actively growing tissue. Insect samples are dominated by less active groups such as Collembola, Psocoptera, and sedentary larvae, e.g. Diptera (Blanton 1990).

The branch bagging method has been modified successfully for use in tall vegetation and even tree canopies where it is used as an alternative to chemical knockdown (Majer & Recher 1988, Schowalter 1995) (see Chapter 7).

Tall vegetation including small trees

Understory shrubs and trees are a structurally significant part of managed and unmanaged forest ecosystems. They are particularly prevalent in tropical biomes, but have been, and still remain, economically important in temperate regions (e.g. coppice with standards).

It is possible and indeed effective to use chemical knockdown (Chapter 7) to sample from tall understory vegetation, provided the plants are screened off from the canopy above (Floren & Linsenmair 1998). Screening can be arranged by constructing a tent above the target species to reduce chemical spread and to prevent insects from the high canopy falling into collection sheets. This technique is likely to produce the most comprehensive and representative samples, although of course erecting the tent will disturb mobile insects and so communities need to be allowed time to settle before treatment begins.

Other sampling techniques such as beating, branch clipping, and bagging will be limited to the space that the investigator can reach from the ground unless access systems such as ladders, ropes, and towers are used. Once the ground is left behind, sampling the understory becomes similar to sampling the canopy (see Chapter 7).

Beating trays

Shrubs and small trees are very difficult to sample quantitatively, and in the past it has been common to use a beating tray. This is almost entirely a qualitative method to be used when constructing species lists or when collecting life-cycle data for particular species. However, some insects are so easily dislodged that using a beating tray can give a relative estimate of population densities (e.g. Geometrid larvae; White 1975).

A tray is held underneath the shrub or small tree and the branches tapped sharply. This causes insects to fall from the branches onto the tray. Insects are then collected up by pooter (aspirator) or fine brush, or funnelled into a collecting jar. The tray is usually constructed from cotton stretched onto a folding wooden frame. White cotton is commonly used since the contrasting color allows even small insects to be seen and removed. If the tray has a standardized area (e.g. 1 m^2) and the number of taps or hits is also standardized, it is possible to make at least some comparisons between vegetation types.

The main biases in this technique lie in the selectivity with which insects will fall from the vegetation, the strength of the beating action, and biases involved in aspiration of samples. It is inevitable that larger insects will be preferentially collected from the sheet and more active ones may escape. If used in surveys of forest understory trees or shrubs then limiting the number of people involved will produced more standardized samples.

Beating trays are often used to collect insects such as Chrysomelidae, Lepidoptera, and Heteroptera. In a comparison between beating trays (drop cloth method), sweep nets, and absolute sampling on cotton for *Lygus lineolaris*, it was found that neither beating nor sweeping captured as many as the absolute method and both were adversely affected by plant height. The beating method was found to be more effective than the sweep net for estimating population densities but was more time-consuming (Snodgrass 1993).

Herms et al. (1990) compared the use of a beating tray in honeylocust trees *Gleditsia triacanthos* with vacuum-sampling. For the most abundant insects (Homoptera and Heteroptera) the beating tray was found to sample early instars more effectively but vacuum sampling was better for adults. Differences were ascribed to the capacity of the beating action to dislodge the small insects from unfolding leaflets. However they were able to use the vacuum sampler over a much larger area of the lower canopy of the trees and therefore samples collected with this technique are likely to be more representative of the whole community.

Vacuum sampling in shrubs and trees

Vacuum sampling can be used quite effectively to collect insects from shrubs and small trees. Three main methods can be employed: holding the nozzle over vege-tation, e.g. slipping the hose over a branch; searching an area of the canopy for a set length of time (Herms et al. 1990); or vacuuming the foliage along a transect or across a 180° arc (Buffington & Redak 1998). The most appropriate method will depend on the density and height of the vegetation, but in all cases the narrow-hosed models will be much easier to use than the heavy wide-hosed types.

Use of secondary characteristics in sampling from the understory

The collection techniques discussed so far in this chapter have been absolute and relative methods of sampling that rely on being able to detect, capture, and remove insects from the habitat. They are generally destructive methods. How-ever, it is possible to make estimates of population densities of insects from secondary characteristics such as spider webs, leaf mines, and even insect frass (Sterling & Hambler 1988, Ozanne & Bell 2003). The use of a feature such as leaf mines will give an accurate measure of population density without calibration, whereas the use of an indicator must be calibrated (Southwood & Henderson 2000). For example, Thorpe and Ridgway (1994) were able to estimate the den-sities of gypsy moth *Lymantria dispar* in oak woodland by measuring the frass drop per unit area and the yield of frass pellets per larva (Table 4.2). A further ad-vantage of using this method to investigate insect populations is that estimates of energy flow can also be made.

Conclusions

It is clear that when they are applied appropriately each of the techniques des-cribed in this chapter can be used successfully to survey insect populations and to investigate community structure and dynamics. To determine which tech-niques are the most appropriate (and in many studies several complementary methods will be needed) some preliminary information should be gathered.

Firstly, data on the structure and composition of the understory habitat should be collected, since the vegetation and the insects that it supports will vary according to forest biome. For example, in tropical rainforests the under-story has a number of structural layers below the high canopy, together with a wide range of vascular epiphytes, whereas in temperate forests understory trees are scarcer, but structurally diverse shrubs and non-vascular epiphytes are com-mon. In addition, the management of primary, secondary, or plantation forest has a significant impact on the structure and composition of the understory community. Information about the vegetation will help the investigator to con-

Table 4.2 Calibration of frass output to larval density for *Lymantria dispar* (gypsy moth) in oak (*Quercus* spp.), following Thorpe and Ridgway (1994).

Frass collection	Frass collection from canopy using 50 cm × 50 cm polyethylene funnels for 12–16 hours from afternoon to morning, recorded as frass density per m²
Frass production	Frass yield per larva recorded for 40 larvae in cups provisioned with oak leaves
Larval density	Mean density of larvae per tree, estimated using the equation: $C = x_d/x_y$ $C = 1/$(area sampled by each frass sampling device) x_d = mean drop (frass per trap) x_y = mean yield (frass per larva) (Leibold & Elkinton 1988)
Extrapolation to whole tree densities	Measure the perimeter of the dripline of each tree, calculate the area and then multiply by larval density

Calculation carried out for fourth instar and again just before pupation.
Frass drop: may be used as a relative population estimate.
Larval density: an absolute population estimate but subject to greater error than frass drop.

sider which techniques are likely to yield samples representative of the habitat. The second step is to gather some baseline information about the insect groups present in the understory and their distribution or the category (see *Introduction*) into which they best fit. These data can be obtained in a pilot study in which several collecting methods are tested. Armed with this background information it will be possible to develop clear experimental hypotheses that will direct the choice of techniques.

The understory is a rich environment and a very productive one for research aiming to answer fundamental ecological questions or to investigate the impact of habitat manipulation and management. With the right tools we can learn much more about the role of this forest layer in ecosystem dynamics.

References

Balogh, J. & Loska, I. (1956) Untersuchungen über die Zoozönose des Luzernenfeldes. *Acta Zoologica Academiae Scientiarium Hungaricae*, **2**, 17–114.
Basset, Y. (1985) Comparaison de quelques méthodes de piégeage de la faune dendrobie. *Bulletin Romand d'Entomologie* 3, 1–14.
Basset, Y., Springate, N.D., Aberlanc, H.P., & Delvare. G. (1997) A review of methods for sampling arthropods in tree canopies. In *Canopy Arthropods* (ed. N. Stork, J. Adis, & R. Didham), pp. 27–52. Chapman & Hall, London.

Belshaw, R. (1992) Tachinid (Diptera) assemblages in habitats of a secondary succession in southern Britain. *The Entomologist*, **111**, 151–161.

Blanton, C.M. (1990) Canopy arthropod sampling: a comparison of collapsible bag and fogging methods. *Journal of Agricultural Entomology*, **7**, 41–50.

Bowden, J., Haines, I.H., & Mercer, D. (1976) Climbing Collembola. *Pedobiologia*, **16**, 298–312.

Bowie, M.H. (1999) Effects of distance from field edge on aphidophagous insects in a wheat crop and observations on trap design and placement. *International Journal of Pest Management*, **45**, 69–73.

Buffington M.L. & Redak, R.A. (1998) A comparison of vacuum sampling versus sweep-netting for arthropod biodiversity measurements in California coastal sage scrub. *Journal of Insect Conservation*, **2**, 99–106.

Burke, D. & Goulet, H. (1998) Landscape and area effects on beetle assemblages in Ontario. *Ecography*, **21**, 472–479.

Byerly, K.F., Gutierrez, A.P., Jones, R., & Luck, R.F. (1978) Comparison of sampling methods for some arthropod populations in cotton. *Hilgardia*, **46**, 257–282.

Canaday, C.L. (1987) Comparison of insect fauna captured in six different trap types in a Douglas-fir forest. *Canadian Entomologist*, **119**, 1101–1108.

Coon, B.F. & Rinicks, H.B. (1962) Cereal aphid capture in yellow baffle trays. *Journal of Economic Entomology*, **55**, 407–408.

Cornelius M.L., Duan, J.J., & Messing, R.H. (1999) Visual stimuli and the response of female oriental fruit flies (Diptera: Tephritidae) to fruit-mimicking traps. *Journal of Economic Entomology*, **92**, 121–129.

Costantini, C., Sagon, N.F., Sanogo, E., & Merzagora, L. (1998) Relationship to human biting collections and influence of light and bednet in CDC light-trap catches of west African malaria vectors. *Bulletin of Entomological Research*, **88**, 503–511.

Darling, D.C. & Packer, L. (1988) Effectiveness of malaise traps in collecting Hymenoptera: the influence of trap design, mesh size and location. *Canadian Entomologist*, **120**, 787–796.

De Barro, P.J. (1991) Attractiveness of four colours of traps to cereal aphids (Hemiptera: Aphididae) in Southern Australia. *Journal of the Australian Entomological Society*, **30**, 263–264.

Dietrick, E.J. (1961) An improved backpack motor fan for suction sampling of insect populations. *Journal of Economic Entomology*, **54**, 394–395.

Disney, R.H.L., Erzinclioglu, Y.Z., Henshaw, D.J. de C., et al. (1982) Collecting methods and the adequacy of attempted fauna surveys, with reference to the Diptera. *Field Studies*, **5**, 607–621.

Dondale, C.D., Nicholls, C.F., Redner, J.H., Semple, R.B., & Turnbull, A.L. (1971) An improved Berlese–Tullgren funnel and flotation separator for extracting grassland arthropods. *Canadian Entomologist*, **103**, 1549–1552.

Ellington, J., Kiser, K., Ferguson, G., & Cardenas, M. (1984) A comparison of sweepnet, absolute and Insectavac sampling methods in cotton ecosystems. *Journal of Economic Entomology*, **77**, 599–605.

Finch, S. (1992) Improving the selectivity of water traps for monitoring populations of cabbage root fly. *Annals of Applied Biology*, **120**, 1–7.

Floren, A. & Linsenmair, K.E. (1998) Diversity and recolonisation of arboreal Formicidae and Coleoptera in a lowland rain forest in Sabah, Malaysia. *Selbyana*, **19**, 155–161.

Gadagkar, R., Chandrashekara, K., & Nair, P (1990) Insect species diversity in the tropics: sampling methods and a case study. *Journal of the Bombay Natural History Society*, **87**, 337–353.

Gray, H. & Treloar, A. (1933) On the enumeration of insect populations by the method of net collection. *Ecology*, **14**, 356–367.

Hand, S.C. (1986) The capture efficiency of the Dietrick vacuum insect net for aphids on grasses and cereals. *Annals of Applied Biology*, **108**, 233–241.

Henderson, I.F. & Whittaker, T.M. (1977) The efficiency of an insect suction sampler in grassland. *Ecological Entomology*, **2**, 57–60.

Herms, D.A., Neilsen, D.G., & Davis Snydor, T. (1990) Comparison of two methods for sampling arboreal insect populations. *Journal of Economic Entomology*, **83**, 869–874.

Hutcheson, J. (1990) Characterization of terrestrial insect communities using quantified, Malaise-trapped Coleoptera. *Ecological Entomology*, **15**, 143–151.

Kirk, W.D.J. (1984) Ecologically selective coloured traps. *Ecological Entomology*, **9**, 35–41.

Lawrence, E. (ed.) (1995) *Henderson's Dictionary of Biological Terms*. 11th edn. Longman, London.

Liebhold, A.M. & Elkinton, J.S. (1988) Techniques for estimating the density of late-instar gypsy moth, *Lymantria dispar* (Lepidoptera: Lymantriidae) populations using frass drop and frass production measurements. Environmental Entomology, **17**, 381–384.

Linders, E.G.A. (1995) Biology of the weevil *Trichosirocalus troglodytes* and impact on its host *Plantago lanceolata*. *Acta Oecologia*, **16**, 703–718.

Lowman, M., Moffet, M., & Rinker, H.B. (1993) A new technique for taxonomic and ecological sampling in rain forest canopies. *Selbyana*, **14**, 75–79.

Macleod, A., Wratten, S.D., & Harwood, R.W.J. (1994) The efficiency of a new lightweight suction sampler for sampling aphids and their predators in arable land. *Annals of Applied Biology*, **124**, 11–17.

Majer, J.D. & Recher, H.F. (1988) Invertebrate communities on Western Australian eucalypts: a comparison of branch clipping and chemical knockdown procedures. *Australian Journal of Ecology*, **13**, 269–278.

Malaise, R. (1937) A new insect trap. *Entomologisk Tidskrift*, **58**, 148–160.

Masner, L. & Goulet, H. (1981) A new model of flight interception trap for some hymenopterous insects. *Entomological News*, **92**, 199–202.

Matthews, R.W. & Matthews, J.R. (1983) Malaise traps: the Townes model catches more insects. *Contributions to the American Entomological Institute*, **20**, 428–432.

Noyes, J.S. (1989) A study of five methods of sampling Hymenoptera (Insecta) in a tropical rainforest, with special reference to the Parasitica. *Journal of Natural History*, **23**, 285–298.

Oliveira, M.L. & Campos, L.A.O. (1996) Preferencia por estratos florestais e por substancias odoriferas em abelhas Euglossinae (Hymenoptera, Apidae). *Revista Brasileira de Zoologia*, **13**, 1075–1085.

Ozanne, C.M.P. & Bell J.R. (2003) Collecting arthropods and arthropod remains for primate studies. In *Field and Laboratory Methods in Primatology: a Practical Guide* (ed. J. Setchell & D. Curtis), pp 214–227. Cambridge University Press, Cambridge.

Parajulee, M.N. & Slosser, J.E. (2003) Potential of yellow sticky traps for lady beetle survey in cotton. *Journal of Economic Entomology*, **96**, 239–245.

Peck, S.B. & Davis, A.E. (1980) Collecting small beetles with large-area "window traps". *Coleopterists' Bulletin*, **34**, 237–239

Roberts, R.H. (1970) Color of malaise trap and the collection of Tabanidae. *Mosquito News*, **30**, 567–571.

Schowalter, T.D. (1995) Canopy invertebrate community response to disturbance and the consequences of herbivory in temperate and tropical forests. *Selbyana*, **16**, 41–48.

Simon, U. (1995) Untersuching der Stratozönosen von Spinnen und Weberknechten (Arach.: Araneae, Opilionida) und der Waldkiefer (*Pinus sylvestris* L.). Diss.FB. Umweltund Planungswissenchaften, TU Berlin. Wissenschaft und Technick Verlag, Berlin.

Snodgrass, G. (1993) Estimating absolute density of nymphs of *Lygus lineolaris* (Heteroptera: Miridae) in cotton using drop cloth and sweep-net sampling methods. *Journal of Economic Entomology*, **86**, 1116–1123.

Southwood, T.R.E. & Henderson, P.A. (2000) *Ecological Methods*. 3rd edn. Blackwell Science, Oxford.

Springate, N.D. & Basset, Y. (1996) Diel activity of arboreal arthropods associated with Papua New Guinea trees. *Journal of Natural History*, **30**, 101–112.

Spurr, S.H. (1980) *Forest Ecology*. 3rd edn. Wiley, New York.

Sterling, P.H. & Hambler, C. (1988) Coppicing for conservation: do hazel communities benefit? In *Woodland Conservation and Research in the Clay Veil of Oxfordshire and Buckinghamshire* (ed. K. Kirby & F.J. Wright), pp. 69–80. Research and Survey in Nature Conservation 15. NCC, Peterborough.

Stewart, A.J.A. & Wright, A.F. (1995) A new inexpensive suction apparatus for sampling arthropods in grassland. *Ecological Entomology*, **20**, 98–102.

Strickler, J.D. & Walker, E.D. (1993) Seasonal abundance and species diversity of adult Tabanidae (Diptera) at Lake Lansing Park-North, Michigan. *Great Lakes Entomologist*, **26**, 107–112.

Sutherland, W.J. (1996) *Ecological Census Techniques: a Handbook*. Cambridge University Press, Cambridge.

Thornhill, E.W. (1978) A motorised insect sampler. *Pest Articles and News Summaries*, **24**, 205–207.

Thorpe, K.W. & Ridgway, R.L. (1994) Effects of trunk barriers on larval gypsy moth (Lepidoptera: Lymantriidae) density in isolated- and contiguous-canopy oak trees. *Environmental Entomology*, **23**, 832–836.

Townes, H. (1962) Design for a Malaise trap. *Proceedings of the Entomological Society of Washington*, **64**, 253–262.

Townes, H. (1972) A light-weight Malaise trap. *Entomological News*, **83**, 239–247.

Usher, M.B. (1990) Assessment for conservation values: the use of water traps to assess the arthropod communities of heather moorland. *Biological Conservation*, **53**, 191–198.

White, T.C.R. (1975) A quantitative method of beating for sampling larvae of *Selidosema suavis* (Lepidoptera; Geometridae) in plantations in New Zealand. *Canadian Entomologist*, **107**, 403–412.

Wigglesworth, V.B. (1972) *The Principles of Insect Physiology*. 7th edn. Cambridge University Press, Cambridge.

Wilson, L.T. & Gutierrez, A.P. (1980) Within-plant distribution of predators on cotton: comments on sampling and predator efficiencies. *Hilgardia*, **48**, 1–11.

Winchester, N.N. & Scudder, G.G.E. (1993) *Methodology for Sampling Terrestrial Arthropods in B.C.* Resources Inventory Committee, B.C. Ministry of Lands and Parks, pp 1–32.

Wise, I.L. & Lamb, R.J. (1998) Sampling plant bugs, *Lygus* spp. (Heteroptera: Miridae), in Canola to make control decisions. *The Canadian Entomologist*, **130**, 837–851.

Index of methods and approaches

Methodology	Topics addressed	Comments
Suction or vacuum sampling	Questions of association with specific plant communities, resources and locations, particularly in habitats such as short grass through to shrubs and small trees. Absolute estimates of population density. Community structure data.	Samples a wide range of taxonomic groups. Particularly effective for aphids and associated predatory beetles such as *Tachyporus* spp (Staphylinidae), medium to large Collembola and Homoptera. More effective than sweeping for collecting Diptera and smaller Hymenoptera. Large heavy species are under-sampled, e.g. large aculeate Hymenoptera. W-type models have been found to collect aphids poorly from soil level and to favor insects in the top sections of plants.
Sweep-net capture	Questions of association with whole plant communities or specific species if there is sufficient foliage volume. Relative estimates of population density.	Samples insects that are known to be present in the top part of the vegetation, e.g. around seed pod. Hymenoptera, Diptera, small Coleoptera, and arachnids are quite well sampled. Lepidoptera are under-sampled.
Malaise traps	Studies of successional change. Detection of the direction of movement of insects through a habitat. Questions of association with the understory environment. Association with specific resources if appropriately located, e.g. gaps and dead wood. Association with specific understory plants if insects are flighted. Relative estimates of population density and species richness.	Small Hymenoptera Hymenoptera, Diptera (especially Tabanidae and Calypterates), Coleoptera, Lepidoptera.

Continued

Methodology	Topics addressed	Comments
Window traps	Association with specific understory resources such as dead wood.	Coleoptera, e.g. dead-wood beetles.
	Relative estimates of density and species richness.	Increased capture rate of small insects such as Microhymenoptera
Color traps	Association with particular plants.	Diptera, Homoptera, Hymenoptera, Coleoptera.
	Resource specialization, e.g. predators and parasitoids.	Brown traps: capture significantly higher densities of Chironomidae.
	Relative estimates of population density and species richness.	
		Yellow traps: more effective for Agromyzidae and cereal aphids; Hymenoptera such as the Chalcidoidea; Coccinellidae.
		White traps: for Phoridae, some species of Syrphidae, non-cereal aphids.
Foliage bagging	Absolute estimates of population density and species richness per unit area and per unit plant material.	Collembola, Psocoptera, and sedentary larvae, e.g. Diptera.
	Association with specific plant species and with specific plant parts.	
Beating tray	Qualitative sampling, relative population estimates, presence absence data, population structure data.	Chrysomelidae, Lepidoptera, Homoptera, Heteroptera, particularly early instars.
Secondary characteristics	Relative or absolute estimates of population density.	Lepidoptera, Coleoptera.

CHAPTER 5

Sampling insects from trees: shoots, stems, and trunks

MARTIN R. SPEIGHT

Introduction

Many insect species can be found living on the outside of twigs, shoots, and bark (Speight & Wylie 2001). So aphids, scale insects, booklice, and thrips are relatively easy to find and identify. However, unlike foliage, where the majority of insects live on the outside of the plant, a large proportion of insects which feed on the shoots, stems, and trunks of trees are to be found inside the plant tissue. Thus borers, tunnelers, and gallers tend to remain hidden for much of their life cycles, and in many cases this concealed habit renders them very difficult to even detect, never mind count. Very often, their presence inside shoots, under bark, or in the timber is only advertised by the effects they have on the host plant. Dead or deformed shoots, holes in bark, or the exudation of frass and resin, are important indicators of the insect within. Another problem often confronts the entomologist even when the insects are found: many active borers are in the larval stage, and their taxonomy, without access to the adult specimen, can be difficult or even impossible. Finally, with such a huge size range of breeding sites and food items, from the smallest twigs to the thickest trunks, both from living trees and also from dead or moribund ones, the sampling procedures employed for these types of insects are many and varied indeed. Table 5.1 summarizes these main habitats, and introduces some of the important insect groups to be found in each one.

Detection

Insects such as aphids, scales, and mealybugs provide relatively obvious indications of their presence, either as the individual insects themselves, or in the wax or so-called "wool" that they produce. Some, such as horse chestnut scale *Pulvinaria regalis*, beech scale *Cryptococcus fagi*, and pine woolly aphid *Pineus pini* are detectable from some distance, since in large densities they coat the stems or trunks of their host trees with white exudations under which they live and/or lay their eggs. However, because so many habitats mentioned in Table 5.1 are internal to the host plant, it may not be at all obvious in many cases that insects are present, and one of the first problems facing more rigorous sampling

77

Table 5.1 Major stem habitats on trees and their associated insect groups.

Habitat	Insect activity	Insect examples
Shoots/stems	Sap feeding	Scale insects (Hemiptera: Coccidae) Woolly aphids (Hemiptera: Adelgidae) Aphids (Hemiptera: Aphididae)
	Boring	Bark beetles (Coleoptera: Scolytidae) Moth larvae (Lepidoptera: Pyralidae) Longhorn beetles (Coleoptera: Cerambycidae)
	Girdling	Longhorn beetles (Coleoptera: Cerambycidae)
	Chewing	Longhorn beetles (Coleoptera: Cerambycidae) Grasshoppers & crickets (Orthoptera: Acrididae & Gryllidae) Moth larvae (Lepidoptera: various)
Bark surface	Sap feeding	Scale insects (Hemiptera: Coccidae) Woolly aphids (Hemiptera: Adelgidae) Aphids (Hemiptera: Aphididae)
	Detritus feeders	Book lice (Psocoptera)
Bark interior	Borers	Moth larvae (Lepidoptera: Pyralidae)
Bark/sapwood interface	Borers	Bark beetles (Coleoptera: Scolytidae) Longhorn beetles (Coleoptera: Cerambycidae)
	Fungivores	Fungus flies (Diptera: Mycetophilidae) Ambrosia beetles (Coleoptera: Platypodidae)
Timber	Borers	"Woodworm" beetles (Coleoptera: Anobiidae) Moth larvae (Lepidoptera: Cossidae & Hepialidae) Longhorn beetles (Coleoptera: Cerambycidae) Jewel beetles (Coleoptera: Buprestidae) Powder post beetles (Coleoptera: Lyctidae)

procedures is deciding whether or not a tree contains anything to sample in the first place. Numerous indications can be observed in forest stands which betray the existence of insect activity within the plant tissues, and Table 5.2 provides a summary of some of the most obvious.

Sampling methods

The sampling methods described in this chapter are arranged basically by the type of habitat or part of the tree where they are normally found (see Table 5.1). Many types of sampling tactic are described, predominantly via the use of examples sourced from the published literature.

Table 5.2 Examples of evidence of insect activity concealed within plant tissues, either currently or in the past.

Part of plant	Evidence	Causal agent
Twigs & shoots	Dead or discolored foliage, general tree dieback	Boring by moth larvae or beetle adult e.g. *Dioryctria cristata* (pine shoot moth), *Tomicus piniperda* (pine shoot beetle)
	Bent, twisted, or deformed shoot, foliage still green	Boring by moth larvae e.g. *Rhyacionia buoliana* (European pine shoot moth)
	Broken or decayed shoot, often with orange-brown sap and resin exudations	Boring by moth larvae e.g. *Hypsipyla grandella* (mahogany shoot borer)
	Periodic swellings along thin stems or twigs	Boring by beetle larvae e.g. *Saperda populnea* (poplar longhorn)
	Swellings or deformed buds (galls)	Gall wasps, gall woolly aphids e.g. *Andricus* spp
External bark	Tan to brown fine granules or dust in bark crevices	Entrance holes of bark beetles in moribund trees e.g. *Ips* spp, *Scolytus* spp
	Heavy sap or resin exudation	Entrance holes of bark beetles in relatively vigorous trees e.g. *Dendroctonus micans* (Spruce bark beetle)
	White to cream powder or dust on bark or at base of tree	Entrance holes of ambrosia beetles e.g. *Trypodendron* spp
	Small circular holes in bark	Exit holes of bark or ambrosia beetles e.g. *Scolytus* spp, *Trypodendron* spp
	Large circular holes in bark	Exit holes of woodwasps e.g. *Sirex* spp
	Medium to large oval holes in bark	Exit holes of longhorn beetles e.g. *Phoracantha* spp
	Medium to large roughly circular holes in bark, accompanied by oozing resin and wood debris	Bore holes of larvae of goat or wood moths e.g. *Cossus cossus*
	Swellings running spirally around trunk	Tunnels of larvae of varicose borers e.g. *Agrilus sexdentatus* (Buprestidae)
	Dry earthern tunnels running mainly up and down trunks and stems	Termites e.g. *Coptotermes* spp
Internal bark/ sapwood surface	Engravings with many branches from a central gallery	Tunnels of bark beetles e.g. *Tomicus, Scolytus, Ips* spp
	Shallow but wide engravings, often containing compacted dust or frass	Tunnels of longhorn or roundhead beetles e.g. *Tetropium* spp

Contd p. 80

Table 5.2 (contd)

Part of plant	Evidence	Causal agent
	Deep oval holes in sapwood	Pupation tunnels of longhorn beetles e.g. *Phymatodes* spp
	Shallow oval holes in sapwood, often surrounded by wood fibers	Pupation chambers of weevils e.g. *Pissodes* spp
	Large circular holes in sapwood	Exit holes of woodwasps e.g. *Sirex* spp
	Small circular holes in sapwood, no blue/black stain, exuding white dust	Entrance/exit holes of powder-post beetles e.g. *Lyctus* spp
	Small circular holes in sapwood, no blue/black stain, no white dust	Entrance/exit holes of wood worm e.g. *Anobium* spp
	Small circular holes in sapwood, surrounded by blue or black stain	Entrance holes of ambrosia beetles e.g. *Trypodendron* spp, *Platypus* spp
Heartwood or inside main trunk	Large random tunnels, often with smooth, sculptured surface texture	Wood-boring termite galleries e.g. *Coptotermes* spp
	One or two wide tunnels, usually circular in cross section	Larval tunnels of goat or wood moths e.g. *Cossus* spp, *Xyleutes* spp
	Small tunnels or chambers, surrounded by blue/black staining in wood tissue	Larval/pupal chambers of ambrosia beetles e.g. *Trypodendron* spp, *Platypus* spp

Shoots and twigs: external

Galls

As with any other part of a tree, sampling of twigs and shoots involves a three-dimensional arena, wherein in order to obtain data representative of the whole tree, the distribution of insect populations within the whole tree has to be considered. This is especially problematic when counts from mature trees are required, since many species of insect are not uniformly distributed throughout the entire height of the host plant. For example, the spruce woolly aphid *Adelges abietis* (Hemiptera: Adelgidae), shows a clumped distribution within crowns of even small trees (Fig. 5.1) (Fidgen et al. 1994). Most adelgid galls occur on lateral shoots of mid-crown branches, so sampling insects at the very top or on the lowest whorls of branches will underestimate overall populations. Very often, pilot surveys, sampling all the of the tree crown, will reveal such dilemmas, so that subsequent, more detailed, assessments can be directed at the regions of the tree supporting highest population densities, and reliable sampling units can be derived. The main snag with this is of course that whole-canopy sampling may well be impossible due to the sheer size of the habitat involved.

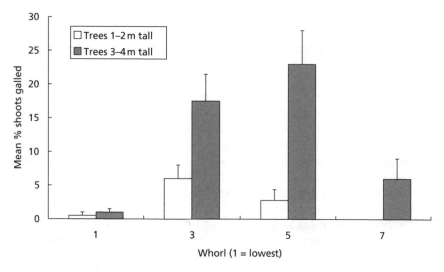

Fig. 5.1 Mean percentage (±s.e.) of shoots of white spruce galled by the adelgid *Adelges abietis*, according to tree height and branch whorl. From Fidgen et al. (1994).

Galls on shoots, twigs, and indeed foliage are relatively easy to sample, since they are large, recognizable, and easy to identify to species from some distance. The insects inside the galls may not be so tractable in terms of abundance or taxonomy, but the effects which they produce on the host plant, such as abnormal shoot or leaf development, leave unmistakable traces. In the case of *Adelges abietis*, once trees in a stand have been selected according to the purpose of the investigation, and the region of the tree where most galls can be found is determined, the proportion of current year shoots per whorl with one or more galls on them is simple to assess by visual counts (Fidgen et al. 1994). However, it is usually impossible to examine all the shoots or twigs on a tree, even a relatively small one, so that sample units have to be established which will provide reliable representations of the whole tree, or, indeed, forest stand. In the case of the gall wasp *Andricus* sp (Hymenoptera: Cynipidae), which stimulates gall formation on oaks in Arizona, USA, it was found that shoot diameter was related to the probability of attack (Pires & Price 2000). Therefore, it was possible in this study to relate the probability of attack by gall wasps to the total number of shoots in diameter categories. In general, the thicker shoots supported more galls than thinner ones, and in fact pilot studies showed that sampling shoots with a diameter of less than 1.5 mm provided underestimates of the overall infestations. So Pires and Price (2000) measured at least 100 shoots growing in the current year, and the number of galls per year counted. In order to provide whole-tree estimates of gall abundance, the total number of shoots for trees taller than 4 m was estimated, based on the number of branches multiplied by the number of estimated shoots per branch. The probability of attack was calculated in relation to the number of sampled shoots in each size (diameter) class.

The incidence of galls themselves does not necessarily provide estimates of insect abundances since in many species of gall-inducing insects, several individuals may inhabit one gall. However, in some cases at least, it is possible to estimate the overall insect population by measuring the size (volume) of the gall—bigger galls contain more insects. Figure 5.2 shows an example from South Australia, which involves the gall wasp *Mesostoa kerri* (Hymenoptera: Braconidae). This insect causes stem galls on *Banksia marginata*, and in order to investigate the population density of wasps in each gall, Austin and Dangerfield (1998) cut terminal branches supporting fresh (i.e. no exit holes) galls from a number of trees. Because galls vary in shape and size, the volume of each one was measured in the laboratory by water displacement in a measuring cylinder, having first removed any attached leaves or twigs. Each gall was then carefully dissected under a stereo-microscope, and adult insects, larvae, or pupae found within the gall chambers counted and preserved. As Figure 5.2 shows, there was a highly significant relationship between gall volume and number of gall wasps contained within. Thus in subsequent sampling routines, a simple measure of gall numbers combined with gall volume would be able to provide accurate estimates of gall wasp population density. Furthermore, it is possible to leave each gall intact so that the numbers of natural enemies versus gall wasps could be assessed by allowing all insects within a gall to emerge naturally, without disturbance.

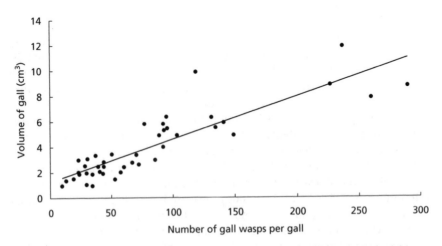

Fig. 5.2 Relationship between the size of galls, measured by water displacement, and the number of gall wasps *Mesostoa kerri* inside. From Austin & Dangerfield (1998).

External feeders

Many insects which live externally on stems and shoots can be collected using "traditional" methods such as beating trays (Wearing & Attfield 2002) or canopy knockdown using insecticidal fogs or sprays (Stork et al. 2001, Speight et al. 2003). However, insects which feed directly on twigs and shoots, such as scales, aphids, or psyllids (all Hemiptera) are usually much more difficult to sample quantitatively, because of their very small size, cryptic appearance, and/or extremely patchy distribution. Very often, microscopic examination is required to count adult (and especially juvenile) populations, so that destructive sampling is frequently required. One basic problem is removing tiny insects from the twigs or shoots on which they reside. As long as they do not stick too tightly to the stems, as in the cases of aphids or psyllids, they may be washed off shoots or stems using soapy water. The resulting liquid can then be strained through wire screens onto muslin filters, from where the numbers of insects of various life stages can be counted under a low-powered microscope. This system of population assessment can work very well. Geiger and Gutierrez (2000) used water-washing as a "rough and ready" tactic for assessing the population density of the leucaena psyllid *Heteropsylla cubana* (Hemiptera: Psyllidae), and compared the procedure with a pilot study where labor-intensive absolute counts where carried out. Water-washing provided very accurate assessments of real densities, with the relationships between the two measurements showing r-squared values from 0.88 to 0.98, and regression slopes near to unity.

Some species of twig and shoot feeders, though small themselves, produce easily recognizable signs of their presence. Woolly aphids (Adelgidae) and certain true aphids (Aphididae) produce waxy secretions under which they live and reproduce, and these secretions provide visual clues about infestation levels and distributions within tree canopies. In addition, some species cause deformation to shoots via feeding damage, which can also be observed without destructive sampling. For example, the balsam twig aphid *Mindarus abietinus* feeds on the buds and elongating shoots of balsam fir *Abies balsamea* in the USA, and these aphids form noticeable aggregations covered with powdery wax and honeydew (Kleintjes 1997). As described above for another species, mid-crown branch tips are known to provide reliable estimates of whole-tree infestation levels. In surveys, 20 host trees were selected randomly from the middle of the plot (to avoid edge effects), and two 25 cm branch tips per tree from the mid-crown region were visually assessed for wax and honeydew, and also the number of distorted shoots per total number of shoots was counted. In this way, gross levels of insect density could be assessed, but in order to relate these general observations to actual numbers of insects, microscopic examination on clipped twigs was required. At the same time, it then proved possible to assess the numbers of predators such as hoverflies and ladybirds associated with the aphids.

The above sampling system may sound simple enough, but obtaining acceptably accurate population densities of insects such as woolly aphids by counting individuals can be difficult because of their very small size and high density. The hemlock woolly aphid *Adelges tsugae* (Hemiptera: Adelgidae), for example, is less than 1 mm long, and can occur in densities over 9 per cm of twig length (Gray et al. 1998). An additional problem involves the frequently highly clumped distributions on host tree twigs. It is important therefore to employ sampling systems which cut down the amount of painstaking and tricky direct counting. Indirect estimates of insect density are desirable, especially for broad-scale population studies, and some researchers have used a so-called binomial sampling plan which uses data collected from two categories. According to Gray et al. (1998), the precision and accuracy of estimated insect density from bino-mial sampling is dependent on the precision and accuracy of equations that de-scribe the relationship between the proportion of samples (in this case twigs) with at least a predetermined insect threshold, and the mean insect density. If such an equation can be found to be reliable, then sampling effort can be much reduced. Likely problems with this type of technique center on the possible in-consistencies over wide geographical areas, and changing insect distributions with life cycle and season.

Gray et al. (1998) sampled 16 to 20 hemlocks in each of several sites over sev-eral generations of the hemlock woolly aphid, selecting trees which would maximize the range of woolly aphid densities. Four equally spaced branch tips around 30 cm long were cut from each of lower and upper tree crowns, and kept cool and humid before they could be examined under a microscope and the adelgids on each twig counted. A twig in this case was defined as the portion of branch extending from a terminal bud to the first node below the terminal. Finally, the relationship between the mean number of adelgids per twig and the proportion of twigs infested with a certain minimum number of adelgids was derived using empirical or theoretical distributions, one example of which is shown in Fig. 5.3.

The empirical model shown in the figure is as follows:

$$\ln x = \ln\alpha + \beta\ln[-\ln(1 - P_T)] \tag{5.1}$$

where x = mean adelgid density, α and β are regression estimates, and P is the proportion of twigs infested with at least T individuals.

In this case, the empirical model only works well for population estimates within a generation, but it means that a rapid assessment of twigs in a sample for the proportion infested with, say, a minimum of three insects provides a reliable estimate of mean numbers overall at that point in time. Hence, as the figure shows, for the combined summer and winter generations from two sites, if 30 percent of twigs in the samples had at least three adelgids on them (P_T=0.3), the model predicts that there will be a mean of 0.66 insects per twig in the whole sample.

Scale insects (Hemiptera: Coccidae) can present even more problems than adelgids and aphids. They are also very small, and many are extremely cryptic

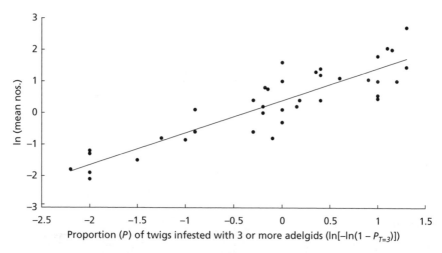

Fig. 5.3 Relationship between mean hemlock woolly adelgid density and proportion of twigs infested with three or more insects, from the empirical distribution model, for three successive generations combined. From Gray et al. (1998).

and hard to recognize, especially in the young stages. An additional problem involves the high mobility of first-instar nymphs (the crawlers), which usually provide the only really dispersive stage of the insect. On a large tree, eggs, crawlers, and later life stages may be patchily distributed on twigs and shoots, with access problems to the high canopy. Clearly, the larger the number of samples that can be taken from all parts of the tree canopy, the more representative the overall mean abundance on twig or shoot may be, but a compromise has to be reached wherein sampling effort is matched with the reliability of the results. Sedentary scales are relatively easily sampled, nevertheless, using clipping tools such as long-armed pruners, and their population densities assessed using a microscope as described above. Crawlers, however, require a static trapping system; citriola scale *Coccus pseudomagnoliarum*, which feeds on the shrub Chinese hackberry, provides an example. Dreistadt (1996) used double-sided transparent sticky tape to trap scale crawlers on the twigs of the host tree. Each twig was encircled with the tape, so that an accurate length of twig was covered, based on the width of the tape employed (in this case, approximately 13 mm). These mini-sticky traps were replaced weekly at the same spot on the branches each time during the active season for the crawlers, though as the crawler stage declined, the tape could be left in position for longer. Dispersing crawlers stuck to the tape, which could be removed and counted under a fairly high-powered microscope (up to 30× magnification). Though in this published example the numbers caught per trap were not corrected for trap (i.e. twig) diameter, it would make between-branch and between-tree comparisons more reliable if the surface area of the sticky trap was measured and the number of crawlers

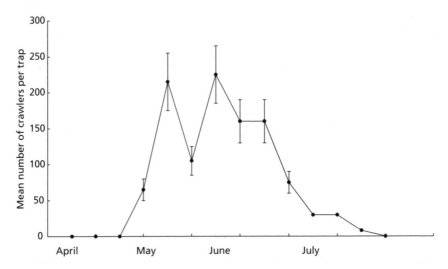

Fig. 5.4 Mean (±s.e.) density of citriola scale crawlers caught on sticky-tape traps on hackberry trees in California. From Dreistadt (1996).

caught on it corrected to a unit area measure, such as density per cm^2 of trap. In this way, seasonal peaks and troughs of scale crawler (and other similarly small but mobile insect) populations can be recorded, as exemplified by Fig. 5.4.

Shoots and twigs: internal

Insects which inhabit shoots and twigs can be grouped under the general title of "borer." They may be adults, as in the case of some bark beetles (Coleoptera: Scolytidae), larvae, as with some longhorn beetles (Coleoptera: Cerambycidae) and many moth larvae (Lepidoptera: Pyralidae and Tortricidae), or both adults and larvae may be found in the same shoot, as with some weevils (Coleoptera: Curculionidae). Whatever the insect involved, most will cause the shoot or twig in which they are living to die, deform or otherwise show fairly clear symptoms that all is not well and attacks are taking place. These symptoms vary with insect species and activity, but many can be easily distinguished one from another.

A very common example from Europe and elsewhere involves the pine shoot beetle *Tomicus piniperda* (Coleoptera: Scolytidae). The larvae of this species live and grow, as with most Scolytidae, under the bark of moribund standing trees and logs (see below), but the young adults have to undergo a process of maturation feeding before they can mature their eggs or sperm (Speight & Wainhouse 1989). This they accomplish by tunneling into the terminal and leader shoots of healthy pine trees. No breeding takes place, merely adult feeding, but the effect on the host tree's shoots can be serious indeed. Damaged shoots eventually fall

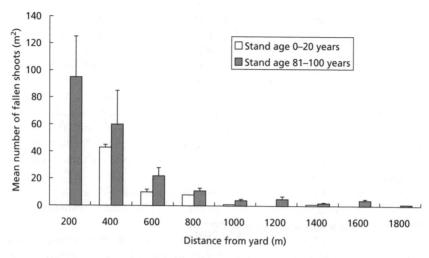

Fig. 5.5 Number of pine shoots (±s.d.) pruned by pine shoot beetles *Tomicus piniperda*, according to distance from a timber yard. From Långström & Hellqvist (1990).

off the trees, especially in high winds, and the easiest way of assessing shoot beetle populations in tree canopies is to count the number of fallen shoots on the ground below. On the assumption that all damaged shoots in a given area of forest have an equal chance of falling to the ground, sample plots can be established wherein all dead shoots are recognized and counted. During studies on *Tomicus* in Sweden, Långström and Hellqvist (1990) set up marker poles along transects leading away from a timber yard (ideal breeding sites for bark beetles), and all fallen pine shoots were collected within 5 m² plots centered on the poles. Any dead shoots not killed by bark beetles are easily recognized, lacking as they do the telltale hollow tunnel up the center of beetle-attacked shoots. In this example, the authors were able to show clear declines in attacks to trees as the distance from the timber yard increased (Fig. 5.5). Any trees within a rough 500 m radius of the yard were clearly at most risk from attack.

However, it may be necessary to carry out more detailed sampling of *T. piniperda* populations and the damage done to standing trees by shoot-boring. In this case the luxury of forest floor sampling is unavailable, and more subtle ways of estimating damage categories may be required. Kauffman et al. (1998) used damage categories based on visual assessments of the continuing activity of adult bark beetles in pine shoots, again from the ground, but this time using whole-tree studies for young pines, or canopy examination using binoculars or ladders for bigger trees. Again, the actual insect was not directly observed at this stage of the sampling. Instead, the structure and especially color of the foliage on shoots was used to categorize damage levels. Changes from dark green, through yellow, to brown indicated increasing intensities of attacks by *Tomicus*.

In order to assess the numbers of beetles emerging from damaged shoots, on their way to lay eggs in logs or stressed trees, these authors had to resort to bagging within pine canopies, followed by destructive sampling. Clumps of three to five shoots in various damaged categories were left attached to the trees, but placed in nylon mesh bags which were sealed around the stems at the bottom so that any beetles emerging from the shoots were unable to escape. The bagged shoots were later cut from the trees and dissected, providing information on (i) the number of shoots with *T. piniperda* tunnels, (ii) the number of dead and alive *T. piniperda* adults in each tunnel, and (iii) the number of dead and alive *T. piniperda* in the bag outside the shoot. Using these procedures, detailed population estimates of young bark beetles could be obtained, and combined with the damage done to the host pine trees.

Unlike most bark beetles, some species of weevil spend their larval and pupal stages within leading shoots of trees. The damage is the same — dead shoots and resultant dieback and deformation, especially of young trees. The white pine weevil *Pissodes strobi* (Coleoptera: Curculionidae) is an extremely common pest of many native and introduced pine and spruce species in North America. Mature adult female weevils puncture feeding pits and lay eggs in the revealed cambium of the top half of the previous year's leader of the host tree. The resulting larvae feed downward beneath the bark, eventually girdling and killing the shoot. Fully grown larvae then pupate in the shoot, and new adults re-emerge from the now dead leaders in summer (Nealis 1998). In order to study the populations dynamics of *P. strobi*, it is necessary to assess population densities within leading shoots not just of the weevil itself, but also of natural enemies such as predators and parasitoids that may also reside within the shoots. According to Nealis (1998), the leader itself is thus a meaningful and convenient sample unit, containing in a single neat "package" most of the agents impacting on weevil survival. In this example, sampling consisted of cutting the entire infested leader at the base of the previous year's growth at regular intervals during the growing season, taking care to spread the destructive load around even-aged tree stands as much as possible, whilst avoiding rows of trees at plantation edges. A census of all leaders in the study area gave the frequency of infested trees in each plantation for the given year. On each sampling occasion, half the collected leaders were dissected immediately and all insects within recorded and counted. The other half were kept intact under controlled conditions (20 °C and 16 : 8 hours light to dark) in 1 m long PVC tubes with plastic collecting funnels at the open bottom end. Any emerging weevils or other insects fell into these funnels and could be preserved and subsequently identified and counted. It is important to note that some insect species may take a considerable time to emerge from wood samples such as these, especially as in this case where parasitic Hymenoptera required a winter diapause before becoming adult. The maintenance of cut samples of trees may be required for many months to ensure all borer individuals have emerged. Shoot-boring Lepidoptera are similar to beetles with the same habits, in that they too usually leave fairly obvious evidence of their presence. The family Tortricidae contains many species which attack leading and lateral shoots of trees, again causing shoot death, tree distortion and considerable dieback. Examples include the notorious temperate and tropical pine shoot moths *Rhyacionia* spp and the mainly North American pine shoot borers *Eucosma* spp. As with the beetles, assessments of attack rates of these insects are simple enough to carry out; systematic walking through young pine stands looking at each tree will show the obvious signs of bent, stunted, brown, or dead leading shoots (Speight & Wylie 2001). Assessing shoot borer damage visually in this manner is easy, quick, non-destructive, and conservative (because some damage may be overlooked) (Prueitt & Ross 1998). In large plantations, where every tree cannot be examined individually, sample grids can be established to sub-sample each stand in a routine manner. Thus, in an example from Oregon, USA, Prueitt and Ross (1998) sampled a total of 120 trees per plot for damage caused by *Eucosma sonomana*, by examining five trees in a row heading north, followed by five trees in the next row heading south, and so on until the full 120 had been studied. More detailed sampling however may be required on occasion, for example when the intensity of shoot attack per tree is of interest. Here, individual whorls from the tops of trees downwards can be sampled, and the levels of shoot borer infestation scored. Clearly, this system will only work well when the tops of the trees are within easy reach of the sampling team.

Some lepidopteran borers are easiest to assess not by the actual damage they cause to shoots (though this may on occasion provide long-range visual clues about the insect's whereabouts), but by the frass and resin exudations which their activities inside the infested shoots cause (Jactel et al. 2002). Perhaps the best example of this is the infamous mahogany shoot borer *Hypsipyla* spp (Lepidoptera: Pyralidae). This pest occurs wherever mahoganies of many genera and species are grown, from Central and South America, through Africa and South Asia, to Southeast Asia and Australia (Speight & Wylie 2001). Adult females lay a single egg on or near terminal shoots and the young larvae tunnel into healthy shoots, eventually killing them. The larva's entrance hole exudes sap, resin, boring debris, and frass (feces), and it is possible to assess the progress of insect development by the nature of these exudations, as Table 5.3 shows (Howard 1995). Using such criteria, the stage of development of the insect in a particular tree can be readily assessed without having to see the larva at all. Because this insect is a typical "low-density" pest, where only one larva per young tree is required to render it worthless for a final timber product, sampling usually takes the form of assessments on a stand scale. For example, in field studies carried out by Newton et al. (1998) in Costa Rica, routine (once a month) inspections of each tree in young stands of two different mahogany species provided, simply and efficiently, cumulative data on the percentage of trees attacked (Fig. 5.6). Sampling the actually larvae within the infested shoots was not, on this occasion, required.

Table 5.3 Stages in the development of resin and frass exudations on leading shoots of mahogany *Swietenia mahagoni* caused by the mahogany shoot borer *Hypsipyla grandella*. From Howard (1995).

Characteristics of exudation	Insect development stage
3 mm diameter ball of cream-colored frass and resin	Early-instar larva in first stages of tunneling in leading shoot
15 to 20 mm frass aggregation, ranging from pale-wood color to orange or light brown	Mature larva, having excavated a tunnel up to 5 or even 10 cm long in leader
Darkened, dissipated, sticky to dry frass aggregation, dark red or brown	Boring ceased, larva either died, pupated, or adult moth emerged

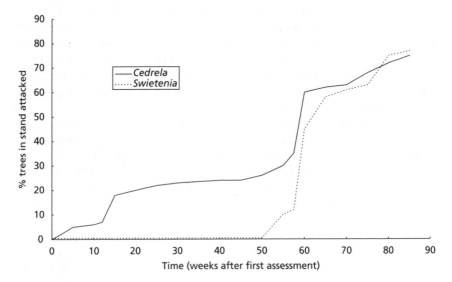

Fig. 5.6 Cumulative number of two mahogany species attacked by *Hypsipyla grandella* in Costa Rica. From Newton et al. (1998).

External bark surface

Certain species of forest insect spend some or all of their life cycle on the outside of the main trunks of trees. Longhorn beetles (Coleoptera: Cerambycidae) lay eggs on the bark surface and the resulting larvae tunnel beneath the bark. Some adult scale insects such as the horse chestnut scale *Pulvinaria regalis* (Hemiptera: Coccidae) also oviposit on main trunks and major branches, whilst other scales, for example the beech scale *Cryptococcus fagisuga* (Hemiptera: Coccoidea) spend all but the active crawler stage on main stem bark.

Counting

Sampling of these populations can involve similar techniques to those described above for insects on the outside of twigs and shoots, but extra problems arise because of the extensive nature of main stems, especially on mature trees. Many external bark insects are not evenly distributed over the whole trunk of trees, and very often tend to be concentrated in certain regions. Figure 5.7 gives an example of the vertical distribution of the maritime pine scale *Matsucoccus feytaudi* (Hemiptera: Margarodidae). The relative tree height in the figure is an indication of the location on the main trunk as a percentage of the total trunk height. Clearly, scales are most abundant in the middle section of the trunk (Jactel et al. 1996). In addition, pine scale, as with so many externally located insects, are very sensitive to bark texture. As Fig. 5.8 shows, scales seem to avoid very thick and very thin bark, but prefer instead thickish bark with many flakes or small cracks. This type of surface provides them with shelter, anchorage, and access to the sap of the bark cambium on which they feed. Any sampling regime for this type of insect which concentrates on very thick or very thin bark, or at the tops or bottoms of trunks, will seriously underestimate scale population density, whilst concentrating on thickish bark in the middle section of tree trunks will not be representative of the whole trunk. Once again, pilot studies should be carried out to decide on the optimal regions of the trunk for detailed sampling.

Some scale insects are readily visible on the outside of the bark, as in the case of the horse chestnut scale *Pulvinaria regalis*. The large adult females lay thousands of eggs in white waxy secretions on the main stems of lime, sycamore, and horse chestnut trees in urban areas of western Europe. The adult herself dies

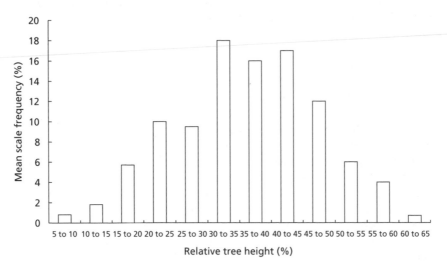

Fig. 5.7 Vertical distribution of the within-tree populations of second-instar pine scale *Matsucoccus feytaudi*. From Jactel et al. (1996).

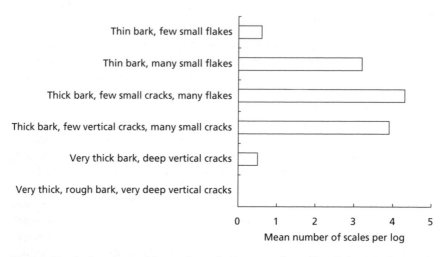

Fig. 5.8 Distribution of second-instar pine scale *Matsucoccus feytaudi* in relation to bark type of host tree. From Jactel et al. (1996).

leaving a very obvious brown body over the wax, providing a readily visible, countable unit. Systematic quadrat sampling using $0.01\,m^2$ wire quadrats placed vertically on tree trunks, using ladders for access, was used by Speight (1994) and each female plus egg mass counted within the quadrat. In an initial calibration exercise, densities of eggs in each egg mass was assessed by microscope counting after removing the wax with a solvent. Subsequently, the size (mean of shortest and longest body dimension) of a full range of small to large dead adults was related to egg numbers per mass, so that in field sampling the mean density of eggs laid per unit bark area could be assessed by estimating the mean size of dead females per quadrat.

Because scales such as *Pulvinaria* are so obvious as adults once the egg-masses have been deposited, widespread surveys of infestation levels can be carried out using visual assessments of whole-tree populations. Speight et al. (1998) developed an "eye-ball" scoring system for *Pulvinaria* egg-masses on town trees, where the estimated percentage cover of trunk and main branches was split into intervals of 10 percent, each being given a "score" from zero to 10. Thus, trees with no egg-masses in evidence at all received a score of zero, whilst trees with the bark completely covered by egg-masses were awarded a score of 10. In order to standardize the sampling assessments, teams of surveyors were trained in a pilot study to ensure that one person's estimate of 30 percent bark coverage (score of 3) was the same as any other's. In the final analysis, these visual assessments of bark insect abundance are only truly of use if they can be related to actual insect density, especially those of the feeding life stages. In the case of *Pulvinaria*, Speight et al. (1998) were able to relate the egg-mass scores per tree to the mean density of scale insect nymphs feeding on the leaves of the

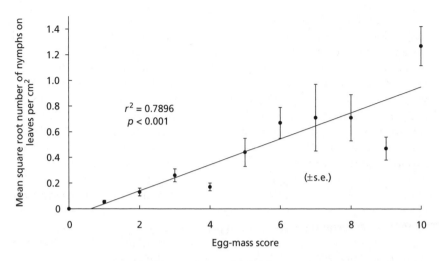

Fig. 5.9 Relationship between egg-mass score on trunks and nymphs per cm² of leaf surface for horse chestnut scale *Pulvinaria regalis*. Data from Speight et al. (1998).

same tree (Fig. 5.9). In this way, the large-scale visual surveys could be used to infer important information for the population densities of feeding herbivores.

Other scale species are much smaller than *Pulvinaria*, and though they leave evidence of their presence via waxy "wool" as before, it is much more difficult to sample insect lifestages accurately. A case in point is the beech scale *Cryptococcus fagisuga* (= *fagi*). This non-armored scale feeds and grows on the bark of main stems of its host trees, and the whole life cycle, bar the dispersing first instars, takes place under the "wool" produced by the insects. As in the previous example, visual estimates of total bark coverage are useful for assessments on a forest-stand level, and they can be taken down to fairly small-scale measurements. For *Cryptococcus*, both Gora et al. (1994) and Lunderstädt (1998) recorded scale density using a five-level scale, from 1 = very sporadically dispersed wax wool "points" indicating the location of scale colonies, up to 5 = wool "points" covering large areas of bark. Thus it was possible to determine the infestation level of each beech tree sampled, measured as the annual mean value of the monthly observed scale density, divided into four classes. These were (a) no or very slight infestation (mean value < 0.2); (b) slight infestation (< 1.0); (c) medium infestation (≤ 2.0); and (d) severe infestation (> 2.0). Finally, Wainhouse and Howell (1983) used a similar scoring system but on a much smaller area of bark. They used small plastic cages with a basal diameter of 13 mm as their sampling unit, firstly assessing the cover of waxy "wool" under each cage using another visual estimation technique where scores of 0, 1, 2, and 3 represented coverage of 0 percent, 1–25 percent, 26–50 percent, and 51–75 percent of the caged area respectively. After some months, the basal area of each cage was then removed from the tree using a boring tool to extract the underlying bark intact. Under a microscope, the waxy coverings of each scale colony were dissected, and classified as follows: (a) established first instars (with wax secretion); (b) second instars; (c) non-fecund adults; (d) fecund adults with eggs either in their bodies or laid; and (e) total number of eggs or hatched first instars. It was possible to calibrate the visual scoring technique with actual insect density, such that there was a positive linear relationship between mean cover score and mean number of adult scales on the bark, providing a validation of the much easier and less time-consuming technique when compared with microscopic examination.

Insects such as longhorn (cerambycid) beetles spend little time on the bark surface, being bark cambium or wood borers as larvae, but they lay their eggs on external bark, and very often this is the only easy part of the life stage for sampling, since the rest is concealed below the bark, making population density assessment much more difficult (see below). One of the most widespread and potentially devastating pests of eucalyptus trees all over the tropical and subtropical world is the eucalyptus longhorn beetle *Phoracantha semipunctata* (Speight & Wylie 2001). Adult female beetles lay their eggs in bark cracks and crevices on trees that are in some way stressed (drought struck, dying, or felled, for example). The simplest way to sample cerambycid eggs laid on bark in this manner is to visually search for them on the bark of cut and sectioned eucalyptus logs, making sure to examine bark cracks and flakes (Way et al. 1992). As in other examples described above, the position on the tree trunks of peak oviposition is not random, but varies with height up the trunk, and also bark texture (Fig. 5.10). From these results it is clear that future intensive sampling could be restricted to the lowest 3 m of trunk in order to study the highest egg densities.

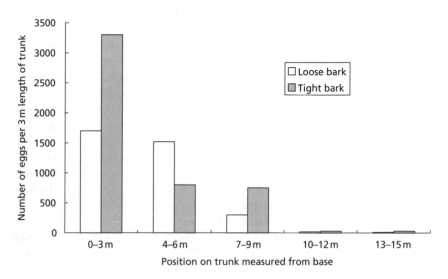

Fig. 5.10 Numbers of eggs laid per tree by *Phoracantha semipunctata* in different oviposition sites on cut trunks of eucalyptus trees. From Way et al. (1992).

Capturing

Insects moving around on the surface of tree bark, either emerging from within the bark or timber (see below) or migrating up or down the trunks, can be captured using a variety of methods and their population densities assessed. On occasion, it has been possible to employ modern technology in the form of 12-volt battery-operated vacuum-cleaners to suck wandering insects off the bark (Jantti et al. 2001), but the time-honored trapping methods are still more widely used. Effectively, there are two basic trap types for this purpose, sticky traps and collecting traps. In the former, a band of sticky material is placed around a tree trunk or limb, and insects that wander onto the surface are retained. In the latter, a physical barrier of some sort attached to the bark guides the insects into a collecting funnel or pot from which they cannot escape.

Sticky bands have been used for many years by horticulturalists to prevent pestiferous insects such as moths and weevils from climbing up trees having emerged from the soil or litter to infest leaves, shoots, or fruits. The technology is basically very simple, and though the actual sticky substance may be a commercially available chemical compound such as Tanglefoot® or Hyvis®, simple grease may suffice for short periods.

One example of the use of sticky bands to monitor insect populations moving up tree trunks concerns the Bruce spanworm *Operophtera brucerata* (Lepidoptera: Geometridae), a close relative of the European winter moth *O. brumata*, in Quebec, Canada (Hébert & St-Antoine 1999). Like the winter moth, Bruce spanworm adult females are flightless, and after emerging from pupae in the soil they crawl up trunks of various deciduous tree species to lay eggs in bark crevices for the winter. In this example, the authors used 15 cm wide strips of different materials coated with Tanglefoot wrapped tightly around tree trunks at about 1.3 m above the ground. Adult moths stuck to the bands were counted, and considerable numbers of insects were caught (Fig. 5.11). Notice however that unless the bands are changed regularly, in this case once a week, medium-term monitoring may underestimate population densities, since there is a limit to the number of individuals that can stick to a band trap, after which it becomes saturated with dead and dying moths and the remainder escape capture. Incidentally, if egg numbers are required, it is possible to use artificial oviposition substrates, again attached to bark, on which insects lay their eggs. In the study by Hébert and St-Antoine, the success of this technique depends on the type of substrate. As can be seen from the figure, polyurethane foam is most successful.

Sticky bands have other practical problems in addition to saturation. The application is messy and time-consuming (especially in cold weather when the sticky glue is hard), and the insects caught are effectively unusable for anything else. Any cracks or crevices in the bark under the bands allow insects to bypass the trap and continue up the tree (Webb et al. 1995).

Collecting pots or traps of various designs are more efficient than sticky bands in the main, though they do involve more expense and setting up. They include

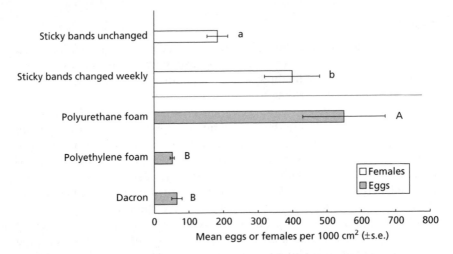

Fig. 5.11 Densities of winter moth eggs and female moths on different artificial substrates and sticky bands, respectively, placed at breast height around the trunks of sugar maple trees. From Hébert & St-Antoine (1999).

the so-called lobster-pot traps, circle traps, or trunk window traps. One of the original "lobster-pot" type traps was designed and used by George Gradwell and George Varley in their classic long-term studies on winter moth in Wytham Wood, Oxfordshire (Agassiz & Gradwell 1977). The trap was made of thin nylon material (actually derived from ladies' stockings!) with a gaping open end pointing downwards held on a wire framework attached securely to the bark. Insects crawling up the bark entered the trap, ascended the nylon funnel, and entered the capture chamber at the top through a small raised opening through which they could not return. Since the width of bark covered by the gape was known, the numbers of winter moth moving up trees per unit surface of tree trunk could be counted over given time periods.

These days, nylon stockings have given way to metal gauze and plastic, but the principle remains the same, and one commercially available trap, the circle trap, is used very extensively, especially in the USA, to monitor numbers of pests such as the plum curculio *Curculio caryae* and the pecan weevil *Conotrachelus nenuphar* (Coleoptera: Curculionidae) (Great Lakes IPM 2000). So-called crawl traps have also been constructed out of modified inverted metal funnels sealed to the bark to catch pine weevils such as *Hylobius pales* and scolytids such as *Ips grandicollis* (Hanula et al. 2002).

More sophisticated "lobster-pot" type traps are now available for collecting tree-trunk insects. One such trap was first designed and used by Moeed and Meads in 1983, and two versions were employed, one to catch insects moving up tree trunks (the "up-trap"), and the other to intercept those moving down trunks (the "down-trap"). Construction details can be obtained from the

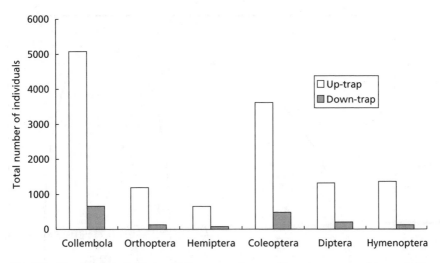

Fig. 5.12 Numbers of individuals within common hexapod/insect groups caught over an 18-month period in trunk traps attached to the bark of four tree species in New Zealand. From Moeed & Meads (1983).

publication; essentially arthropods crawling up or down tree trunks come into contact with the mesh "girdle" around the tree and are channeled into the collecting jar. As can be seen from Fig. 5.12, large numbers of individuals can be caught with this system. It is significant that the majority are moving up the tree, with smaller numbers moving down.

A final type of trunk trap is the trunk window trap, which has been used to collect insects emerging from the bark or wood of standing trees, and also from fungal polypores growing on tree trunks. A vertical pane of clear Perspex is mounted at 90 degrees to the tree trunk, with a collecting funnel and bottle below. Insects attempting to walk over the vertical pane, or indeed those that fly into it having just taken off from the bark surface, fall into the mesh funnel and are then collected in the jar below where they are preserved (Kaila et al. 1997). Trunk window traps have been used by these authors to compare the populations of saproxylic Coleoptera in boreal forests in Finland. Figure 5.13 illustrates some of their results. Using detrended correspondence analysis (DCA), it was clear that beetle populations from dead trees left standing in clear-cut forests were distinctly different from those from the same habitats in mature, uncut, forests. Forest management has important consequences for dead-wood insects, and the conservation status of such material cannot be overemphasized. Indeed, some authors recommend that trunk window traps, whilst not collecting all insect species equally, are simple and effective tools for sampling saproxylic beetles in the tropics as well as in temperate regions (Grove 2000).

Without considerable modification, trunk traps are not suitable for sampling arthropods walking or crawling on horizontal branches. Instead, it is possible to

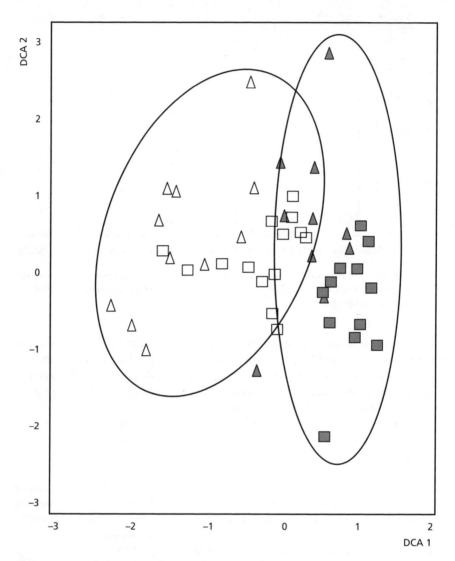

Fig. 5.13 Detrended correspondence analysis (DCA) of beetle samples from trunk window traps in clear-cut (open symbols) and mature (closed symbols) forests in Finland. Triangles and squares denote two sites of each type. From Kaila et al. (1997).

use a form of pitfall trap, a capture system used the world over for sampling mobile animals on the forest floor. Koponen et al. (1997) designed an arboreal pitfall trap which they employed to sample arthropods moving on large horizontal branches of old oak trees in Finland. The trap consists of a plastic collar constructed from a water pipe, coated with the non-stick substance, Fluon®, which is fitted tightly around the branch. Animals encountering the collar fall into a plastic funnel and thus into a collecting bottle below. Since the funnel is substantially below the branch, rainwater tends not to fill the bottle; drain holes prevent the trap becoming waterlogged in very bad weather. In their study using this trap, Koponen et al. caught a total of 32,938 arthropods from seven study sites with five traps per site, over a 4–5 week interval.

Bark/sapwood interface

Sampling insects living inside trees, especially main branches and trunks, can be very problematic, and almost always involves some destruction. If the trees are already dead, and the bark can be removed easily, then insects within can be fairly readily collected and counted. The difficulty is of course that this type of sampling will destroy the habitats for these insects, and if, as so many species are, they are rare or endangered, then the very sampling procedure may make matters worse.

Sampling techniques for insects that spend at least part of their life cycles (usually larvae and pupae) beneath tree bark can be arranged into several categories, including hand/eye searching, trap-logging, log dissection, bark removal, emergence trapping or caging, and externally trapping for flying adults. Each will be considered in turn.

Hand-searching

Direct searching is probably the most effective sampling method when the aim is to find as many scarce insect (and other animal) species under bark or in timber as possible within a short time (Siitonen & Martikainen 1994). Adult individuals may be found walking about on the surface of logs, dead trees, and stumps, and such material is usually an irresistible magnet for entomologists! Loose bark can be pulled off, and adults, larvae, and pupae may be collected with ease. Experienced people may on occasion be able to rear young stages in controlled environments, though once removed from their natural habitats bark-dwelling insect pupae, and larvae especially, are notoriously difficult to care for until adulthood. As with many other examples of this type of exploration, the intensity of sampling, normally in these cases equated to the area of bark searched, can be related to the success of discovering new species. Figure 5.14 shows how species accumulation curves can be produced for bark beetles (Coleoptera: Scolytidae) in old pine stumps in British Columbia, Canada (Safranyik et al. 1999). The curve only begins to level out as bark surface sampled exceeds 10,000 cm^2 (1 m^2). Clearly, perfunctory, rapid hand-searching will not reveal the presence of all species in a timber sample.

Old, dead trees soon loose their tightly attached bark, revealing evidence of insect infestations in the past in the form of tunnel engravings and patterns on the sapwood surface. Though the actual insects responsible have long gone,

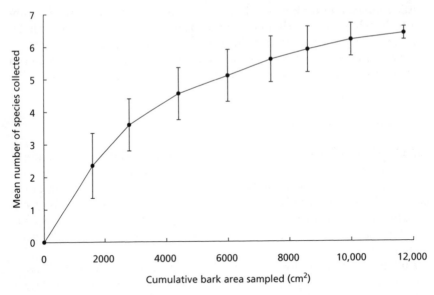

Fig. 5.14 Mean species accumulation curves of bark beetles attacking pine stumps according to area of bark sampled. From Safranyik et al. (1999).

numbers, sizes, and distributions can still be sampled. Furthermore, since bark beetles for example produce species-specific gallery patterns, taxonomic work can also be accomplished by examining such material. Macías-Sámano and Borden (2000) investigated the interactions of two scolytid species in grand fir in Canada, by examining whole trunks of fallen fir trees from which the bark had sloughed off naturally. A string marked off in meter intervals was pinned along the entire length of the tree trunks, and at each 1 m point along the string, the numbers of galleries of each beetle species were counted in order to study interspecific competitive interactions.

Trap-logging

Most species of bark- or wood-inhabiting insects are attracted to stressed, moribund, or dead timber; healthy trees are normally defended against borer attack by chemical and/or physical means (Speight et al. 1999). Hence, if logs are left out in a forest in a susceptible state, ripe for colonization, then adult borers from the surrounding habitats can be expected to oviposit in this material, eventually producing new generations of themselves which can then be collected and counted. In this way, for example, it may be possible to census populations of insects in a forest which are otherwise very difficult to find or identify. This can be especially problematic in species-rich areas such as tropical rainforests. Tavakilian et al. (1997) set out to investigate the host–plant relationships of tropical longhorn beetles (Coleoptera: Cerambycidae) in French Guiana. The only way to provide incontrovertible evidence of host–plant associations is to rear an adult insect from an accurately identified host plant. In this example, the authors felled 690 trees and lianas in the dry season, when it was thought most longhorn beetles would be active. Rare tree species were favored in the hope that poorly known beetle species would be encountered. The felled trees were left on the forest floor for about four months, during which time naturally oc-curring longhorn beetles had the opportunity to lay their eggs in the moribund material. After this time, each log was examined for evidence of insect attack, such as oviposition scars or frass exudations. Logs with such evidence were cut into 80 cm lengths, and placed in cages (see also below). The cages were moni-tored twice per week, and emerging insects collected, preserved, and identified where possible. As a result of all this effort, around 350 cerambycid species were collected, 90 of which were undescribed.

Bark removal and log dissection

In situations where the larvae or pupae of bark borers need to be sampled di-rectly and quantitatively, there is usually no recourse but to take the tree or log apart, or at least to remove sections of its bark. A whole mature tree is impossi-ble in practical terms to sample in this way, so it is important before intensive sampling begins to establish where in the trunk is the highest likelihood of find-ing the target insects. Bark-boring insect species are not uniformly distributed over the whole trunk of trees, but instead tend to congregate in certain areas, re-lated to stem girth, height above ground, and especially bark thickness. Put sim-ply, small species of bark beetles for instance tend to be found under thinner bark, as compared with large species that need the extra space for their tunnels and galleries that thick bark provides. Figure 5.15 provides an example of three species of bark beetle (Coleoptera: Scolytidae) all of which have larval stages in galleries under the bark of damaged spruce trees. As can be seen, the small *Xylechinus* is normally restricted to the tops of trees where the bark is thinner, and the other, larger species cannot grow to maturity (Jakuš 1998). Obviously there would be no point in sampling this species in areas of trunk from near the bottom of mature trees. These data were collected in a fairly standard fashion that involves a lot of unavoidable destruction. Spruce tree trunks first had their branches removed, and were then divided into 2 m long sections (generally called billets). From the middle of each billet, a 50 cm long strip of bark was peeled, with the width of the bark strip being equal to the girth of the trunk at the peeling position. This strip was then divided into 4 equal quarters, each of which was the sampling unit. Beneath each sampling unit, bark beetle galleries were identified (scolytids have very distinct species-specific gallery patterns) and broods (assemblages of larvae and pupae) counted. The density of bark bee-tle attack was then calculated by dividing the number of brood systems by the area of the sample unit, in this case measured in dm^2.

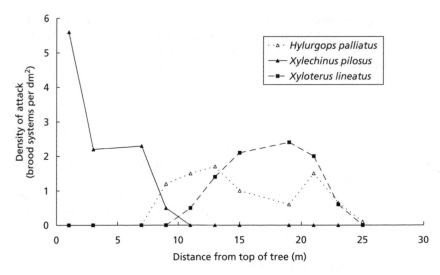

Fig. 5.15 Attack density distribution of three species of bark beetle on bark from snapped or uprooted spruce trees according to position on trunk. From Jakuš (1998).

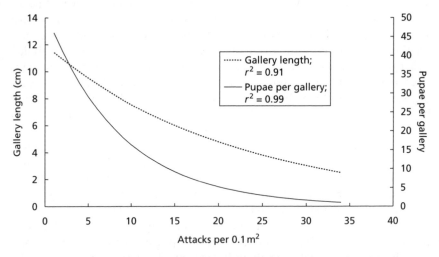

Fig. 5.16 Relationships between attacks of *Ips cembrae* and (a) gallery length and (b) pupae per gallery (fitted curves). From Zhang et al. (1992).

Table 5.4 suggests general rules for sampling of beetles under bark, which involve sectioning the main trunk, removing the bark from selected parts of each section, and sampling bark beetle populations by counting individuals and measuring the dimensions of their galleries or brood chambers. In this way, for example, Zhang et al. (1992) were able to demonstrate competitive interactions within one beetle population as shown in Fig. 5.16. As can be seen, as attack

Table 5.4 Examples from the literature of methods of sampling scolytids under bark.

Beetle species	Common name	Tree species & country	Material sampled	Bark removal technique	Population measurements	Reference
Ips grandicollis	Five-spined engraver beetle	Pine, Australia	Billets 35 cm long by 12–20 cm diameter	All bark stripped from each billet	Numbers of beetles, alive or dead, counted in each life stage, plus enemy species. Attack density assessed as egg galleries per 100 cm², egg gallery length (cm) per 100 cm² bark area in 10 replicates of 4 billets each	Lawson 1993
Ips typographus	Spruce bark beetle	Spruce, Belgium	Standing trees	Circular samples hammered from bark with sharpened metal tubing, 1 dm² cross-section	Numbers of individuals of each development stage. Longitudinal samples taken at 4 cardinal positions around bark circumference, at 1 m intervals. Circular samples taken at 2 to 4 m intervals, all around circumference (numbers varied according to trunk diameter)	Gonzalez et al. 1996
Ips cembrae	Larch bark beetle	Larch, China	Felled, fire-damaged trees	One rectangular bark sample (20 × 50 cm) stripped per tree at breast height	Attack density (attacks per 0.1 m²), number of egg galleries per attack, egg gallery length, number of egg niches, and pupal chambers per gallery	Zhang et al. 1992
Ips grandicollis	Five-spined engraver beetle	Pine, Australia	Billets 70 cm long by c. 15 cm diameter – 2 per felled tree	From 4 equal quadrants of billet, 7 beetle entrance holes selected at random, and a 4 × 4 cm piece of bark chiseled out	Ranked categories (1 = maternal gallery only, no eggs → 6 = most new adults emerged)	Stone & Simpson 1990
Ips spp *Hylurgops* spp *Pityogenes* spp *Polygraphus* spp *Xyloterus* spp	Spruce bark beetles	Spruce, Slovakia	Snapped or uprooted trees cut into 2 m long sections	50 cm strip of bark peeled from middle of each section. Bark strips divided into 4 quarters (upper and lower, left and right quarters of trunk) – each quarter a sample unit	Brood systems for each beetle species counted, position on trunk recorded. Density of beetle attack = number of beetle brood systems divided by area of sample, in dm²	Jakuš 1998

frequency or density increased, so the gallery length decreased, as did the number of larvae successfully reaching the pupal stage, indicative of increasing competition for limited resources (space and food).

Entrance or emergence hole sampling

If data on bark beetle survival and emergence success are required, it may not be necessary to laboriously remove bark; the numbers of holes in the bark made either by ovipositing females entering the bark or, more frequently, by new adults leaving their pupal chambers may be all that is needed. Cut logs or branches are the usual sample section as before. Entrance holes can be difficult to spot, but assessing their density has been used to investigate the relationships between tree stress and bark beetle attack (Kelsey & Gladwin 2001). For emergence studies, the cut ends of the logs should be covered with paraffin wax or plastic to avoid the logs drying out, and this material can then be stored under controlled environment conditions (or on occasion simply left in the forest), until all adult emergence is thought to be over. In the situations where only one species of scolytid is known to be inhabiting the billets, their uniform exit holes are easily recognized and counted. Any different size or shape holes in the bark are likely to be caused by other insects, such as bark beetle parasitoids. For example, Lozano et al. (1997) investigated the interactions between the olive bark beetle *Phloeotribus scarabaeoides* and its wasp parasitoid *Cheiropachus quadram* (Hymenoptera: Pteromalidae) in southern Spain. The emergence holes in olive bark made by the two insects were readily distinguishable, by virtue of their different sizes (the parasitoid holes being significantly smaller in diameter). Cut logs measuring 40–60 cm long and 4–8 cm in diameter were placed in an olive grove in March, and left there all summer to allow bark beetle infestation and parasitoid attack to proceed. Once adults of both species had emerged in autumn, the logs were retrieved and the densities of attack (in numbers per dm^2) calculated from their respective emergence holes in each log. A lot of useful information can be obtained over relatively large areas with little effort.

Emergence trapping

In addition to merely counting emergence holes, it may be useful on occasion actually to collect the insects which emerge from tree bark, or, indeed, from deeper inside timber. The actual species residing under bark may not in fact have been identified, especially when species complexes, natural enemies, or rare species are under investigation. Many workers have used some from of emergence traps or cages to collect emerging insects for later preservation and study. Emergence traps usually consist of some form of bag made from cotton or fine mesh nylon, in which a billet of timber is placed. The bags containing billets may either be left in the field or, more conveniently, placed in a rearing room maintained at around 25 °C and 50 percent relative humidity (Reid & Robb

1999). To facilitate the collection of insect specimens, bags can be suspended with plastic funnels at the bottom end into which insects fall. Below the funnels small vials of 70 percent alcohol catch and preserve the collection. If the vials are emptied in a regular basis, say once or twice per week, then information on the timing of emergences can also acquired. Once emergence is thought to be complete, the logs can be dissected and beetle galleries measured as described above. As might be expected, there are fairly close relationships to be found between the number of emergence holes in bark and the actual number of beetles emerging (Fig. 5.17) (Turchin & Odendaal 1996).

It might be considered that there is little extra information to be gained in a sampling program that merely collects one species of insect. However, this type of sampling has been used successfully to study the colonization of new, exposed logs by beetles such as scolytids, and the interactions with their naturally occurring enemies in the forest. For example, Weslien (1992) cut billets of Norway spruce from a Swedish forest, and exposed them for a varying number of weeks in the forest to attack from the spruce bark beetle *Ips typographus*, as well as any other insect species which might also colonize the bark or timber. After the allotted weeks of exposure, each log was caged and emerging insects collected as described above. Figure 5.18 presents some of the results. *Ips* was found to colonize logs in the first week of exposure, whereas predators and parasitoids did not take up residence in any numbers until logs had been exposed for four or even eight weeks. These enemies would appear therefore to be responding to the growing larvae of *Ips* under the bark before being stimulated to enter the bark themselves. Competitors (other scolytid species) also did not appear until exposure had lasted for several weeks; it is likely that they are less aggressive bark colonizers, and require more moribund bark before attacking it.

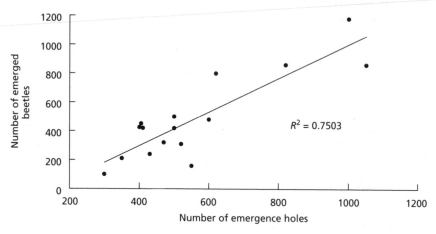

Fig. 5.17 Relationship between number of *Dendroctonus frontalis* adults emerging from log and number of new emergence holes. From Turchin & Odendaal (1996).

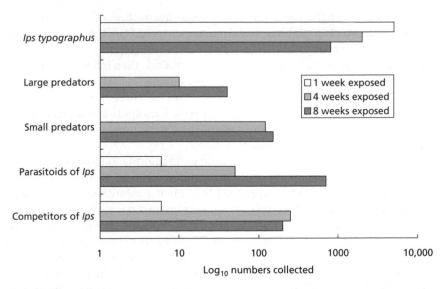

Fig. 5.18 Numbers of insects emerging from caged spruce logs exposed for varying lengths of time prior to caging. From Weslien (1992).

Flight trapping

Once adult bark- and wood-boring insects have emerged from bark or timber, their role is to find a mate, and especially of course to find new, suitable tree hosts to colonize. This is usually accomplished by flight, and since suitable breeding material may be patchily and irregularly distributed throughout the forest, its discovery may involve repeated flight over several days (Madoffe & Bakke 1995). Because of this behavior, it is possible to sample bark- and wood-boring insect populations using flight traps, with or without pheromone baits. Flight and pheromone traps are discussed elsewhere in this book (see Chapter 6), and so only their use for wood and bark borers will be briefly described here. Window flight traps consist in general of perpendicular plates of transparent plastic mounted over a funnel leading into a collecting jar (see also *External bark surface*, above). Lower-stem flight traps are variations on the same theme, and can be constructed out of very basic materials, including cut-down plastic milk bottles (Erbilgin & Raffa 2002). Unless these traps are baited with specific pheromones (Erbilgin et al. 2001), beetles and other bark or wood borers flying about hit the plastic panes by chance and fall into the jar. The very randomness of the system, coupled with the fact that overall abundance is not measured directly, rather activity, means that this type of trap is not used regularly for bark- or wood-borer sampling. Window flight traps have been used to study assemblages of bark beetle species in old-growth forests in Finland, for example (Martikainen et al. 1999), and as Fig. 5.19 illustrates, many sample plots may be required before

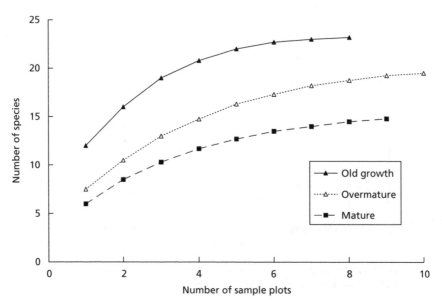

Fig. 5.19 Cumulative number of insect species associated with bark beetles caught in window flight traps in Finnish spruce forests. From Martikainen et al. (1999).

there is some degree of certainty that all species in the habitat under investigation have been found. When more precise information is available on the flight patterns and relative abundance of a single species of beetle, pheromone traps are rather more useful.

Synthetic pheromones are now available commercially for a wide range of scolytid species, and traps baited with them are used routinely to monitor pest levels, especially in high-risk areas such as dockyards where imported timber may be infested by unwanted exotic pests. Most commonly, bark beetle baits are employed in funnel traps, tree-trunk traps or drainpipe traps, all variations on the same theme. As sampling tools in forest stands, one major question involves the range, or effective sampling area, of a pheromone trap. Both the trap design and the concentration of the chosen pheromone bait come into play here. Figure 5.20 shows one example of pheromone trapping for the spruce bark beetle *Ips typographus*, a species with which so much pheromone work has been done over the years. Clearly, pheromone-baited window and pipe traps show a similar attraction radius, whilst sticky traps with the same bait are much less effective. Note that the dose of the chemical attractant has a vital influence on the trap's efficiency, but only up to a certain concentration. Perhaps most surprising, for this example at least, is the very small attraction radius exhibited: an individual *Ips* flying beyond about 2 m from a particular trap would appear not to be influenced at all. This problem thus has important implications for the spacing and distribution of pheromone traps for bark beetles in a forest stand.

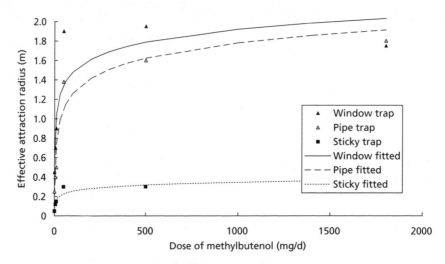

Fig. 5.20 Effective attraction ratio for three types of pheromone trap for *Ips typographus* according to pheromone dose. From Schlyter (1992).

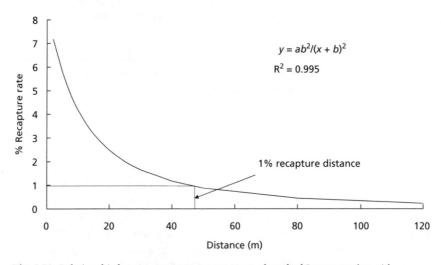

Fig. 5.21 Relationship between percentage recapture of marked *Ips typographus* with distance of pheromone trap from release site (fitted model). From Zolubas & Byers (1995).

A further problem with pheromone traps for scolytids and other flying beetles concerns the fact that beetles tend to disperse widely from their point of emergence from tree or bark material, and the capture rate of pheromone traps declines dramatically as the distance from the source increases (Fig. 5.21). In this examples, it would appear that traps have to be located really very close to

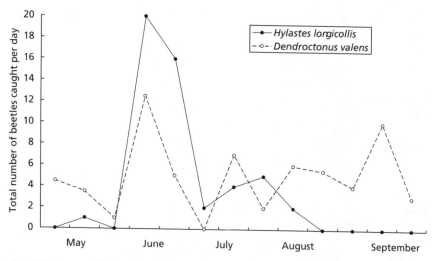

Fig. 5.22 Seasonal flight patterns for two species of scolytid assessed by pheromone trap captures in Oregon. From Peck et al. (1997).

the origin of bark beetle flight patterns—a few tens of meters at most is the maximum range for reliable estimations of flying population density. Possibly one of the most useful ways of employing pheromone traps in the sampling of bark beetle populations is to describe peak flight patterns for various species. Assuming that each species of insect reacts similarly to pheromone trap location and distribution, comparisons of seasonal flight activity can be made to identify when pest species are most active, and hence when forest stands or individual trees are most at risk. In Oregon, USA, Peck et al. (1997) set up multiple funnel (Lindgren) traps, baited with a cocktail of synthetic pheromones for a variety of bark beetle species, mixed with ethanol which is a general stimulus for bark beetles trying to locate suitable host trees. Figure 5.22 depicts the seasonal flight patterns of two scolytid species sampled from May to September. Species such as *Hylastes longicollis* have an obvious and fairly tight peak flight period in June, whilst others such as *Dendroctonus valens* are active through much of the summer season and hence much less predictable in terms of risk evaluation (Peck et al. 1997).

Wood borers

Probably the most difficult group of insects to sample in trees are those species which bore deep into the timber itself, frequently all the way into the heartwood. It is often well-nigh impossible even to detect their presence, never mind to assess their abundance, and apart from the wholesale felling of forests followed by laborious dissection with a chainsaw, quantitative sampling of populations in the timber is just about impossible. Emergence traps and cages containing logs or billets will eventually reveal some of these species, mainly moths and beetles, though some take literally years to complete larval development and eventually appear as adults. One or two species do give their presence away by producing active holes in bark from which frass and resin exude. One example is the teak beehole borer *Xyleutes ceramica* (Lepidoptera: Cossidae). The larvae of this so-called wood moth tunnel deep into the heartwood of standing, seemingly healthy trees, excavating large and extensive tunnels. Rarely do trees exhibit obvious symptoms of infestation; only the final timber product becomes severely degraded. The hole maintained by the larva on the outside of the bark is the only real external sign of the insect within, but this does allow for the possibility of inspections in tropical forest stands, scoring trees for usually no more than presence or absence of the pest. Smaller wood-destroying insects such as the well-known group of beetles called woodworm (Coleoptera: Anobiidae) can be sampled quantitatively, but again only via assessing emerging adult populations. Seybold and Tupy (1993) used laboratory cages containing billets of Norway spruce in California to investigate infestations of the anobiid *Ernobius mollis*. This species excavates extensive tunnels in the phloem and xylem of moribund, especially fire-damaged, timber. Emergences from the cages were found in this study to continue for over 13 months, suggesting both delayed emergence and, more problematically, possible re-infestation of wood material with bark remaining actually within the cages. In the latter case, population density sampling runs the risk of rather severe over-estimation.

Conclusions

Sampling insects on the twigs, shoots, and trunks of trees presents numerous problems. These include uneven population distributions over external and internal surfaces, the frequently concealed and inaccessible nature of the life stage under investigation, and the general magnitude of the sampling procedures required to provide reliable and representative results. Pilot studies are frequently required to narrow down sampling areas and to select the best protocols. Destruction of the host trees, and the insects that associate with them, is often unavoidable. Until the advent of technologies that enable us to see inside bark and timber, such as X-rays or ultrasound, sampling these groups of insects will always be difficult.

References

Agassiz, D. & Gradwell, G. (1977) A trap for wingless female moths. *Proceedings and Transactions of the British Entomological and Natural History Society*, **10**, 3–4, 69–70

Austin, A.D. & Dangerfield, P.C. (1998) Biology *of Mesostoa kerri* (Insecta: Hymenoptera: Braconidae: Mesostoinae), an endemic Australian wasp that causes stem galls on *Banksia marginata. Australian Journal of Botany*, **46**, 559–569.

Dreistadt, S.H. (1996) Citricola scale (Homoptera: Coccidae) abundance on Chinese hackberry and scale control with spray oil or acephate trunk implants. *Journal of Economic Entomology*, **89**, 481–487.

Erbilgin, N. & Raffa, K.F. (2002) Association of declining red pine stands with reduced populations of bark beetle predators, seasonal increases in root colonizing insects, and incidence of root pathogens. *Forest Ecology and Management* **164**, 221–236.

Erbilgin, N., Szele, A., Klepzig, K.D., & Raffa, K.F. (2001) Trap type, chirality of alpha-pinene, and geographic region affect sampling efficiency of root and lower stem insects in pine. *Journal of Economic Entomology* **94**, 1113–1121.

Fidgen, J.G., Teerling, C.R., & McKinnon, M.L. (1994) Intra- and inter-crown distribution of the eastern spruce gall adelgid, *Adelges abietis* (L.), on young white spruce. *Canadian Entomologist*, **126**, 1105–1110.

Geiger, C.A. & Gutierrez, A.P. (2000) Ecology of *Heteropsylla cubana* (Homoptera: Psyllidae): Psyllid damage, tree phenology, thermal relations, and parasitism in the field. *Environmental Entomology* **29**, 76–86.

Gonzalez, R., Gregoire, J.C., Drumont, A., & de Windt, N. (1996) A sampling technique to estimate within-tree populations of pre-emergent *Ips typographus* (Col., Scolytidae). *Journal of Applied Entomology*, **120**, 569–576.

Gora, V., Koenig, J., & Lunderstaedt, J. (1994) Physiological defence reactions of young beech trees (*Fagus sylvatica*) to attack by *Phyllaphis fagi. Forest Ecology and Management*, **70**, 245–254.

Gray, D.R., Evans, R.A., & Salom, S.M. (1998) Hemlock woolly adelgid (Homoptera: Adelgidae) dispersion and the failure of binomial sampling to estimate population density. *Environmental Entomology*, **27**, 564–571.

Great Lakes IPM (2000) Insect and animal traps. http://www.greatlakesipm.com [accessed May 12, 2004].

Grove, S. (2000) Trunk window trapping: an effective technique for sampling tropical saproxylic beetles. *Memoirs of the Queensland Museum*, **46**(1), 149–160.

Hanula, J.L., Meeker, J.R., Miller, D.R., & Barnard, E.L. (2002) Association of wildfire with tree health and numbers of pine bark beetles, reproduction weevils and their associates in Florida. *Forest Ecology and Management* **170**, 233–247

Hébert, C. & St-Antoine, L. (1999) Oviposition trap to sample eggs of *Operophtera bruceata* (Lepidoptera: Geometridae) and other wingless geometrid species. *Canadian Entomologist*, **131**, 557–565.

Howard, F.W. (1995) Reduction of damage to mahogany by mahogany shoot borer and mahogany leaf miner by use of azadirachtin. *Journal of Tropical Forest Science*, **7**, 454–461.

Jactel, H., Perthuisot, N., Menassieu, P., Raise, G., & Burban, C. (1996) A sampling design for within-tree larval populations of the maritime pine scale, *Matsucoccus feytaudi* Duc. (Homoptera: Margarodidae), and the relationship between larval population estimates and male catch in pheromone traps. *Canadian Entomologist*, **128**, 1143–1156.

Jactel, H., Goulard, M., Menassieu, P., & Goujon, G. (2002) Habitat diversity in forest plantations reduces infestations of the pine stem borer *Dioryctria sylvestrella. Journal of Applied Ecology*, **39**, 618–628.

Jakuš, R. (1998) Patch level variation of bark beetle attack (Col., Scolytidae) on snapped and uprooted trees in Norway spruce primeval natural forest in endemic condition: proportions of colonized surface and variability of ecological conditions. *Journal of Applied Entomology*, **122**, 543–546.

Jantti, A., Aho, T., Hakkarainen, H., Kuitunen, M., & Jukka, S. (2001) Prey depletion by the foraging of the Eurasian treecreeper, *Certhia familiaris*, on tree-trunk arthropods. *Oecologia (Berlin)*, **128**, 488–491.

Kaila, L., Martikainen, P., & Punttila, P. (1997) Dead trees left in clear-cuts benefit saproxylic Coleoptera adapted to natural disturbances in boreal forest. *Biodiversity and Conservation*, **6**, 1–18.

Kauffman, W.C., Waltz, R.D., & Cummings, R.B. (1998) Shoot feeding and overwintering behavior of *Tomicus piniperda* (Coleoptera: Scolytidae): implications for management and regulation. *Journal of Economic Entomology*, **91**, 182–190.

Kelsey, R. & Gladwin, J. (2001) Attraction of *Scolytus unispinosus* bark beetles to ethanol in water-stressed Douglas-fir branches. *Forest Ecology and Management*, **144**, 229–238

Kleintjes, P.K. (1997) Midseason insecticide treatment of balsam twig aphids (Homoptera: Aphididae) and their aphidophagous predators in a Wisconsin Christmas tree plantation. *Environmental Entomology*, **26**, 1393–1397.

Koponen, S., Rinne, V., & Clayhills, T. (1997) Arthropods on oak branches in SW Finland, collected by a new trap type. *Entomologica Fennica*, **8**, 177–183.

Långström, B. & Hellqvist, C. (1990) Spatial distribution of crown damage and growth losses caused by recurrent attacks of pine shoot beetles in pine stands surrounding a pulp mill in southern Sweden. *Journal of Applied Entomology*, **110**, 261–269.

Lawson, S.A. (1993) Overwintering mortality of *Ips grandicollis* Eichh. (Col., Scolytidae) and its parasitoid, *Roptrocerus xylophagorum* Ratz. (Hym. Pteromalidae), in South Australia. *Journal of Applied Entomology*, **115**, 240–245.

Lozano, C., Kidd, N.A.C., Jervis, M.A., & Campos, M. (1997) Effects of parasitoid spatial heterogeneity, sex ratio and mutual interference on the interaction between the olive bark beetle *Phloeotribus scarabaeoides* (Col., Scolytidae) and the pteromalid parasitoid *Cheiropachus quadrum* (Hum., Pteromalidae). *Journal of Applied Entomology*, **121**, 521–528.

Lunderstädt, J. (1998) Impact of external factors on the population dynamics of beech scale (*Cryptococcus fagisuga*) (Hom., Pseudococcidae) in beech (*Fagus sylvatica*) stands during the latency stage. *Journal of Applied Entomology*, **122**, 319–322.

Macías-Sámano, J.E. & Borden, J.H. (2000) Interactions between *Scolytus ventralis* and *Pityokteines elegans* (Coleoptera: Scolytidae) in *Abies grandis*. *Environmental Entomology*, **29**, 28–34.

Madoffe, S.S. & Bakke, A. (1995) Seasonal fluctuations and diversity of bark and wood-boring beetles in lowland forest: implications for management practices. *South African Forestry Journal*, **173**, 9–15.

Martikainen, P., Siitonen, J., Kaila, L., Punttila, P., & Rauh, J. (1999) Bark beetles (Coleoptera, Scolytidae) and associated beetle species in mature managed and old-growth boreal forests in southern Finland. *Forest Ecology and Management*, **116**, 233–245.

Moeed, A. & Meads, M.J. (1983) Invertebrate fauna of four tree species in Orongorongo Valley, New Zealand, as revealed by trunk traps. *New Zealand Journal of Ecology*, **6**, 39–53.

Nealis, V.G. (1998) Population dynamics of the white pine weevil, *Pissodes strobi*, infesting jack pine, *Pinus banksiana*, in Ontario, Canada. *Ecological Entomology*, **23**, 305–313.

Newton, A.C., Cornelius, J.P., Mesen, J.F., Corea, E.A., & Watt, A.D. (1998) Variation in attack by the mahogany shoot borer, *Hypsipyla grandella* (Lepidoptera: Pyralidae), in relation to host growth and phenology. *Bulletin of Entomological Research*, **88**, 319–326.

Peck, R.W., Equihua, M.A., & Ross, D.W. (1997) Seasonal flight patterns of bark and ambrosia beetles (Coleoptera: Scolytidae) in northeastern Oregon. *Pan Pacific Entomologist*, **73**, 204–212.

Pires, C.S.S. & Price, P.W. (2000) Patterns of host plant growth and attack and establishment of gall-inducing wasp (Hymenoptera: Cynipidae). *Environmental Entomology*, **29**, 49–54.

Prueitt, S.C. & Ross, D.W. (1998) Effects of environment and host genetics on *Eucosma sonomana* (Lepidoptera: Tortricidae) infestation levels. *Environmental Entomology*, **27**, 1469–1472.

Reid, M.L. & Robb, T. (1999) Death of vigorous trees benefits bark beetles. *Oecologia (Berlin)*, **120**, 555–562.

Safranyik, L., Shore, T.L., & Linton, D.A. (1999) Attack by bark beetles (Coleoptera: Scolytidae) following spacing of mature lodgepole pine (Pinaceae) stands. *Canadian Entomologist*, **131**, 671–685.

Schlyter, F. (1992) Sampling range, attraction range, and effective attraction radius: estimates of trap efficiency and communication distance in coleopteran pheromone and host attractant systems. *Journal of Applied Entomology*, **114**, 439–454.

Seybold, S.J. & Tupy, J.L. (1993) *Ernobius mollis* (L.) (Coleoptera: Anobiidae) established in California. *Pan Pacific Entomologist*, **69**, 36–40.

Siitonen, J. & Martikainen, P. (1994) Occurrence of rare and threatened insects living on decaying *Populus tremula*: a comparison between Finnish and Russian Karelia. *Scandinavian Journal of Forest Research*, **9**, 185–191.

Speight, M.R. (1994) Reproductive capacity of the horse chestnut scale insect, *Pulvinaria regalis* Canard (Hom., Coccidae). *Journal of Applied Entomology*, **118**, 59–67.

Speight, M.R. & Wainhouse, D. (1989) *Ecology and Management of Forest Insects*. Oxford University Press, Oxford.

Speight, M.R. & Wylie, F.R. (2001) *Insect Pests in Tropical Forestry*. CABI, Wallingford.

Speight, M.R., Hails, R.S., Gilbert, M., & Foggo, A. (1998) Horse chestnut scale (*Pulvinaria regalis*) (Homoptera: Coccidae) and urban host tree environment. *Ecology (Washington DC)*, **79**, 1503–1513.

Speight, M.R., Hunter, M.D, & Watt, A.D. (1999) *Ecology of Insects: Concepts and Applications*. Blackwell, Oxford.

Speight, M.R., Intachat, J., Chey, V.K., & Chung, A.Y.C. (2003) Insects in managed canopies. In *Arthropods of Tropical Forests: Spatio-Temporal Dynamics and Resource Use in the Canopy* (ed. Y. Basset, R. Kitching, S. Miller, & V. Novotny). Cambridge University Press, Cambridge.

Stone, C. & Simpson, J.A. (1990) Species associations in *Ips grandicollis* galleries in *Pinus taeda*. *New Zealand Journal of Forestry Science*, **20**, 75–96.

Stork, N.E., Hammond, P.M., Russell, B.L., & Hadwen, W.L. (2001) The spatial distribution of beetles within the canopies of oak trees in Richmond Park, UK. *Ecological Entomology*, **26**, 302–311.

Tavakilian, G., Berkov, A., Meurer, G.B., & Mori, S. (1997) Neotropical tree species and their faunas of xylophagous longicorns (Coleoptera: Cerambycidae) in French Guiana. *Botanical Review*, **63**, 303–355.

Turchin, P. & Odendaal, F.J. (1996) Measuring the effective sampling area of a pheromone trap for monitoring population density of southern pine beetle (Coleoptera: Scolytidae). *Environmental Entomology*, **25**, 582–588.

Wainhouse, D. & Howell, R.S. (1983) Intraspecific variation in beech scale populations and in susceptibility of their host *Fagus sylvatica*. *Ecological Entomology*, **8**, 351–359

Way, M.J., Cammell, M.E., & Paiva, M.R. (1992) Studies on egg predation by ants (Hymenoptera: Formicidae) especially on the eucalyptus borer *Phoracantha semipunctata* (Coleoptera: Cerambycidae) in Portugal. *Bulletin of Entomological Research*, **82**, 425–432.

Wearing, C.H. & Attfield, B. (2002) Phenology of the predatory bugs *Orius vicinus* (Heteroptera: Anthocoridae) and *Sejanus albisignata* (Heteroptera: Miridae) in Otago, New Zealand, apple orchards. *Biocontrol Science and Technology*, **12**, 481–492.

Webb, R.E., White, G.B., & Thorpe, K.W. (1995) Response of gypsy moth (Lepidoptera: Lymantriidae) larvae to sticky barrier bands on simulated trees. *Proceedings of the Entomological Society of Washington*, **97**, 695–700.

Weslien, J. (1992) The arthropod complex associated with *Ips typographus* (L.) (Coleoptera, Scolytidae): species composition, phenology, and impact on bark beetle productivity. *Entomologica Fennica*, **3**, 205–213.

Zhang, Q.H., Byers, J.A., & Schlyter, F. (1992) Optimal attack density in the larch bark beetle, *Ips cembrae* (Coleoptera: Scolytidae). *Journal of Applied Ecology*, **29**, 672–678.

Zolubas, P. & Byers, J.A. (1995) Recapture of dispersing bark beetle *Ips typographus* L. (Col., Scolytidae) in pheromone-baited traps: regression models. *Journal of Applied Entomology*, **119**, 285–289.

Index of methods and approaches

Methodology	Topics addressed	Comments
External examination for insects	Assessments of which trees in a stand may be attacked.	Only efficient for external feeders, especially somewhat sessile forms (e.g. sap-feeders, gall formers etc).
	Assessment of attack severity.	
	Presence of insects on the surface.	
	Identity of insects.	
	Absolute estimates of population density and species richness.	
	Assessment of community structure, e.g. guild.	
	Collection of live specimens for subsequent experimental work on population dynamics and feeding strategies.	
External examination for evidence of damage	Assessments of which trees in a stand may be attacked.	Shapes and sizes of holes, as well as extruded frass or bore dust, provide good indication of pest type (e.g. woodwasps, longhorn beetles, bark beetles, ambrosia beetles, powderpost beetles).
	Assessment of attack severity.	
	Presence of insects within shoots and stems.	
	Density of emerged populations.	
	Emergence of herbivores and enemies via relative sizes and shapes of emergence holes.	
	Region of stem or trunk infested.	
Removing bark or splitting shoots and logs	Location of boring species.	Mainly collects larvae which need to be reared to adulthood (difficult) to identify accurately.
	Identity of boring species.	
	Extent of attack.	
	Effects on the tree.	

Continued

Methodology	Topics addressed	Comments
Sticky traps, lobster-pot traps; baited traps	Identifies and assesses densities of mobile species, larvae nymphs or adults. Identifies timing of lifecycle changes such as adult emergence and oviposition.	
Trap logging	Estimates of population density and species richness of flying adults looking for new host trees to colonize.	May be used to manage pests such as bark and ambrosia beetles.
Flight and pheromone traps	Estimates of population density and species richness of flying adults. Early warning of pest invasion in stands, log yards and ports.	A more sophisticated and species-specific version of trap logging.

CHAPTER 6

Insects in flight

MARK YOUNG

Introduction

The ability to fly, that makes insects such interesting animals for study, also has a great impact on the ways that they can be caught. Their high mobility both helps them to approach traps and helps them avoid them, so specialized trapping strategies are needed. Essentially such traps either rely on actively attracting insects, or they merely intercept their flight. Examples of both are included below.

Light traps

The legendary attraction of insects to light has long been used in ecological studies. Although it is still not certain why the attraction takes place, the results of standardized light trapping have produced vital ecological insights in many different places and conditions.

Basic principle of use

Even if it is not yet clear why light sources attract insects, it is certain that the effect depends on the degree of contrast between the light source and its surroundings, with reduced catches when contrast is low (Nag & Nath 1991). The basic design of a light trap therefore includes a single bright source and a method of trapping the insects that arrive. The behavior of insects when very close to the light may change, however, in that they tend to avoid the very close, bright light by flying into the dark area around it. In fast-flying species, such as large Lepidoptera, their momentum may carry them right up to the light trap, whereas slow-flying small Diptera may avoid being trapped by diverting from the light into the dark adjacent area (Muirhead-Thomson 1991). Most of the variation in trap design arises from the need to increase trap efficiency for the chosen type of insect and to achieve practicality, portability, and economy, for what is necessarily a cumbersome and expensive trap type.

Type of light source

It is believed that different types of light source are optimal for different insect orders. For each type of source, however, catch is increased as the intensity of light is increased. Tungsten filaments, emitting largely in the visible spectrum, are apparently most attractive to many Diptera, whereas Lepidoptera and Trichoptera respond best to ultraviolet-rich sources, such as mercury-vapor bulbs, including even "black" lights, which emit only in the ultraviolet range (Southwood & Henderson 2000). However, the addition of suction traps to mercury-vapor traps increases the catch of small Diptera (Downey 1962), and this implies that the apparent preference for visible sources by these species may actually be because the extra intensity of the ultraviolet traps leads to an increased "escape" behavior close to the trap.

Waring (1980, 1990) compared the catch of moths in traps fitted with bulbs emitting different ultraviolet levels. He found that a 6 W fluorescent tube attracts only 15–40 percent of the number of individuals and 50 percent of the number of species, compared with a 125 W "MB" bulb, and it is well known that on light nights the 6 W tube is almost ineffective. Bowden and Church (1973) showed that catches of Lepidoptera may be 3–10 times greater on clear, moonless nights than on clear nights when the moon is full, a finding confirmed by Taylor (1986). Although this may be because fewer moths fly at full moon, it is more likely that the reduced catch reflects the lower light source contrast on a moonlit night. Bowden and Morris (1975) showed that some insects fly more at a full moon, but are caught less often then. Haufe and Burgess (1960) found light traps to be ineffective for mosquitoes on light Arctic nights, and substituted trap "sources" which were merely high-contrast black and white striped cylinders.

Comparative catchability

Light traps vary greatly in their effectiveness for different types of insect. Diptera, Lepidoptera, and Trichoptera are orders frequently trapped, whereas only some species of Coleoptera and Hymenoptera appear. However, this may merely reflect the relative abundance of night-flying representatives of these groups. The selectivity also applies within orders. Moth trappers know that male moths almost always outnumber females and that some species may be abundant at nearby sugar baits but do not enter light traps (Young 1997). Part of the reason may just be that some species (and females) fly less than others (and males) but some residual selectivity does remain and interpretation of catches must reflect this.

Influences on catch efficiency and effective catching distance

Gaydecki (1984), McGeachie (1987), and others have observed that high catches of nocturnal insects are made on relatively warm, still, dark, and humid nights. Gentle rain may not reduce catches but heavy rain and moderate or strong winds do so. Gregg et al. (1994) found reduced catches of migratory moths when wind speed increased, as did Edwards et al. (1987) for *Culicoides* midges. Wind direction is also important, insofar as it reflects meteorological conditions. In Britain southwesterly winds are associated with mild, humid airstreams, and high insect catches, whereas northeasterlies are colder and drier and catches are low.

Direct observation of small Diptera shows that they are often unaffected by light traps beyond a range of 5 meters, whereas large Lepidoptera are sometimes seen to veer towards the light when up to 20 or more meters distant. Roberts (1996) released marked moths of the family Noctuidae at various distances from a standard 125 W "Robinson pattern" trap and showed a rapid drop-off in recapture, such that beyond 15 m the trap was almost completely ineffective. However, Graham et al. (1961) estimated the trapping radius of their trap as 60+ m for the pink bollworm *Pectinophora gossypiella* in the USA. Roberts (1996) observed that flying moths often settled well short of the trap, and Hartstack et al. (1968) found a similar behavior in *Heliothis* species, whereas *Trichoplusia ni* tended to fly up into the trap.

Trap designs

A basic design for Lepidoptera is that of the "Robinson" pattern trap. The light is held above a downward-pointing, open-ended cone, within which are 2–4 vertical baffles. Moths drop down through the cone and are trapped in the box-like base. The upper portion of the base is transparent, so that moths trying to escape fly to this instead of finding their way back through the narrow cone opening. This trap resembles a lobster pot, and many minor modifications exist (Southwood & Henderson 2000). It usually uses a generator or mains-powered high-intensity UV source, whereas a low-wattage fluorescent tube, run from a car battery, is used in the more portable but essentially similar "Heath" trap.

In the 1930s C.B. Williams designed a standardized but relatively inefficient "Rothamsted" trap for Lepidoptera at Rothamsted Experimental Station (Williams 1948) (Fig. 6.1). A high-power tungsten bulb is held under an opaque square top and glass baffles lead moths into a killing jar. This trap is very resistant to wet, windy weather and is designed to be used every night of the year in a standardized trapping sequence. The use of the killing jar prevents escape and will increase catches in all trap designs. By the use of timing devices it is also possible to separate the trap into components, so as to investigate catches in each part of the night.

Light traps for Diptera tend to incorporate a suction device, so as to complete the catching of individuals that approach the trap. A common example of this type is the "New Jersey" trap (Mulhern 1942), which was designed to catch mosquitoes, but many variants exist and some are illustrated in detail

Fig. 6.1 Rothamsted Insect Survey standard design light trap. These are used in long-term monitoring of moth populations. Picture courtesy of Ian Woiwod, Rothamsted Experimental Station.

by Southwood and Henderson (2000). Rawlings et al. (2003) used up to 40 standardized 8 W black light traps, incorporating a suction trap held below the light, scattered across all of South Africa, to sample *Culicoides* midges as part of a faunal and veterinary study. The lights attract the midges but very small flying insects tend not to enter the actual trap unless they are sucked into it by the suction device.

Examples of use

The large influence of variable weather conditions, plus the variation in contrast and exposure at different sites, makes it very difficult to compare light catches from different places and/or different nights. Consequently, light traps are difficult to standardize and so their use has to be carefully controlled. They are excellent for surveys or general assessment of species richness and they have played a large part in quantifying the species richness of different tropical forest areas (e.g. Robinson & Tuck 1993). They are also used to detect the emergence of pest species, prior to control measures.

If used over long periods, however, the nightly variation is "averaged out" and useful data accrue. The Rothamsted trap data series, for some of their array of traps now running for nearly 40 years, has allowed extensive analysis of ecological patterns of diversity and population dynamics. Early work by Taylor et al. (1976) established the validity of measures of diversity in insect communities, and more recently Woiwod and Hanski (1992) were able to use long-term data runs to demonstrate density-dependent effects, which had proved highly elusive in the more short-term studies. Woiwod and Harrington (1994) provide an overview of the work of the Rothamsted Insect Survey, illustrating the crucial importance of their light-trap series.

Chapman et al. (2002) provide an interesting example where the use of a light trap has validated another method of assessing the abundance of a flying insect. They were interested in the abundance of *Plutella xylostella* as a high-altitude migrant, and were primarily using vertically orientated radar to assess this. Although reasonably confident that they could discriminate *P. xylostella* on the basis of the size of the radar reflections and the wing beat frequency, they used an MV light trap (and also a net deployed from a tethered balloon) to confirm the presence and general abundance of the moth.

Suction traps

Suction traps have been used very extensively to trap small flying insects, with a particular emphasis on biting Diptera and on aphids. These are not caught easily by light traps or netting methods, although many species now have pheromone or other baits available for them. Experience has shown that suction traps can provide a very effective catching efficiency, although they may be heavily influenced by weather conditions and precise trap placement. Some standardized traps have been running for many years, providing an unrivaled dataset for detecting seasonal and annual patterns in the populations of the target species. They have almost completely replaced "trawl" traps, where a net is towed through the air for a given distance, at a set speed.

Basic designs and simple modifications

The basic design of a suction trap is that a motorized fan draws or forces air through a filter, on which the insects are caught. Modifications affect the power of the fan (and therefore the volume of air processed in a standard time); the direction, placement, dimension, and precise shape of the entry port; and the nature of the filter and trapping chamber. Frequently, the suction device is associated with another trapping method, usually a light trap, a bait trap, or a pheromone trap. The time of use, the height at which used, and the trap surroundings also influence the trapping efficiency.

Since all of the factors that are varied have an undoubted effect on the size and composition of the catch, it is essential in any study to standardize these factors and to define them very clearly in all reports.

Standard Rothamsted suction trap

In 1964 the first of what is now a European series of standardized suction traps for aphids was set up at Rothamsted Experimental Station (Fig. 6.2). This was designed by L.R. Taylor, as a modification of previously used traps at the research station (Taylor 1951), and is characterized by a 12-meter-high entry tube. This is designed to ensure that the trap catches aphids which are moving significant distances in the air column, rather than just ones in close proximity to a host plant, and this reflects the proposed use for the traps. As well as providing information on aphid species richness and diversity, and data on year-to-year population changes, the traps are used to provide early warning of the arrival of pest species. The spring movements of crop pest aphids can be plotted by the network of traps and farmers given appropriate warning of the need to use insecticide spray. An *Aphid Bulletin* is published to provide this advice and is circulated within one week of each capture event (Woiwod & Harrington 1994).

The currently used fan draws in $45\,m^3$ per minute through a pipe of 244 mm internal diameter, resulting in high "extraction" efficiency but also a very high wind velocity in the pipe (Muirhead-Thompson 1991). To avoid damaging the insects, the wind velocity is reduced by having an expansion chamber just above the filter. This results in a "capture" velocity of just over 1 m per s, which keeps the catch in good condition. A jar with preservative fluid is the actual receptacle in which the catch is stored. To run such a powerful fan, it is necessary to use a mains power supply, so that this design of trap is not portable. The traps are shaped and colored so that they are neither attractive nor repellent to insects but merely catch the individuals present in a volume of air.

In 1994 15 standard Rothamsted suction traps were in operation in Britain, with 14 more in France and several others in the rest of Europe (Woiwod & Harrington 1994). These provide a wide-ranging picture of the pattern of aphid

Fig. 6.2 Rothamsted Insect Survey standard design suction trap. These are used to provide warning of increasing aphid numbers in spring and early summer. Picture courtesy of Ian Woiwod, Rothamsted Experimental Station.

movement across Europe and are the model which is also being adopted worldwide for national schemes.

The use of data from these standardized, continuous catches has allowed rigorous testing of the relationships between climatic variables and aphid abundance. For pest species, such as *Aphis fabae* or *Myzus persicae*, this has also allowed development of predictions of aphid abundance based on winter weather (Bale et al. 1992). It has been found that the temperatures recorded in January/February, which affect mortality of overwintering aphids, are well correlated with numbers the following season, with time of migration to the host crop depending largely on accumulated day-degrees from midwinter. This has allowed good prediction of the likely time of arrival, and abundance, of the pest species, so helping farmers plan preventative spraying. Using these models it has also been possible to predict the likely impact of climatic warming, with alarmingly high increases proposed for an overall 3°C rise in temperatures (Anon. 1991).

The Rothamsted dataset for aphids (and also some moths from their light traps) have also been used successfully to test for the presence of "chaotic" dynamics in insect populations, a procedure only possible where long-term data are available (Zhou et al. 1997).

As well as for small Diptera, suction traps have been used for many other small aerial invertebrates, for example ballooning spiders (Dean & Stirling 1985) and beetles (Leos Martinez et al. 1986).

Modified designs

Suction traps began to be used extensively in the 1950s and early 1960s and a very commonly used type of fan was the 9-inch (230 mm) diameter commercially available Vent-Axia® (Johnson & Taylor 1955). This moves around $10 m^3$ per min, considerably less than the Rothamsted design, but sufficient to make large catches of small Diptera in favorable conditions. In a very common design (the "exposed cone" design), this fan is set at the top of a long mesh cone and the insects are gradually sieved down to the bottom of the cone, into a collecting tube. Passage past the fan blades does not usually harm small insects. Often there is a series of plates which can be dropped into the collecting tube at pre-set intervals so as to segregate the catch. This allows investigation of the time of flight of the species being caught.

More recently smaller fans, powered by batteries, have been used in traps specifically designed for a particular type of insect, often in association with an attractive "lure". In this case the insects are attracted from the surroundings by the lure and then trapped by being sucked into the collecting tube. This prevents calculations of absolute insect density in the air but maximizes numbers caught. These traps are frequently fully portable and can even be hauled up into the canopy of a forest.

It is now routine to modify the precise design of the trap to suit the study insect. Some of the slower flying mosquitoes or midges (e.g. Ceratopogonidae) need only low-powered fans but are very sensitive to the placement and shape of the trap opening. Some are caught only in exposed locations and others only within shelter; many are only caught if a CO_2 or pheromone bait is also used to draw the insects to the vicinity of the trap. A further consideration is whether the shape and coloration of the trap is itself attractive or repulsive, for if the purpose of the trap is to produce a measure of the density of individuals in the location, then the trap itself must be neutral in attractive effect. If the trap appears to be a vertical shape then it may be used as a cue, above which form swarms of male Diptera. These may then be caught in huge abundance, distorting the true abundance of the species.

Calibration of suction trap catches and the effect of differing wind speed

If a suction trap is neither attractive nor repulsive, then the catch may represent the true number of insects that were present in the volume of air that has been filtered. This will only apply to individuals that fly so weakly that they cannot take evasive action, of course, and progressively stronger flying species will be caught less and less effectively. If the trap really does provide a proper estimate of the numbers of insects present then it becomes possible to make full estimates of abundance by considering the volume of air sampled. Since wind speed both affects the efficiency of capture and also physically moves the air and its insects past the trap opening, it is also necessary to consider wind speed in any calculations. Taylor (1962) investigated the relationship between these factors and derived an equation for the efficiency of capture of each of a series of commonly used traps. He then produced a logged conversion factor for each size of insect, at various wind speeds, so as to arrive at the following formula:

$$\text{log catch per hour} + \text{conversion factor} = \text{log density} \cdot 10^6 \text{ cu ft of air} \qquad (6.1)$$

There is no doubt, however, that this formula is too simple to account fully for the many factors that influence catches, and so results from it (and more recent modifications) must be interpreted critically.

Taylor (1962) also noted that there was some variation in the efficiency of any one design, and so a difference of at least 6 percent in catch for a 9-inch (230 mm) fan, or 10 percent for an 18-inch (460 mm) fan, is necessary before a real difference in the level of catch can be claimed.

Factors influencing catches by suction traps

It has been found that the precise surroundings of a suction trap can alter the catch dramatically, as shown by a study of mosquitoes in Florida which used moveable mesh barriers and potted shrubs to vary the degree of exposure (Bidlingmayer 1975). Different species were caught in open, sheltered, and partially sheltered conditions. It has also been found that the height of the trap inlet is of considerable importance. Some woodland mosquitoes in the UK fly only below 30 cm, and so in one study these were caught using a trap which had been sunk into the ground (Service 1971).

This relationship between height of inlet and species caught has always been considered (*vide* the 12-meter-high Rothamsted traps), and Southwood and Henderson (2000) record a procedure to develop an equation which relates catch abundance to outlet height for any given species. In The Gambia it was similarly found that different species flew at different heights, although there was also a difference associated with time of night.

Four main factors influence the size and composition of the catch, aside from height or placement. These are wind speed, insect size, suction rate and pipe diameter, and precise shape of the inlet tube. Many years ago comparisons were made between the efficiency of suction traps and of sticky traps and tow nets at different wind speeds (Johnson 1950). In general wind speeds below 3–5 miles per hour (5–8 km·per h) favor suction traps, whereas above 7 miles per hour (11 km per h) tow nets still work but suction traps are ineffective. The relationship also depends on insect size, however, for large Diptera are predominantly caught on sticky traps, rather than by suction traps.

The wind speed disrupts the suction gradient, as can be detected by using smoke particles to visualize it, but even in still air small traps may only trap insects that approach within 20 cm of the inlet. It has also been found that the "exhaust" airflow can disrupt the suction gradient, so that, if more than one trap is used, care must be taken in their mutual placement (Bidlingmayer & Hem 1981). The influence of wind speed is indirect, as well as direct, in that fewer individuals, especially of small species, may fly in high wind speeds.

The area of the inlet diameter influences the catch, as does its precise shape and direction, and the use of directional traps and gauze baffles have shown that different species respond differently to these factors. Often mosquitoes fly upwind in a gentle breeze and so are caught more when the inlet is directed downwind. However, the overall shape of the trap is also important, especially for biting species, and some species of *Simulium* approach and are caught by a suction trap resembling a standing human, whereas for others a longitudinal "cow" shape is most attractive (Coupland 1990). Many large species are barely trapped at all and presumably fly against the suction action; this can be substantiated by using radar that shows the presence near a trap of species such as medium to large moths, which nevertheless do not appear in the trap contents (Schaefer et al. 1985).

The use of "baits" with suction traps

It is now commonplace to use either light, or attractively colored shapes, or volatile baits, with suction traps (Fig. 6.3). For example, researchers at the Institute of Animal Health at Pirbright have designed and used extensively a portable combined light and suction trap for various small Diptera (the "Pirbright" trap). These traps are now widely used, as by Linton (1998) and others, who caught minute *Culicoides* midges in the UK and Spain. (The use of volatile-bait traps is described below.)

If light traps have become an almost universal type of trap for large flying insects, then suction traps have certainly become similarly generally used for small flying species. As with light traps, however, they are so easily affected by precise trapping conditions that great care must be used in interpreting their catches.

Water (or pan) traps

Many flying insects settle onto surfaces whose tone and/or color contrast clearly with the background. Such behavior is obviously related to the taxes that lead to recognition of, and settling onto, such things as flowers or ponds

Fig. 6.3 Onderstepoort Veterinary Institute design combined light and suction trap. Small insects are attracted by the light and then sucked down into the collecting chamber.

and so, as expected, different species and sexes of insects have different responses. This behavior is taken advantage of in the use of "water" or "pan" traps, which are nothing more than a contrasting surface and water to trap the insects that settle onto it.

Basic design and use

Simple water traps (Fig. 6.4) are merely shallow dishes, usually around 15 cm diameter, which are held at a standardized height above the ground, often 1 m. They contain water, to which has been added a drop of detergent to reduce surface tension, and sometimes an added preservative, such as formalin. The traps are left in place for a set time, often 2–4 days, after which their catches are filtered through a sieve and stored. Such traps catch a very wide range and

Fig. 6.4 Simple water trap. Insects are attracted to settle on the bright surface and become trapped in the water.

abundance of flying insects and so can be used in general faunal surveys, as well as in more focused studies. The catch is very dependent on weather conditions, for in windy or cool weather few insects fly and are caught. Nevertheless, the trap cheapness allows easy replication and if sufficient traps are used then the catch data can often be regarded as semi-quantitative.

Modifications

A frequent modification is to use upright, mutually perpendicular baffles, set above the trap so as to intercept flying insects (Coon & Rinicks 1962). However, the biggest influence on the selectivity and efficiency of the traps is the color of the pan. Harper and Story (1962) showed that different colored traps attract varying numbers of the sugar beet fly *Tetanops myopaeformis*, and numerous authors have repeated this experience, including Leong and Thorp (1999), who studied bees that visit white-flowered *Limanthes douglasii rosea*. They found that female *Andrena* were attracted to white and blue pans, whereas males were caught more in white and less in blue and yellow ones, differences that might be related to natural flower visiting. The effect of different colored dishes, as well as of differently patterned baffles, has also been investigated by researchers attempting to monitor numbers of tsetse flies (Deansfield et al. 1982). Blue was highly attractive to *Glossina tachinoides*, whereas white was best for *G. morsitans*, and the catch was increased by the use of black baffles. Other authors have found that red traps are favored by some beetles (Dafni et al. 1990), whereas white and yellow pans are visited most by Diptera in general (Disney et al. 1982). Following such experience, most general studies have used one of these two pale colors (Kirk 1984), whereas focused studies should use a preliminary catching period to select the most efficacious color.

Types of use

Water traps have been used for widely different studies. The fact that most flying insects are attracted at least to some extent has allowed their use in faunal surveys and comparisons. For example, Young and Armstrong (1994) used white traps in various stand types of native pinewoods to make comparisons between the stands. They found that the catches included some of all orders of flying insects present in the forests, but that Diptera, Coleoptera, and Hymenoptera predominated. Furthermore, the trap catches allowed easy differentiation between stand type. However, the catches were greatly affected by precise location, so that sheltered traps caught best during windy episodes, although easily visible traps in open forest were generally to be preferred. They and others (e.g. Disney et al. 1982) found many "tourist" species in the catches and these reduce the site specificity of the results and make them less suitable for focused site comparisons than results from (for example) pitfall traps.

McGeachie (1987) used a circular array of water traps at varying distances from light sources to assess the position at which insects alighted, when they approached the light from different directions. In this example the catches were merely of settling insects, rather than ones attracted by the pans, but they proved very effective.

Leong and Thorp's (1999) study was more specific. They were catching insects that pollinated a particular plant, and the color and size of the pans used were designed to act as "super-flowers". However, as well as catching the actual pollinators, they also caught many other insects and this "wastage" may not be widely acceptable.

The catching power and simplicity of the water trap has often been combined with pheromone traps to produce highly effective designs. Many pest species have been monitored using such combined traps. For example, Thompson et al. (1987) compared the catch of the European corn borer *Heliothis zea* in light traps and pheromone traps, with either a sticky surface or a water trap. At times light traps worked best, but water traps improved the success of the pheromone lures.

In summary, water traps will provide either a wide-ranging and abundant catch, which reflects the general composition of flying fauna of an area, or a more focused assessment of one taxon. However, they cannot achieve better than a semi-quantitative result and so interpretation of catch data must be cautious. Nevertheless, their simplicity and economy make them useful in many circumstances.

Sticky traps

The basic trapping style of sticky traps is similar to that of water traps. Insects settle onto a surface and are caught there. This process may either be passive, where the sticky surfaces merely intercept insects that are blown or fly inadvertently onto them, or active, where the insects choose to settle onto the surfaces. The shape, position, and color of the traps all influence the trapping efficiency, and they are often used in conjunction with other trap types.

The major practical differences between water and sticky traps are twofold. Water traps have to be set facing upwards, whereas sticky traps can be angled and shaped to whatever design is required, which can be a major advantage. However, insects caught on sticky traps are frequently badly affected by the process. The usual practice is to use a sticky substance that can be dissolved, usually chemically but in some cases by warmth, so that the insects can then be sieved out and preserved in alcohol. However carefully this is carried out, delicate insects often lose scales and/or appendages in the process and may also be distorted, so reducing ease of identification. This matters less for beetles or robust wasps, but moths and small flies may be beyond use. This is a substantial problem, which greatly reduces the usefulness of sticky traps for general purposes. Their main use is therefore in focused studies, where another attractant is also used, so that only one species or taxon is being caught, and where the condition of the catch is unimportant.

Basic design and use

A basic sticky trap is a surface coated with the trapping gum. Various substances are available commercially and are of varying holding strength. For tiny insects a rather fluid grease, or even castor oil, may suffice, and this will cause less damage, whereas for larger insects a thick resin or gum is better, and polyisobutylene is now widely used. The gums themselves act to preserve the insects for a limited period and a standard practice is to cover the sticky surface, with its trapped insects, in the field with "cling film," so that sheets can be stored together for later analysis. As soon as possible thereafter the gum is dissolved (Murphy 1985) and the catch transferred to fluid preservative.

If an upright cylinder or flat surface is used, then the catch will include insects that have merely been "blown" onto the surface. In most circumstances, however, the surface is angled and shaped to produce a more selective catch, often in conjunction with a specific attractant.

The cost of each sticky trap is small and so many can be included in a sampling program. This allows extensive replication, so that confidence can be assigned to results, and this is a major advantage. However, the trapping surface can become fully saturated, or the gum can lose its stickiness (when cold or coated with dust), so that there may not be a linear catch rate over time. Catches are also highly dependent on weather conditions, insofar as these affect flight, so that interpretation of results must allow for this.

Modifications

The most common use of sticky traps is to provide the catching power in traps using attractive lures, frequently pheromones. However, the color, shape, and size of the sticky surfaces also affect the basic efficacy and these will be discussed first.

Different species of insect fly at different heights and respond to colors in different positions. Finch and Collier (1989) provided flat sticky squares, angled from facing directly upwards, by 45° intervals through directly downwards back to upwards, and recorded the catch type in different agricultural fields. Syrphidae were almost exclusively caught on vertical surfaces, *Psila rosea* (the turnip root fly) on faces pointing 45° downwards, and *Delia* species on upwards or upwardly angled plates.

Vertically placed cylinders with sticky surfaces are claimed to catch freely flying insects without bias, but it was realized very early on (e.g. Taylor 1962) that, even if the wind speed and direction were recorded, it was not possible to relate catch density directly to actual abundance. Furthermore, the color of the surface influences the catch. (For this reason, suction traps were developed to trap aphids in realistic numbers.) Many researchers have used flat, square surfaces, but others have mimicked the shape of the natural target of their insects. Kring (1970) compared the efficiency of catch of red spheres and yellow panels for apple maggot flies *Rhagoletis pomonella* and found that a red hemisphere set on a yellow background out-performed a flat red circle on the same yellow background. Later workers, such as Jones (1988), have also found that red spheres are the optimum shape for this species of pest. However, Meyerdirk et al. (1979) found no difference in catches between triangular, square, circular, and rectangular yellow surfaces for the citrus blackfly *Aleurocanthus woglumi*.

The effect of different colors of sticky traps has been studied often, including the preferential choice of red spheres reported above. Katsoyannos (1987) found that the preferential sequence of colors for Mediterranean fruit flies *Ceratitis capitata* was yellow > orange > black > red = green > white = blue. Webb et al. (1985) showed that greenhouse whitefly *Trialeurodes vaporarium* showed a rather similar preference spectrum, namely yellow > green = orange > white = violet = blue = red = black. They showed that the bright yellow traps even out-competed leaves as a landing surface.

Comparative trap efficiency

The crucial question is how well sticky trap catches compare with real abundances of target species. Heathcote (1957) made an early attempt to compare the real efficiency of different trap types by noting the numbers of aphids caught by water, flat sticky, and cylindrical sticky traps, as a ratio to catches from adjacent suction traps. He found that results were very variable, with high catches for some species from water traps (e.g. ratio = 3.91 for *Tuberculoides annulatus*), and on cylindrical sticky traps (e.g. 1.92 for *Aphis fabae*), but generally low catches on flat sticky traps (ratios from 0.01 to 0.84). In a greenhouse environment Gillespie and Quiring (1987) compared the numbers of *Trialeurodes vaporariorum* found on plants with those caught on yellow sticky traps. They found that the density of traps influenced the answer. It was possible to swamp the system with traps, so that almost all insects were on traps, rather than on plants. At low trap densities, however, there was a reasonable relationship between insects on leaves and on traps. Traps also proved to be very sensitive and detected whiteflies at levels that were barely perceptible on plants.

In field conditions Ramaswamy and Cardé (1982) compared the efficiency of traps under different conditions for the spruce budworm *Choristoneura fumiferana*. Sticky surfaces, associated with pheromone lures, greatly increased the sensitivity of the traps, but only if they were changed sufficiently often to prevent surface saturation (but not so often as to disturb recently arrived moths).

This last example re-emphasizes the point that sticky traps are most often used in association with another lure. They can be highly effective, and are cheap and easily replicated; however, they only represent real population levels in certain closely defined circumstances.

Baited traps

Certain insects are attracted powerfully to volatile chemicals, and these chemicals can be used in traps to attract and kill the insects, either so as to monitor their numbers or to actually reduce their populations. Blood-sucking species, such as tsetse flies *Glossina* spp or mosquitoes (e.g. *Aedes* spp), for example, are attracted by host odors (as well as host shapes), whereas male moths of many forest pest species are attracted by sex pheromones released by females.

Basic trap designs and common modifications

The elements of a bait trap are the source of attractant and a trapping surface or container. Although these basic elements are similar for "bait" and "pheromone" traps, there are sufficient differences for them to be discussed separately.

"Bait" traps

Bait traps divide roughly into those which use host odors to attract female flies which are hoping to feed, and those which use carrion or dung in which the female flies hope to oviposit (carrion or dung traps). Vessby (2001) used naturally collected dung placed in suspended "bucket" traps to collect samples of the dung beetles *Aphodius* spp, and showed clearly that different catches were made if the dung was allowed to dry out, rather than if it was watered to keep it moist.

Early carrion traps typically used liver as a bait and flies were caught if they flew from the bait up into a conical catching chamber above the bait. These can be highly effective for some species, such as blow-flies (e.g. *Lucilia* spp) that cause "strike" in sheep. However, the traps suffer from operational problems and are influenced by various environmental conditions (Gillies 1974).

First of all the age of the liver bait influences the species caught, so that it is difficult to compare different catches. However, one response to this, to discard baits after three days (Vogt et al. 1983), leads to the traps being very inefficient when used against screw-worms (*Cochliomyra* spp.) in the USA. It was found that these are only attracted efficiently by baits over 5–7 days old (Coppedge et al. 1978). Secondly many non-target species may be caught, including large and disruptive carrion beetles and ants. To prevent this, baits are now typically presented on poles incorporating ant "baffles" and with a coarse screen in front of the catching chamber. The liver bait is also usually shielded by a gauze cover, to prevent excessive egg laying and subsequent changes due to maggot feeding. As an alternative to liver, commercially available chicken legs were used by Smith and Merrick (2001) to attract carrion-feeding *Nicrophorus* beetles.

Inevitably more female than male flies are caught, although some males also arrive both to feed and to find mates. It has also been found that the proportion of males varies through the day and with weather conditions (Vogt et al. 1983).

Generally catches of both males and females increase directly with temperature, up to a threshold level, but males are caught more in bright sunlight. Increasing wind speed reduces catches but some species are found mainly in exposed traps, whereas others come mainly to sheltered sites.

Recently the principal development in "carrion" types has been the use of specific chemicals in place of meat baits. For screw-worm controls, it has been found that a cocktail of the volatile chemicals that are released as the liver decomposes make a very effective attractant, originally called Swormlure (Coppedge et al. 1978). A series of improved recipes has since been used, varying both the chemicals used and their relative concentrations.

Whole animals as baits

Traps which use "host" baits were originally literally that. Whole animals were used to attract the biting flies (Phelps 1968, McCreadie et al. 1984). Even now an effective way to collect individual tsetse flies is to "poot" them off the surface of a human or animal bait, and Coupland (1994) used a similar method to assess the activity and behavior of *Simulium* in Scotland. This catches small numbers of flies in good condition, and early attempts to make larger catches involved suddenly dropping nets over tethered oxen. The practical problems of using such traps are easily imagined and they are not now widely used, except in conjunction with small bait animals such as rabbits. However a commonly used variant keeps the whole animal as a bait but uses a series of electrocuting surfaces or nets around the animal on which the flies are killed and caught (e.g. Vale 1974, Rogers & Smith 1977). Alternatively a suction trap may be placed directly above a small animal bait, as has been used to trap various nuisance flies in Trinidad (Davies 1978).

Greater efficiency of capture for low-density biting flies is provided by the use of moving baits, often over a set transect, with stops at specified catching stations. However, as Glasgow (1961) found, it is still only possible to detect gross population changes using such methods.

The attractant nature of an animal is often a combination of its size, shape, heat output, exhaled breath components including CO_2, and body odor. It is difficult to mimic all of these, hence the continued use of whole animals, but various trap designs use combinations. For example Coupland (1990) used a cow-sized, rectangular black Malaise-style trap to catch *Simulium* spp, but catches were enormously enhanced by adding a slow leak of CO_2 from the underside of the trap. This was provided cheaply by using inner tires as pressurized reservoirs of CO_2. Anderson and Yee (1995) describe similar combined devices used for *Simulium* spp in America.

More recent improvements have come principally from the use of improved formulations of attractant chemicals. These range from CO_2 to aldehydes, ketones, and octenols. Octenols have proved especially useful, partly because they are only slowly volatile and so a long release is possible from an impregnated lure (Hall et al. 1984). Blackwell et al. (1996) used 1-octen-3-ol, a component of the body odor of ruminant mammals, to attract *Culicoides* midges. They combined field trials with laboratory Y-tube preference tests and electro-antennograms, to provide a full picture of the effectiveness of this chemical. Typically this chemical is attractive at low to moderate concentrations but becomes repellent at high concentrations.

In this study the chemicals were used in "Delta" traps, which are open, triangular sticky traps, but the effectiveness of different traps is discussed below.

Pheromone traps

The attractiveness of virgin female insects to males, especially in Lepidoptera, has long been known, and even in the 1930s attempts were made to use crude extracts of female gypsy moth *Lymantria dispar* abdomens to attract males. These were largely unsuccessful, but living female moths have often been used as lures (e.g. for attracting male spruce budworm moths *Choristoneura fumiferana*; Miller & McDougall 1973). Once the attractive chemicals were identified and could be synthesized, as began in the 1960s and 1970s, it became possible to make effective pheromone traps (Fig. 6.5). These are now used routinely to attract males to killing lures, so as to reduce pest populations; to swamp wild female attractiveness, so as to disrupt mating; and to monitor pest populations by detecting males as soon as they appear (Cardé & Minks 1995).

An early discovery was that the female attractants are usually a subtle blend of varying concentrations of several chemicals; only rarely, as in the gypsy moth, are single chemicals used (Cardé & Baker 1984). This complexity has made the production of effective lures much more difficult, and, although super-effective formulations have occasionally been produced, a wild female still generally outperforms the chemical lure.

Typically the chemical lure is impregnated into a "plug," which is then held close to sticky trap surfaces, or above a collecting box (e.g. Sanders 1988). A common problem of sticky traps is that they become saturated, so that the catch is only proportional to numbers of males present in the local environment when these numbers are low. To overcome this problem large collecting chambers with an insecticide, or water-filled chambers, are sometimes used, but these are more expensive and more complicated to use than simple sticky surfaces (Kendall et al. 1982).

The study of Keil et al. (2001) illustrates the problem that, whereas males can be caught at pheromone lures, the capture of females needs a different technique. They marked the orchard pest *Cydia pomonella* by incorporation of a red dye in the larval diet, released the marked moths and then re-caught the males in pheromone traps and the females using light traps.

The design of the trap has been very varied, for it has been found that even minor variations can influence the catch of a species significantly. This has resulted in a plethora of commercial designs. For some pest species it is now

Fig. 6.5 Simple pheromone trap. Pheromone is released from the "wicked" tube and insects are caught on the sticky interior surface of the trap.

known what design is to be preferred, but often a sampling program merely has to make use of an easily available and cheap trap such as the Delta or Pherocon® design. The precise height and location of the trap also influences catches significantly, so preliminary trials are recommended to investigate optimal trapping conditions. Some of this variation is associated with the way that the volatile pheromone spreads away from the trap in a plume. These plumes can be visualized by releasing visible substances, such as smoke or soap bubbles, from the lure site and assuming that these behave in the same way as the pheromone chemical. The wind speed and its constancy, as well as the presence of obstacles, affects the plume and this alters the attractiveness of the lure (Elkinton et al. 1987).

The concentration of the attractive chemical also affects the behavior of male moths, and hence whether they are ultimately trapped. Typically very low concentrations lead to increased alertness, slightly higher concentrations lead to walking activity, higher still induce flight, but atypically high concentrations may be repellent (Cardé & Charlton 1984). The flight tends to be upwind in the pheromone plume but erratically cross-wind if the plume is lost. The distance of attraction can be tens of meters in favorable conditions for large moths but only a few meters for small Diptera (David et al. 1983).

The use of pheromone traps

Pheromone traps are used to determine whether a pest species is present in a geographical area, to confirm whether immigration or emergence has happened, to trap and kill pests, and/or to monitor population levels. For the latter use particularly it is essential to know how trap catches relate to actual population levels.

Various studies have related numbers caught at pheromone lures with either direct counts of larvae or pupae, or with light-trap catches. In general very good relationships have been found between catches and actual numbers at low population levels (Speight & Wainhouse 1989). Frequently males are caught even when population levels are too low to be detected in any other way. However, as natural population levels increase, pheromone trap catches begin to level off and they become non-representative at moderate to high levels. This is because of a range of factors, including physical trap saturation and principally the swamping effect of many wild females (Croft et al. 1986). At high population levels light traps perform significantly better than pheromone traps. Of course this also applies at times when the insects are not mating, perhaps in an initial immature period or before hibernation. Barbour (1987) compared the relationship between pheromone trap catches of the pine beauty moth *Panolis flammea* and absolute counts of pupae in standardized soil areas. He found that he had to make an allowance for the patchy nature of the pupal distribution and for the mobility of the moths but that, after these adjustments, regression analysis yielded relatively high r^2 values (e.g. 61 percent), indicating that the pheromone traps were providing reasonable representation of the actual moth numbers.

Numerous studies have now been carried out using pheromone lures, and many successful control programs depend on them (Ridgway et al. 1990). Well-worked examples include that of the pink bollworm *Pectinophora gossypiella* in the USA, Egypt, and Pakistan (Campion et al. 1989) and the spruce budworm *Choristoneura fumiferana* in the USA and Canada (Silk & Kuenen 1988).

In studies where initial detection of a pest's presence, or the attainment of a threshold level requiring a pest control program, is what is required, then pheromone traps are ideal, and their importance in trapping insects cannot be over-stated. If a representation of the full range of population levels is required then they are less useful.

Interception traps

There is an essential difference between those traps that are undetected by insects and act by literally intercepting their flight and those that are detected but act merely as a screen, trapping insects that land and crawl on them. True interception traps do not attract insects and may provide an unbiased estimate of the true population flying in an area. In this they differ fundamentally from attractive traps and are more analogous to suction traps or direct netting.

Undetected traps

Undetected interception traps are usually a series of sheets of a fully transparent material, these days often Perspex or acrylic, that are set out either at random, or close to a reference point. Such traps are called window traps and have been used to catch a wide range of insects, especially including bark beetles (Canaday 1987, Young & Armstrong 1994) and dispersing ground beetles (van Huizen 1977). The sheets are placed above a water reservoir, into which the insects drop and are retained, especially if a drop of detergent or oil is added to the water. By using angled Perspex sheets, often in a cross shape, it is possible to detect the direction of flight. Sometimes an object of investigation, such as a dead tree, may be surrounded by window traps, so allowing the relationship between wind direction and insect approach to be determined (Tunset et al. 1988).

In practice raindrops and/or dust soon adhere to window traps, however, rendering them visible and so reducing their unbiased activity.

Other undetected interception traps are fast moving nets, including tow nets and sweep nets. Sweep nets are very widely used to dislodge and collect insects from relatively long ground vegetation. They are of limited comparative use, because of the factors discussed below, but they do provide a quick, simple, and sometimes acceptable indication of the relative abundance of some of the insects flying close to the ground layer. For example, Banks and Brown (1962) found less than 10 percent variation in catch between replicated sets of sweeps in wheat fields. The efficacy of a sweep varies with many factors, however, including the height, size, and power of the netting stroke; length, nature, and density of the vegetation; whether the vegetation is wet; weather conditions; time of day; and the "holding" power of different insects. Despite this, sweep-netting is still widely used to provide quick estimates of abundance, and studies to determine the influence of the confounding factors have been carried out over many years (e.g. Gray & Treloar 1933, Rudd & Jensen 1977, Cherrill & Sanderson 1994).

Tow nets may be towed by planes (e.g. Reling & Taylor 1984) or trucks (e.g. Bidlingmayer 1974), or may be rotated on long arms, although the last hardly acts as a tow net, since the airspeed is usually too low. The advantage of these tow nets is that very large volumes of air are sampled, leading to relatively large catches of even scarce insects, but the nets are too cumbersome to be restricted to individual components of the environment. The exception is that airborne nets can provide an estimate of insects that are dispersing well above the vegetation layer. Although studies using such nets are frequently directed at specific types of insects, nevertheless they do catch a wide range of species, fitting them to faunal surveys.

Visible interception traps

Most interception traps are detected by the insects they catch, but are supposedly neutral in their effect, not being strongly attractive. This assumption is easily challenged, however, even in the case of the most widely used of all such traps, the Malaise trap (Malaise 1937), and Roberts (1972) showed that the color, size, and placement of the trap all influence catches.

The Malaise trap (see Chapter 4) is a "tent" of various types of material, arranged so that insects are led up into the trap's inner corner, at which is placed a non-return collecting jar. The original design has been modified several times and is often used in conjunction with an attractant bait, such as CO_2 (e.g. Coupland 1990). Studies have shown that Malaise traps are far from random in their catch, and that even different genera of flies within one family show different responses. Tallamy et al. (1976) found that *Chrysops* horse flies were not caught easily, whereas *Tabanus* were frequent visitors. However, such traps are still frequently used to make generalized catches where this is appropriate. For example, Petersen et al. (1999) successfully studied the dispersion of Plecoptera and Trichoptera from a stream using a series of Malaise traps.

Other more directional "funnel" traps have been designed to detect movement patterns in various insects, including mosquitoes in Africa (Gillies et al. 1978).

Overall, interception traps do have a role in specific, planned studies, where their apparent lack of bias can be tested, and also in the production of faunal lists. However, it is often difficult to interpret their catches. Recently they have frequently been combined with an attractant lure, to produce a more focused result.

Conclusions

This chapter reviews some commonly used techniques, but an essential message is that virtually every study has to use a modified technique. In a review of the last 30 papers published in a representative journal, namely *Ecological Entomology*, it was found that not one merely used a "standard" widely used method. Most incorporated elements of a general technique, such as sweep-netting, but all field work had been designed specifically for its study. In all cases, a preliminary period had been used to develop an appropriate method. Furthermore, methods are becoming ever more specific in their catches, except when they are deliberately chosen to be wide-ranging, and the improved chemical formulae used in pheromone traps illustrate this trend. It seems likely that advances in trapping techniques will come from better observation of the behavioral responses of the target insect, rather than from more modern technology, although miniaturization of components and improved battery life are bound to be important. As Southwood and Henderson (2000) comment in their preface, most of the trapping techniques were designed years ago and have stood the test of time, essentially unaltered, apart from the fine-tuning needed for each individual study.

References

Anderson, J.R. & Yee, W.C. (1995) Trapping blackflies (Diptera: Simuliidae) in northern California.1. Species composition and seasonal abundances on horses, host models and in insect flight traps. *Journal of Vector Ecology* **20**, 7–25.

Anon. (1991) *The Potential Effects of Climate Change in the United Kingdom*. First Report of the United Kingdom Climate Change Impacts Review Group. HMSO, London.

Bale, J.S., Harrington, R., & Howling, G.G. (1992) Aphids and winter weather. I. Aphids and climate change. In *Proceedings of the Fourth European Congress of Entomology and the XIII Internationale Symposium für die Entomofaunistik Mitteleuropas, Volume 1*, pp 139–143. Hungarian Natural History Museum, Budapest.

Banks, C.J. & Brown, F.S. (1962) A comparison of methods of estimating population density of adult sunn pest, *Eurygaster integriceps* Put. (Hemiptera, Scutelleridae) in wheat fields. *Entomologia Experimentalis et Applicata*, **5**, 255–260.

Barbour, D.A. (1987) Monitoring pine beauty moth by means of pheromone traps: the effect of moth dispersal. In *Population Biology and Control of the Pine Beauty Moth* (Panolis flammea) (ed. S.R. Leather, J.T. Stoakley, & H.F. Evans), pp. 49–56. Forestry Commission Bulletin 67, Edinburgh.

Bidlingmayer, W.L. (1974) The influence of environmental factors and physiological stage on flight patterns of mosquitoes taken in the vehicle aspirator and truck, suction, bait and New Jersey light traps. *Journal of Medical Entomology*, **11**, 119–146.

Bidlingmayer, W.L. (1975) Mosquito flight paths in relation to the environment: effect of vertical and horizontal visual barriers. *Annals of the Entomological Society of America*, **68**, 51–57.

Bidlingmayer, W.L. & Hem, D.G. (1981) Mosquito flight paths in relation to the environment: effects of forest edge upon trap catches in the field. *Mosquito News*, **41**, 55–59.

Blackwell, A., Dyer, C., Mordue (Luntz), A.J., Wadhams, L.J., & Mordue, W. (1996) The role of 1-octen-3-ol as a host-odour attractant for the biting midge, *Culicoides impunctatus* Goetghebuer, and interactions of 1-octen-3-ol with a volatile pheromone produced by parous female midges. *Physiological Entomology*, **21**, 15–19.

Bowden, J. & Church, B.M. (1973) The influence of moonlight on catches of insects in light traps in Africa. II. The effect of moon phase on light-trap catches. *Bulletin of Entomological Research*, **63**, 129–142.

Bowden, J. & Morris, M. (1975) The influence of moonlight on catches of insects in light-traps in Africa. III. The effective radius of a mercury-vapour light-trap rid the analysis of catches using effective radius. *Bulletin of Entomological Research*, **65**, 303–348.

Campion, D.G., Critchley, B.R., & McVeigh, L.J. (1989) Mating disruption. In *Insect Pheromones in Plant Protection* (ed. A.R. Jatsum & R.F.S. Gordon). Wiley, Chichester.

Canaday, C.L. (1987) Comparison of insect fauna captured in six different trap types in a Douglas-fir forest. *Canadian Entomologist*, **119**, 1101–1108.

Cardé, R.T. & Baker, T.C. (1984) Sexual communication with pheromones. In *Chemical Ecology of Insects* (ed. W.J. Bell & R.T. Cardé), pp. 355–383. Chapman & Hall, London.

Cardé, R.T. & Charlton, R.E. (1984) Olfactory sexual communication in Lepidoptera. Strategy, sensitivity and selectivity. In *Insect Communication* (ed. T. Lewis), pp. 241–265. Academic Press, London.

Cardé, R.T. & Minks, A.K. (1995) Control of moth pests by mating disruption: successes and constraints. *Annual Reviews of Entomology*, **40**, 559–585.

Chapman, J.W., Reynolds, D.R., Smith, A.D., Riley, J.R., Padgley, D.E., & Woiwod, I.P. (2002) High altitude migration of the diamondback moth *Plutella xylostella* to the UK: a study using radar, aerial netting, and ground trapping. *Ecological Entomology*, **27**, 641–650.

Cherrill, A.J. & Sanderson, R.A. (1994) Comparison of sweep-net and pitfall trap samples of moorland Hemiptera: evidence for vertical stratification within vegetation. *Entomologist*, **113**, 70–81.

Coon, B.F. & Rinicks, H.B. (1962) Cereal aphid capture in yellow baffle trays. *Journal of Economic Entomology*, **55**, 407–408.

Coppedge, J.R., Ahrens, E.H., & Snow, J.W. (1978) Swormlure-2 baited traps for detection of native screwworm flies. *Journal of Economic Entomology*, **71**, 573–575.

Coupland, J.B. (1990) The ecology of black flies (Diptera: Simuliidae) in the Scottish Highlands in relation to control. PhD thesis, University of Aberdeen.

Coupland, J.B. (1994) Factors influencing nuisance blackfly (Diptera: Simuliidae) activity in the Scottish Highlands. *Medical and Veterinary Entomology*, **8**, 125–132.

Croft, B.A., Knight, A.L., Flexner, J.L., & Miller, R.W. (1986) Competition between caged virgin female *Argyrotaenia citrana* (Lepidoptera: Tortricidae) and pheromone traps for capture of released males in semi-enclosed courtyard. *Environmental Entomology*, **15**, 232–239.

Dafni, A., Bernhardt, P., Shmida, A., Iruri, Y., Greenbaum, S., & O'Toole, C. (1990) Red bowl-shaped flowers: convergence for beetle pollination in the Mediterranean region. *Israel Journal of Botany*, **39**, 81–92.

David, C.T., Kennedy, J.S., & Ludlow, A.R. (1983) Finding a sex pheromone source by gypsy moths released in the field. *Nature*, **303**, 804–806.

Davies, J.B. (1978) Attraction of *Culex portesi* Senevet & Abonnenc and *Culex taeniopus* Dyar & Knab (Diptera: Culicidae) to 20 animal species exposed in a Trinidad forest. *Bulletin of Entomological Research*, **68**, 707–719.

Dean, D.A. & Sterling, W.L. (1985) Size and phenology of ballooning spiders at 2 locations in eastern Texas (USA). *Journal of Arachnology*, **13**, 111–120.

Deansfield, R.D., Brightwell, R., Onah, J., & Okolo, C.J. (1982) Population dynamics of *Glossina morsitans submorsitans* Newstead, and *G. tachinoides* Westwood (Diptera: Glossinidae) in sub-Sudan savanna in northern Nigeria. I. Sampling methodology for adults and seasonal changes in numbers caught in different vegetation types. *Bulletin of Entomological Research*, **72**, 175–192.

Disney, R.H.L., Erzinclioglu, Y.Z., Henshaw, D.J. de C., et al. (1982) Collecting methods and the adequacy of attempted fauna surveys, with reference to the Diptera. *Field Studies*, **5**, 607–621.

Downey, J.E. (1962) Mosquito catches in New Jersey Mosquito traps and ultra-violet light traps. *Bulletin of the Brooklyn Entomological Society*, **57**, 61–63.

Edwards, P.B., Kettle, D.S., & Barnes, A. (1987) Factors affecting the numbers of *Culicoides* (Diptera: Ceratopogonidae) in traps in coastal South-East Queensland (Australia) with particular reference to collections of *Culicoides subimmaculatus* in light traps. *Australian Journal of Zoology*, **35**, 469–486.

Elkinton, J.S., Schal, C., Ono, T., & Cardé, R.T. (1987) Pheromone puff trajectory upwind flight of male gypsy moth in the forest. *Physiological Entomology*, **12**, 399–406.

Finch, S. & Collier, R.N. (1989) Diptera caught on sticky boards in certain vegetable crops. *Entomologia Experimentalis et Applicata*, **52**, 23–27.

Gaydecki, P.A. (1984) A quantification of the behavioural dynamics of certain Lepidoptera in response to light. PhD thesis, Cranfield Institute of Technology.

Gillespie, D.R. & Quiring, R. (1987) Yellow sticky traps for detecting and monitoring greenhouse whitefly (Homoptera: Aleyrodidae) adults on greenhouse tomato crops. *Journal of Economic Entomology*, **80**, 675–679.

Gillies, M.T. (1974) Methods for assessing the density and survival of blood sucking Diptera. *Annual Review of Entomology*, **19**, 345–362.

Gillies, M.T., Jones, M.D.R., & Wilkes, T.J. (1978) Evaluation of a new technique for recording the direction of flight of mosquitoes (Diptera: Culcidae) in the field. *Bulletin of Entomological Research*, **68**, 145–152.

Glasgow, J.P. (1961) The variability of fly-round catches in field studies of *Glossina*. *Bulletin of Entomological Research*, **51**, 781–788.

Graham, H.M., Glick, F.A., & Hollingsworth, J.P. (1961) Effective range of argon glow lamp survey traps for pink bollworm adults. *Journal of Economic Entomology*, **54**, 788–789.

Gray, H. & Treloar, A. (1993) On the enumeration of insect populations by the method of net collection. *Ecology*, **14**, 356–367.

Gregg, P.C., Fit, G.P., Coombs M., & Henderson, G.S. (1994) Migrating moths collected in tower-mounted light traps in northern New South Wales, Australia: influence of local and synoptic weather. *Bulletin of Entomological Research*, **84**, 17–30.

Hall, D.R., Beevor, P.S., & Cork, A. (1984) 1-Octen-3-ol; a potent olfactory stimulant and attractant for tsetse isolated from cattle odours. *Insect Science and its Application*, **5**, 335–339.

Harper, A.M. & Story, T.P. (1962) Reliability of trapping in determining the emergence period and sex ratio of the sugar-beet root maggot *Tetanops myopaeformis* Röder (Diptera: Otitidae). *Canadian Entomologist*, **94**, 268–271.

Hartstack, A.W., Hollingsworth, J.P., & Lindquist, D.A. (1968) A technique for measuring trapping efficiency of electric insect traps. *Journal of Economic Entomology*, **61**, 546–552.

Haufe, W.O. & Burgess, L. (1960) Design and efficiency of mosquito traps based on visual response to patterns. *Canadian Entomologist*, **92**, 124–140.

Heathcote, G.D. (1957) The comparison of yellow cylindrical, flat and water traps and of Johnson suction traps, for sampling aphids. *Annals of Applied Biology*, **45**, 133–139.

Johnson, C.G. (1950) The comparison of suction trap, sticky trap and townet for the quantitative sampling of small airborne insects. *Annals of Applied Biology*, **37**, 268–285.

Johnson, C.G. & Taylor, L.R. (1955) The development of large suction traps for airborne insects. *Annals of Applied Biology*, **43**, 51–61.

Jones, V.P. (1988) Longevity of apple maggot (Diptera: Tephritidae) lures under laboratory and field conditions in Utah. *Environmental Entomology*, **17**, 704–708.

Katsoyannos, B.L. (1987) Effect of color properties of spheres on their attractiveness for *Ceratitis capitata* (Wiedmann) in the field. *Journal of Applied Entomology*, **104**, 79–85.

Keil, S., Gu, H., & Dorn, S. (2001) Response of *Cydia pomonella* to selection on mobility: laboratory evaluation and field verification. *Ecological Entomology*, **26**, 495–501.

Kendall, D.M., Jennings, D.T., & Houseweart, M.W. (1982) A large capacity pheromone trap for spruce budworm moths (Lepidoptera: Tortricidae). *Canadian Entomologist*, **114**, 461–463.

Kirk, W.D. (1984) Ecologically selective coloured traps. *Ecological Entomology*, **9**, 35–41.

Kring, J.B. (1970) Red spheres and yellow panels combined to attract apple maggot flies. *Journal of Economic Entomology*, **63**, 466–469.

Leong, J.M. & Thorp, R.W. (1999) Colour-coded sampling: the pan trap colour preferences of oligolectic and nonoligolectic bees associated with a vernal pool plant. *Ecological Entomology*, **24**, 329–335.

Leos Martinez, J., Granovsky, T., Williams, H.J., Vinson, S.B., & Burkholder, W.E. (1986) Estimation of aerial density of the lesser grain borer (*Rhyzopertha dominica*) (Coleoptera: Bostrichidae) in a warehouse using dominicalure traps. *Journal of Econonomic Entomology*, **79**, 1134–1138.

Linton, Y.M. (1998) Characterisation of the South African *Culicoides imicola* (Kieffer, 1913) species complex, and its phyllogenetic status in Europe. PhD thesis, University of Aberdeen.

Malaise, R. (1937) A new insect trap. *Entomologisk Tidskrift*, **58**, 148–160.

McCreadie, J.W., Colbo, M.H., & Bennett, G.F. (1984) A trap design for the collection of haematophagous Diptera from cattle. *Mosquito News*, **44**, 212–216.

McGeachie, W.J. (1987) The effect of air temperature, wind vectors and nocturnal illumination on the behaviour of moths at mercury vapour light traps. PhD thesis, Cranfield Institute of Technology.

Meyerdirk, D.E., Hart, W.G., & Burnside, J. (1979) Evaluation of a trap for the citrus blackfly *Aleurocanthus woglumi* (Homoptera: Aleyrodidae). *Canadian Entomologist*, **111**, 1127–1129.

Miller, C.K. & McDougall, G.A. (1973) Spruce budworm moth trapping using virgin females. *Canadian Journal of Zoology*, **51**, 853–858.

Muirhead-Thomson, R.C. (1991) *Trap Responses of Flying Insects*. Academic Press, London.

Mulhern, T.D. (1942) New Jersey mechanical trap for mosquito surveys. *New Jersey Agricultural College Experimental Station Report*, **421**, 1–8.

Murphy, W.L. (1985) Procedure for the removal of insect specimens from sticky-trap material. *Annals of the Entomological Society of America*, **78**, 881.

Nag, A. & Nath, P. (1991) Effect of moonlight and lunar periodicity on the light trap catches of cutworm *Agrotis ipsilon* (Hufn.) moths. *Journal of Applied Entomology*, **111**, 358–360.

Petersen, I., Winterbottom, J.H., Orton, S., et al. (1999) Emergence and lateral dispersal of adult Plecoptera and Trichoptera from Broadstone Stream, UK. *Freshwater Biology*, **42**, 401–416.

Phelps, R.J. (1968) A falling cage for sampling tsetse flies (Glossina: Diptera). *Rhodesia Journal of Agricultural Research*, **6**, 47–53.

Ramaswamy, S.B. & Cardé, R.T. (1982) Nonsaturating traps and long-life attractant lures for monitoring spruce budworm males. *Journal of Economic Entomology*, **75**, 126–129.

Rawlings, P., Meiswinkel, R., Labuschange, K., Welton, N., Baylis, M., & Mellor, P.S. (2003) The distribution and species characteristics of the *Culicoides* biting midge fauna of South Africa. *Ecological Entomology*, **28**, 559–566.

Reling, D. & Taylor, R.A.J. (1984) A collapsible tow net used for sampling arthropods by airplane. *Journal of Economic Entomology*, **77**, 1615–1617.

Ridgway, R.L., Silverstein, R.M. & Inscoe, M.N. (eds) (1990) *Behaviour-Modifying Chemicals for Pest Management: Applications of Pheromones and other Attractants*. Dekker, New York.

Roberts, I. (1996) The efficiency of light traps for moths in relation to meteorological conditions. BSc Thesis, Aberdeen University.

Roberts, R.H. (1972) The effectiveness of several types of Malaise traps for the collection of Tabanidae and Culicidae. *Mosquito News*, **32**, 542–547.

Robinson, G.S. & Tuck, K.R. (1993) Diversity and faunistics of small moths (Microlepidoptera) in Bornean rainforest. *Ecological Entomology*, **18**, 385–393.

Rogers, D.J. & Smith, D.T. (1977) A new electric trap for tsetse flies. *Bulletin of Entomological Research*, **67**, 153–159.

Rudd, W.G. & Jensen, R.L. (1977) Sweep net and ground cloth sampling for insects in soybeans. *Journal of Economic Entomology*, **70**, 301–304.

Sanders, C.J. (1988) Monitoring spruce budworm population density with six pheromone traps. *Canadian Entomologist*, **120**, 175–183.

Schaefer, G.W., Bent, G.A., & Allsopp, K. (1985) Radar and opto-electronic measurements of the effectiveness of Rothamsted Insect Survey suction traps. *Bulletin of Entomological Research*, **75**, 701–715.

Service, M.W. (1971) Flight periodicities and vertical distribution of *Aedes cantans* (Mg), *Aedes geniculatus* (01), *Anopheles plumbeus* Steph, and *Culex pipiens* L. (Diptera: Culicidae) in southern England. *Bulletin of Entomological Research*, **60**, 639–651.

Silk, P.J. & Kuenen, L.P.S. (1988) Sex pheromones and behavioural biology of the coniferophagous *Choristoneura*. *Annual Reviews of Entomology*, **33**, 83–101.

Smith, R.J. & Merrick, M.J. (2001) Resource availability and population dynamics of *Nicrophorus investigator*, an obligate carrion breeder. *Ecological Entomology*, **26**, 173–180.

Southwood, T.R.E. and Henderson, P.A. (2000) *Ecological Methods*. 3rd edn. Blackwell, Oxford.

Speight, M.R. & Wainhouse, D. (1989) *Ecology and Management of Forest Insects*. Oxford University Press, Oxford.

Tallamy, D.W., Hansens, E.J., & Denno, R.F. (1976) A comparison of Malaise trapping and serial netting for sampling a horsefly and deerfly community. *Environmental Entomology*, **5**, 788–792.

Taylor, L.R. (1951) An improved suction trap for insects. *Annals of Applied Biology*, **38**, 582–591.

Taylor, L.R. (1962) The absolute efficiency of insect suction traps. *Annals of Applied Biology*, **50**, 405–421.

Taylor, L.R., Kempton, R.A., & Woiwod, I.P. (1976) Diversity statistics and the log-series model. *Journal of Animal Ecology*. **45**, 255–272.

Taylor, R.A.J. (1986) Time series analysis of numbers of Lepidoptera caught at light traps in East Africa, and the effect of moonlight on trap efficiency. *Bulletin of Entomological Research*, **76**, 593–606.

Thompson, D.V., Capinera, J.L., & Pilcher, S.D. (1987) Comparison of an aerial water-pan pheromone trap with traditional trapping techniques for the European cornborer (Lepidoptera: Pyralidae). *Environmental Entomology*, **16**, 154–158.

Tunset, K., Nilssen, A.C., & Anderson, J. (1988) A new trap design for primary attraction of bark beetles and bark weevils (Coleoptera, Scolytidae and Curculionidae). *Journal of Applied Entomology*, **106**, 266–269.

Vale, G.A. (1974) The responses of tsetse flies (Diptera: Glossinidae) to mobile and stationary baits. *Bulletin of Entomological Research*, **64**, 545–588.

Van Huizen, T.H.P (1977) The significance of flight activity in the life cycle of *Amara plebeja* Gyll. (Coleoptera, Carabidae). *Oecologia*, **29**, 27–41.

Vessby, K. (2001) Habitat and weather affect reproduction and size of the dung beetle, *Aphodius fossor*. *Ecological Entomology*, **26**, 430–435.

Vogt, W.G., Woodburn, T.L., Morton, R., & Ellem, B.A. (1983) The analysis and standardisation of trap catches of *Lucilia cuprina* (Widemann) (Diptera: Calliphoridae). *Bulletin of Entomological Research*, **73**, 609–617.

Waring, P. (1980) A comparison of the Heath and Robinson M.V. moth traps. *Entomologists' Record and Journal of Variation*, **92**, 283–289.

Waring, P. (1990) Abundance and diversity of moths in woodland habitats. PhD thesis, Oxford Brookes University.

Webb, R.E., Smith, F.F., & Affeld, H. (1985) Trapping greenhouse whitefly with coloured surfaces; variables affecting efficacy. *Crop Protection*, **4**, 381–393.

Williams, C.B. (1948) The Rothamsted light trap. *Proceedings of the Royal Entomological Society of London A*, **23**, 80–85.

Woiwod, I.P. & Hanski, I. (1992) Patterns of density dependence in moths and aphids. *Journal of Animal Ecology*, **61**, 619–629.

Woiwod, I.P. & Harrington, R. (1994) Flying in the face of change: the Rothamsted Insect Survey. In *Long-term Experiments in Agricultural and Ecological Sciences* (ed. R.A. Leigh & A.E. Johnston), pp. 321–342. CAB International, London.

Young, M.R. (1997) *The Natural History of Moths*. T. & A.D. Poyser, London.

Young, M.R. & Armstrong, G. (1994) The effect of age, stand density and variability on insect communities in native pine woodlands. In *Our Pinewood Heritage* (ed. J.R. Aldhous), pp. 201–221. FC, RSPB, SNH, Edinburgh.

Zhou, X., Perry, J.N., Woiwod, I.P., & Harrington, R. (1997) Detecting chaotic dynamics of insect populations from long-term survey data. *Ecological Entomology*, **22**, 231–241.

Index of methods and approaches

Methodology	Topics addressed	Comments
Light traps	Faunal surveys and general biodiversity assessment. Accumulation of generalized long-term data sets. Partly focused survey of selected groups.	Catches only nocturnal insects. Different light sources are attractive to different insect types. Greater catches result from high contrast between light and background. Catches affected by moon phase and by some meteorological factors. Catches of small insects increased when suction trap is added. Catches at best semi-quantitative.
Suction traps	General faunal surveys, especially of smaller insects. Partly focused surveys of selected groups. Standardized, often long term catches.	Power of fan and detailed design of nozzle affects catch type and size. Often used in conjunction with other traps. Standardized designs have produced well-replicated catches. Catches can be calibrated and predicted to some extent. Catch much affected by local conditions and weather.
Water (or pan) traps	Faunal surveys and general biodiversity assessment. Wide-ranging, easily replicated sampling designs. Partly focused surveys (only by careful design of trap).	Catch much affected by color of trap. Catch much affected by local conditions and weather. Difficult to design so that catch is focused on particular groups. Catches many "tourist" species and has high "by-catch."

Continued

Methodology	Topics addressed	Comments
Sticky traps	General faunal surveys and biodiversity assessment, especially of small species. Partly focused sampling, by careful trap design. Frequently aimed at pest species, aimed to reduce population or detect presence.	Very easily and cheaply replicated. Very difficult to calibrate catches. Easily and cheaply replicated, often commercially available. Precise trap design affects catch greatly. Often used in association with other trap types. Catch is often damaged by sticky substance used. Traps can become saturated or clogged with dust.
Baited traps	Single species may be caught by use of correct pheromone bait. Functional group (e.g. carrion feeders) caught by correct bait. Often aimed at pest species, aimed to reduce population or detect presence.	Very difficult to calibrate catches. Species specificity possible with correct bait (often pheromone or other chemical attractant). May be sex-specific (e.g. males attracted by female pheromone). Precise condition of bait may affect catch type and quantity. Baits may be whole animals. Often used against pests, so traps may be commercially available. Often used in conjunction with another trap type to ensure efficient catch.
Interception traps	Faunal surveys or generalized biodiversity assessment. At best may be unbiased collection of all flying insects.	May catch at densities well below detection level of other traps. May be partially calibrated. If clean and well designed may offer truly general catch. Transparent sheets quickly lose effectiveness if wet or dirty. Malaise traps may be biased in catch. Difficult to calibrate effectively.

CHAPTER 7

Techniques and methods for sampling canopy insects

CLAIRE M.P. OZANNE

Introduction

The forest canopy has been described as the last biological frontier (Erwin 1983, Lowman & Wittman 1995). Whether this designation is justified or not, much remains to be discovered about this significant terrestrial habitat. Forest biomes currently extend across 25 percent of the world's land surface (FAO 1999). Tropical moist forests cover approximately 7 percent of this area yet are estimated to support 75 to 85 percent of all insect species, described and undescribed (Hammond 1992). Recent research suggests that up to 50 percent of forest insect species can be found in the canopy, with the percentage of true canopy specialists lying between 7 and 13 percent (Collembola: Rodgers & Kitching 1998; Coleoptera: Hammond et al. 1997; the figure is slightly higher for mites at 17–18%: Winchester 1997, Walter et al. 1998). This stratum of the forest habitat is therefore of great importance to global biodiversity (Ozanne et al. 2003).

More importantly, however, canopy insects contribute significantly to a number of fundamental forest ecosystem processes. These processes include nutrient cycling by herbivores (Schowalter et al. 1981), decomposition by leaf-surface and suspended-soil arthropods (Nadkarni & Longino 1990), predator–prey interactions (Winchester 1997, McGeoch & Gaston 2000), and the pollination and dispersal of forest plants and epiphytes (Aizen & Feinsinger 1994, Marini-Filho 1999). Forest canopies also provide us with the opportunity to test an array of hypotheses that attempt to explain population and community dynamics at a wide range of scales—from that of the individual leaf or needle (centimeters), through branch and tree (meters), to plantation stand, catchment area, and whole forest (kilometers).

Until Erwin's pioneering work in the late 1970s in the tropics, Crossley and Schowalter's work in the USA, and Southwood's work in Europe in the early 1980s, little attempt had been made to investigate canopy arthropods in a quantitative manner (Erwin 1982, Crossley et al. 1976, Schowalter et al. 1981, Southwood et al. 1982; but see Martin 1966). The exceptions to this were a few individual species regarded as a threat to timber production (see Speight & Wainhouse 1989). The establishment of several canopy science networks since 1994 (e.g. the International Canopy Network (ICAN), the European Science

Foundation Tropical Canopy Research Programme, and the Global Canopy Programme) have stimulated a notable rise in research (Nadkarni & Parker 1994, Stork & Best 1994, Mitchell 2001), which is now being carried out in a range of tropical and temperate managed and planted forests. Work on canopy arthropods in now being carried out in western Canada (Winchester & Ring 1996a), through the USA (Schowalter & Ganio 1998) and South America (Adis et al. 1998b, Basset & Charles 2000), across Europe (Schubert & Ammer 1998, Ozanne 1999), Africa (Moran et al. 1994, Wagner 2000, Winchester & Behan-Pelletier 2003), and India (Devy 1998), to Japan (Watanabe 1997), Southeast Asia (Guilbert 1998, Floren & Linsenmair 2000), Australia, and New Zealand (Lowman et al. 1996, Didham 1997, Kitching et al. 2000a, Majer et al. 2000).

Much of the research published in the literature is aimed at reporting basic information on canopy communities—addressing issues of species distribution, population densities, and community structures. Only very recently have more complex questions about the role of arthropods in ecosystem function, spatial distribution, and response to human impact begun to be addressed. There are three reasons that account for the historical lack of good data on canopy arthropods: firstly the challenge of accessing the canopy without significant disturbance, secondly the sheer species richness and complexity of many canopy communities, and thirdly the difficulties of sampling with appropriate replication and experimental design (partly due to accessibility difficulties, richness, and heterogeneity).

Access to the canopy

There is a growing literature on canopy access techniques (Heatwole & Higgins 1993, Dial & Tobin 1994, Moffet & Lowman 1995, Mitchell et al. 2002) which often divides the methods into "high tech" and "low tech" according to the equipment and cost (Barker 1997). Some of the sampling methods that are discussed in this chapter can be utilized from the ground and therefore do not strictly require access (e.g. chemical knockdown and branch clipping). However, in high canopies (e.g. over 20 m) they may be made more effective by operating them from within the canopy. Other sampling methods described here specifically require the operator to be in the canopy itself (Basset et al. 2003a).

One of the simplest access techniques allows the researcher to climb directly into the canopy using a single rope and harness—the single rope technique, or SRT (Barker & Standridge 2002). This method of access has the benefit that a climber can get into the trees with the minimum of disturbance, although it is essential that appropriate training is undertaken to ensure safety. The equipment required is less expensive and more transportable than in other access techniques and it has the significant advantage that a greater number of trees can be used in the course of a study, thus permitting proper replication of samples. Trees can be rigged with ropes in several ways. Most methods involve attaching the rope to a cord that can be used to pull the rope over a stable branch. Fishing line that has been shot (or thrown; Dial & Tobin 1994) over the branch pulls up the cord. Placing the line up into the right location in the canopy requires some expertise and a great deal of practice.

More permanent access techniques include platforms and walkways, towers, and cranes. Platforms and walkways have now been erected in a number of temperate and tropical forests (Reynolds & Crossley 1995, Inoue et al. 1995, Ring & Winchester 1996) (Fig. 7.1). These can act as foci for multi-faceted research projects, and they are useful for detailed studies of a small area of canopy and for work in which a chronological sequence of samples is required from the same location. Economically, walkways have the added attraction that they can be used to combine research and eco-tourism.

The most permanent of the canopy access structures is the canopy crane. Currently there are eight crane projects worldwide, based in the Pacific Northwestern USA (Wind River), Panama (two cranes, Panama City), Australia (Cape Tribulation, Queensland), Sarawak (Lambir Hills), Japan (Tomakomai, Hokkaido), Germany (Burgaue near Leipzig, Solling), and Switzerland (Hofstetten). These are large construction cranes consisting of a tower, boom, and suspended gondola (Mitchell et al. 2002).

Cranes have the advantage over some other access methods that, after the installation process is complete, non-destructive research can be carried out with

Fig. 7.1 Canopy walkway at the Forest Research Institute Malaysia (FRIM), Kepong, Kuala Lumpur, Malaysia. Photo: M.V. Graham.

minimal impact on canopy organisms. The cranes also allow researchers to study processes *in situ*, often requiring bulky analytical equipment which in the past could only have been used in the laboratory (Bauerle et al. 1999). The tower of the crane can be used to fix monitoring and trapping equipment, whilst the gondola can move the researcher horizontally and vertically through and above the canopy. The disadvantage of a crane lies in the fixed location, which reduces the opportunity for replication of samples. Additionally, because the cost of purchase and installation necessitates preservation of the site for future work, destructive or removal sampling is often discouraged, restricting the scope for entomological studies.

The most dramatic method of accessing the canopy is undoubtedly via the canopy raft ("radeau des cimes") and its smaller companion the sledge or luge (Hallé & Blanc 1990). These large red inflatable and netting platforms (raft — hexagonal structure 580 m^2; luge — triangular 16 m^2) are carried over the forest by a dirigible and can be placed down onto the upper canopy surface. Researchers can climb to the raft using ropes or be carried over the canopy surface to required points in the forest. A number of sampling techniques including branch clipping, Malaise traps, and sweeping can be used from the platforms, and researchers can walk about freely on the raft, enabling direct observation of animal–plant interactions — providing the researchers have a good head for heights!

Sampling issues

There are a number of sampling issues to be considered when designing a study that involves collecting insects from the forest canopy. Firstly, canopy communities may be composed of organisms drawn from a wide range of taxa, and the insect assemblages may be particularly species-rich. The highly diverse insect samples collected from tropical canopies have been used as a basis for estimates of global species richness made initially by Erwin (1982), then May (1990), and more recently Stork (1993). In addition, insects and other arthropods are often very numerous (Table 7.1).

Because of this diversity, a collecting event can yield samples that require considerable time to clean out debris and to sort, even to a low taxonomic level such as order. Since canopy insect communities are poorly known, sampling frequently yields species new to science. The use of morpho-taxa (Oliver & Beattie 1996) to assist with identification is well established in tropical work, but even this approach requires considerable taxonomic expertise.

The second issue is that the distribution of insects within the forest canopy is heterogeneous. Insects exhibit vertical, horizontal, and temporal variation in location & density (Costa & Crossley 1991, Hollier & Belshaw 1993, Springate & Basset 1996, Ozanne et al. 1997, Rodgers & Kitching 1998, Foggo et al. 2001, Basset et al. 2003b). This heterogeneity may be the focus of

Table 7.1 Typical canopy insect densities for a range of tree species.

Tree species	Collection method	Density	Reference
Quercus robur	Pyrethrum knockdown	591.3 per m^2	Southwood et al. 1982
Eucalyptus marginata	Branch clipping	247.1 per kg leaf biomass	Abbott et al. 1992
Pinus sylvestris	Pyrethrum knockdown	1046.31 per m^2	Ozanne 1996
Aporusa lagenocarpa	Pyrethrum knockdown	220.3 per m^2	Floren and Linsenmair 1997
Tropical forest, Brunei	Pyrethrum knockdown	117.4 per m^2	Stork 1991
Rain forest canopy, Cameroon	Branch clipping	16.56 per sample (approx. 0.85 m^2 leaf area)	Basset et al. 1992

investigation, but if it is not then steps need to be taken in the study design to take account of it. Thus the issues of representative sampling, adequate replication, and the avoidance of pseudoreplication need to be addressed (Hurlbert 1984, Guilbert 1998).

Finally, the range of taxa present and the complexity of the habitat mean that a study may require the use of several sampling techniques to collect the appropriate data. No one technique can collect all groups of insects equally, and indeed many sampling methods (e.g. activity traps) have strong biases (Basset et al. 1997). The most effective technique or combination of techniques must be chosen in the light of the research questions that the study is seeking to answer or the hypotheses to be tested.

Chemical knockdown

Chemical knockdown is arguably the most effective, comprehensive, and replicable of canopy sampling techniques (Stork & Hammond 1997, Majer et al. 2000). Knockdown can be used to collect insects and other arthropods from vegetation that spans the canopy height range from understory saplings and shrubs (Hill et al. 1990), to the upper sections of tropical emergent trees which may be 40–50 m from the ground (Adis et al. 1997, Paarmann & Stork 1987). Knockdown has been used successfully to collect insects from the complete canopy of individual trees (Floren & Linsenmair 1997), to investigate within canopy variation (Kitching et al. 1993), and to study the spatial distribution of organisms across forests (Ozanne et al. 2000). The technique can be used to gather information on insect population densities and community structure, e.g. guild proportions and trophic structure (Kitching et al. 2002), and to collect live specimens for subsequent experimental work on population dynamics and feeding strategies (Paarmann & Kerck 1997).

Knockdown is a passive sampling method and involves the delivery of a contact chemical that affects the insect nervous system—often temporarily. The most commonly used chemicals induce repetitive axon firing, resulting in a loss of coordinated movement which causes insects to fall from the vegetation or from flight. Knockdown can be rapid, occurring in a matter of minutes, although differential rates of absorption of the chemical through the cuticle can extend this time to more than an hour (Paarmann & Kerck 1997, Ozanne unpublished data).

Two main techniques are employed to deliver the chemical to the canopy: fogging and misting. Both collect insects that are in flight through the canopy or are surface-dwellers on the leaves, flowers, fruit, twigs, branches, and trunk of the tree. Insects on the outer surfaces of epiphytes (vascular and non-vascular) may also be collected, as could those on the surface of suspended soils. Knockdown is not an entirely comprehensive sampling method. The technique will not reliably collect insects that spin leaves together or that inhabit leaf domatia and epiphytes or that bore into bark (Stork & Hammond 1997, Walter & Behan-Pelletier 1999). Thus, as with other techniques, its use should be appropriate to the research questions being asked.

Fogging

Fogging was first used for collecting arboreal arthropods by Roberts (1973). Fogging machines produce a hot cloud of chemical droplets that rises upwards and outwards in a still air column, allowing the chemical to reach the heights required to sample rainforest emergents. The thermal fog is produced by allowing the chemical to drip in a controlled manner onto a hot surface generated by the exhaust from the petrol-driven engine. There is a range of machines but the most commonly used in insect sampling are the Swingfog® and Dyna-Fog® versions.

The fog can be delivered more reliably to the upper canopy by hoisting the machine into the canopy on a system of ropes and pulleys. However the fogger may be difficult to control once suspended and so this requires careful rope rigging. Some research groups have developed a radio-control mechanism that turns on the flow of chemical once the fogger has reached the appropriate location in the canopy (Adis et al. 1998a). Fogging is typically carried out for 5–10 minutes in one location.

The efficiency of collection is dependent on the environmental conditions. The inability to control the movement of the fog even in relatively calm conditions is one of the greatest disadvantages of this technique. In turbulent air the fog may not reach the canopy above the collecting trays and thus although insects may be knocked down they will not be collected and the wind may sweep away falling insects. Fogging should therefore be carried out at dawn or dusk when the air is still. If it is raining then fog will not rise and disperse in the required manner and if the foliage is wet the chemical tends to pool on the leaves, reducing its effectiveness; insects stick to the foliage rather than falling.

Mistblowing

The second method used in knockdown sampling is mistblowing. The principles involved here are quite different from those of fogging. The mistblower consists of a 2-stroke engine driving a fan that blows a strong air current along the delivery pipe. The chemical is allowed to drip into the air current at a rate controlled by the nozzle aperture, and as it hits the air stream the liquid is sheared into small droplets and carried up into the canopy (Fig. 7.2).

The height to which the mist reaches is determined in part by the power of the engine and fan and in part by the density of the foliage. Typically the mist reaches between 6 and 12 m (Southwood et al. 1982, Ozanne et al. 1988), although the mistblower can be hoisted up into the canopy in the same manner as a fogger to increase its range (Kitching et al. 2000b). The volume of chemical used and the droplet size spectrum can be controlled such that mistblowers can be set up for low-volume (LV: 20–300 l/ha) or ultra-low-volume delivery (ULV:

Fig. 7.2 Hurricane Major mistblower (Cooper Pegler). Photo: M.R. Speight.

5–20 l/ha). Misting is usually carried out for only a few minutes (0.5–5 minutes) depending on the canopy volume and chemical flow rates. Machines most commonly used are the Hurricane-Major® and the Stihl® backpack mistblowers.

Similar factors to those affecting fogging influence the efficiency of this method. Wind or rain will reduce the effectiveness of sampling. However, a short shower during the knockdown period after spraying can, ironically, increase the catch as insects are washed off the foliage into the collecting trays (personal observation). The structure of the foliage influences the dispersal of the chemical within the canopy, affecting, for example, the amount of active ingredient reaching the upper and lower surfaces of leaves and or needles (Ozanne et al. 1988).

Comparison of misting and fogging suggests that fogging results in knockdown over a much wider area, particularly downwind of the sample point. This is an important disadvantage in sensitive habitats and in areas where other trees or proximate locations are going to be sampled. The proportion of insects in different groups in the sample can vary with technique (M.R. Speight et al. unpublished data). This may be attributed partly to the method of chemical dispersal and partly to the different chemicals commonly used with the different techniques (natural pyrethrum vs. synthetic pyrethroids). Misting seems to be the most effective (M.R. Speight et al. unpublished data).

The chemicals used in knockdown are mixtures of either natural pyrethrins or synthetic pyrethroids. These are usually carried in an oil (e.g. kerosene), and in ULV delivery this is used undiluted, but in LV delivery an emulsion is made with water. The chemical may be synergized by piperonyl butoxide if the aim is to kill the insects rapidly, but larger species are capable of recovering from a knockdown event.

Natural pyrethrum has the advantage that it is inactivated by ultraviolet light more rapidly than synthetic equivalents (probably within 24–48 hours), a significant factor when sampling in sensitive sites or when conducting recolonization studies which require repetitive sampling (e.g. Floren & Linsenmair 1997). Natural pyrethrum should be used if live specimens are required (Adis et al. 1997), but these are much more expensive than synthetics. In Europe and North America it is necessary to observe pesticide handling and application procedures including health and safety regulations when using these chemicals. The time taken for insects to fall from the tree varies with their location, susceptibility, and size, but the majority of animals can be collected up after two hours (Ozanne 1991, Stork & Hammond 1997).

The second element of the knockdown sampling system is the collecting tray or mat used to capture fallen insects. These have become more sophisticated and therefore more efficient over the last 20 years, moving from large plastic sheets spread on the ground (Yamashita & Ishii 1976), through cloth trays stretched on wooden frames (Southwood et al. 1982), to conical hoops made from vinyl or tenting material (Ozanne 1996, Adis et al. 1998a, Kitching et al. 2000b) (Fig. 7.3). Vinyl hoops work particularly well because they are robust

Fig. 7.3 Collecting hoops (Natural History Museum UK design). Photo: I.P. Palmer.

and the surface is very shiny, allowing insects to roll down into the jar at the apex. Remaining insects can be washed or gently brushed into the jar, which should contain a small amount of preserving fluid (e.g. 70% ethanol).

Current collecting hoops have been developed to reduce the handling of specimens, which are easily damaged (although the knockdown chemical seems to produce autotomy in some long-legged insects anyway; Paarmann & Kerck 1997), and to prevent small insects and mites from being left behind. Hoops are of a standard surface area (0.5 or 1 m^2) to allow densities of insects to be quantified per unit ground area, and they are hung under the canopy by clipping to branches, to a network of cords, or to a tower. Strong cord and large clips allow the hoops to be easily handled without tangling. Problems may arise if hoops are hung too early and catch debris from the canopy, or if the jar fills with rainwater during the collection period; some hoops have a built in storm vent. The specific placement of trays depends on the study design. In plantations they can be set up under the canopy of several trees to reduce the effect of between-tree variation (Ozanne 1996), while trays near to the trunk may have different catches from those at the crown margins. Trays can be attached to a tower at different heights to investigate vertical distribution of species in the canopy (M.R. Speight, personal communication).

Overall, knockdown compares favorably with other canopy sampling techniques.

1 Compared with *beating*, it collects higher densities (Fig. 7.4, Lowman et al. 1996).

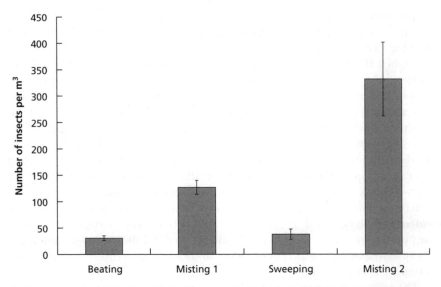

Fig. 7.4 Mean densities of insects sampled in Australian sub-tropical rainforest using three collection techniques: beating, sweeping, and pyrethrum misting (misting 1 = misting after beating; misting 2 = misting after sweeping) ±s.e. (from Lowman et al. 1996).

2 Compared with *sweeping*, it collects higher densities (Fig. 7.4, Lowman et al. 1996), and better estimates of collembolan and dipteran densities (Lowman et al. 1996).
3 Compared with *branch clipping*, it produces better estimates of the richness of parasitic Hymenoptera (Blanton 1990); it estimates the density of large mobile insects and cryptic insects less well (Majer & Recher 1988); it underestimates sessile insects e.g. Psyllidae (Majer & Recher 1988); it is not as good for biomass estimation (Blanton 1990).

Branch bagging and clipping

A viable alternative sampling strategy to chemical knockdown, and one preferred by a number of research groups, is branch bagging and clipping. This method may be used at a wide range of canopy levels, limited only by the height to which the mechanism can be operated accurately from the ground, or by the canopy access technique used. The technique was first reported for canopy sampling by Crossley et al. in 1976, and has been used in temperate forests to investigate vertical stratification of communities (Schowalter & Ganio 1998) and the diel movement of arthropods within the canopy (Ohmart et al. 1983), and in tropical forest to study plant–herbivore relationships (Basset & Höft 1994). Branch bagging and clipping can be used to measure a number of population and community attributes, including presence and absence of species, population density, guild structure, and heterogeneity of distribution. The method can be used to standardize insect densities to units of plant biomass and surface area (Basset et al. 1992, Ohmart et al. 1983).

Branch clipping involves passing a mesh, cloth, or plastic bag over the end portion of a branch and then drawing the bag closed to prevent the escape of mobile insects (although see Ohmart et al. (1983) where branches were clipped first and dropped into a calico bag). The branch section is then cut off and the sample brought to the ground. The bag and clippers are usually attached to long poles or arms that can be fixed in length or telescopic, allowing them to be pushed up into the canopy from the ground (Basset & Höft 1994), along branches from a platform (Winchester & Ring 1996b, Winchester 1997), or from a cherry-picker (Majer et al. 1990). Shorter poles, affording more control, can be used when the clipping is carried out from a walkway or from the canopy raft or luge (Basset et al. 1992). The amount of branch and foliage cut down in any one sample varies from 20 to 60 leaves (Johnson 2000), through samples of 2–5 g in weight (Schowalter et al. 1988), to larger samples up to 120 g in weight (Majer & Recher 1988).

The selection of branches to be clipped is a key step in designing an effective investigation, and samples can be taken at random or from specified locations to investigate particular microhabitats (Johnson 2000). Insects shaken off the foliage into the bag can be counted live *in situ* (see Johnson (2000)), but frequently the samples are removed from the site for further study. In some studies the bag is filled with CO_2 or other chemical (e.g. pyrethroid spray or ethyl acetate; Basset et al. 1992) before closure and in others the bag is chilled (Schowalter et al. 1981) to prevent escape on opening. Storage of the clipped samples (perhaps in a cool environment to reduce mould growth) can allow insects that are difficult to sample, e.g. dipteran larvae and pupae, to emerge as adults (I.P. Palmer, personal communication).

Branch clipping is an excellent method for sampling sedentary insects on branch and leaf surfaces. Comparative work indicates that the technique is able to capture insects from all orders. Several studies suggest that large mobile insects, e.g. Odonata, are under-sampled (Cooper & Whitmore 1990, Johnson 2000) but other groups do not have time to avoid the bag as it is drawn over the foliage. The technique is not effective for sampling aerial components of the canopy fauna such as midge clouds (Chironomidae) (Johnson 2000).

Branch clipping has the advantage over many other canopy sampling techniques that the species richness or density of insects can be converted directly to units of plant biomass and/or leaf and branch surface area (Schowalter et al. 1981, Abbott et al. 1992, Winchester & Ring 1996b). This can provide valuable data on herbivore loads and microhabitat preferences of insects in the canopy. The technique can also be used to investigate epiphyte communities — particularly those of non-vascular epiphytes such as moss mats and lichens — and to collect insects that are leaf-spinners or that hide in deep bark crevices. The most important disadvantage is that it is the branch tips that are usually clipped. This introduces a bias towards insects that are attracted to rapidly grow-ing tissue often found at the apices of branches and a bias away from inver-tebrates inhabiting large branches and the trunk.

Aerial and arboreal traps: Malaise, interception, emergence and light

Active and passive trap systems can be used to investigate insects moving with-in the forest and in the spaces in and around the canopy, such as above the canopy surface, in gaps, or at edges. Traps that have been designed for use in other habitats (described in other chapters in this book), can be employed effec-tively above the ground, although sampling efficiency may be affected by the location in which the traps are placed in the three-dimensional spaces of tree crowns. Appropriate placement will depend on the research question. Traps can be used to answer general questions about canopy community structure or to test specific hypotheses about the use of particular strata of the canopy. For ex-ample, Compton et al. (2000) used sticky traps to sample the location and movement of fig wasps within and above the canopy surface.

With some modification, flight interception traps have also been used in the canopy environment. Hill and Cermak (1997) describe a plastic window trap fitted with collecting trays at the base and a roof to keep out the rain. They used this apparatus to compare ground and canopy insect catches by hoisting some traps up into the canopy, securing them with guy ropes to prevent twisting in the wind. Traps were installed in locations that ensured foliage did not interfere with the capture surface. A wide range of arthropod groups was collected, with Coleoptera, Diptera, and Hymenoptera dominating the samples.

The most effective canopy traps can be built by combining the best features of different ground-based mechanisms. For example, combination Malaise and interception traps have been designed for use in the canopy (Fig. 7.5). Springate and Basset (1996) used such a trap to investigate diel movement of insects with-in tree crowns. The apparatus consisted of a rectangular cross-panel of black netting (see Chapter 4 for Malaise trap design and discussion of effects of netting color) with a white netting roof connected to a collecting jar. This part of the trap intercepts a range of insect groups including Diptera and Hymenoptera. A clear plastic funnel was also attached below the main body of the trap and connected to a large collecting jar containing ethanol. This part of the apparatus acts as a window trap, capturing those insects that close their wings and drop down-wards on alighting such as the Coleoptera. In order to allow the trap to be left out for long periods of time an overflow grid was inserted in the middle of the lower jar to cope with heavy rainfall.

Combination traps are essentially activity-dependent and therefore under-estimate the contribution of sedentary and flightless arthropods to the com-munity (Springate & Basset 1996). Their effectiveness is influenced by crown

Fig. 7.5 Combined Malaise and interception trap at the MASS site, Vancouver Island, BC, Canada. Photo: I.P. Palmer.

structure and by their location in relation to insect flight paths. However, Behan-Pelletier and Winchester (1998) found that traps set in the canopy of temperate rainforest in British Colombia caught significant numbers of flightless arthropods (e.g. oribatid mites, Acarina), perhaps because they were carried through the canopy by air currents or because they are actively moving about within tree crowns.

Light traps can also be used very effectively in the canopy to collect actively flying insects. They are particularly efficient at sampling Lepidoptera and Coleoptera, but also capture Hemiptera and other insect groups. Light traps have been used in forests to investigate the impact of fragmentation on communities (Kitching et al. 2000b) and to investigate vertical distribution of moths (Intachat & Holloway 2000). Light traps generate three kinds of data:

presence/absence data for individual species, qualitative data for comparative work between sites, and relative estimates of population densities (Southwood & Henderson 2000). With careful calibration these relative estimates can be made absolute. Estimates of population density are generated per unit trapping effort, for example, per trap night (TNI: trap night index).

The mostly commonly used light traps are Rothamsted tungsten-filament and Robinson mercury-vapor light-traps (Intachat & Woiwood 1999), and a Pennsylvania trap modified for wet environments and canopy suspension (Kitching et al. 2000b). The traps can be hoisted up to the required height in the canopy using ropes and pulleys adjusted so that they can be let down to be emptied and then re-hoisted. Alternatively they can be fixed to canopy access towers. Efficiency of trap operation in the canopy is dependent on moonlight, weather conditions (e.g. cloudy and clear nights may produce quite different data sets), temperature, and vegetation density (which affects penetration of the light source) (Bowden 1982). Where multiple traps are used they should be spaced such that the light cannot be seen from any other trap to avoid interference. Light-trap catches complement those from other sampling techniques such as chemical knockdown, which seems to be less effective at capturing Lepidoptera. The main disadvantage of light-trap catches is the difficulty of determining where the insects have come from within the forest.

Insects in fruit, seeds, and silk: moss cores, suspended soils, and bark sprays

Most of the collecting techniques described in this chapter have been designed to capture insects that are free-living on the surface of the vegetation, or flying through the air spaces in the canopy. There are, of course, a number of insect groups that make a significant contribution to the canopy community but live within the plant tissue (stems, leaves, seeds, and fruit), within epiphytes, in silk cocoons, bark fissures, and suspended soils, and that are rarely represented in more general canopy samples.

In temperate rainforest and tropical cloud forest, trees support large moss mats and a considerable quantity of suspended soil (Nadkarni & Longino 1990, Winchester & Ring 1996a). These diverse micro-/mesohabitats contribute significantly to ecosystem processes in the canopy and support rich, diverse, and distinctive invertebrate communities. They present an interesting sampling challenge because organisms can only be collected from them by accessing the canopy directly using one of the techniques discussed at the start of this chapter.

Invertebrates are collected by taking samples of the habitat (soil, leaf litter, or moss) in the canopy and then removing the animals in the laboratory either by active extraction (e.g. Winkler extraction, Tullgren funnel) or by washing (Behan-Pelletier et al. 1996). Habitat samples should be of a known weight or volume so that the density of animals can be standardized to habitat unit (e.g. biomass of moss). This is usually achieved by taking a core sample (e.g. moss mat cores of 3×5 cm; Winchester and Ring 1996b, Winchester 2002). Samples may be collected from particular locations along branches, from different heights in the canopy, or from the center of epiphytes (Rodgers & Kitching 1998, Walter et al. 1998), depending on the research question. Core samples are very effective since it is clear where the animals are located in the canopy, and therefore the technique lends itself to answering questions about key ecosystem processes. Other micro-/mesohabitats within the canopy that can yield insects are seeds and fruit, which may be sampled by clipping from the vegetation or by collecting fallen fruit from the ground (e.g. figs; W. Paarmann, personal communication).

Conclusions

The canopy is a spatially and architecturally complex environment, supporting a host of insects. Some canopy insects are tourists (*sensu* Moran & Southwood 1982), others are habitat generalists that move between forest strata, whilst some are canopy specialists well adapted to the particular niches available in tree crowns (e.g. leaves, bark crevices, epiphyte surfaces, and suspended soils). In order to collect data that can answer the kinds of questions entomologists might wish to ask about these insects, a range and often a combination of sampling techniques is required.

Sampling in the canopy is distinct from sampling ground vegetation in the challenges it poses in terms of access, community richness, and spatial heterogeneity. Several studies mentioned in this chapter suggest that the canopy fauna may indeed have a composition that is distinct from that of the understory, ground, or soil. In order to gain a fuller understanding of insect ecology and to conduct hypothesis testing in a range of globally representative habitats, we have to continue to rise to, and overcome, the challenges presented by this frontier between the biosphere and the atmosphere.

References

Abbott, I., Burbidge, T. Williams, M., & Van Heurck, P. (1992) Arthropod fauna of jarrah (*Eucalyptus marginata*) foliage in Mediterranean forest of Western Australia: spatial and temporal variation in abundance, biomass, guild structure and species composition. *Australian Journal of Ecology*, **17**, 263–274.

Adis, J., Paarmann, W., da Fonseca, C.R.V., & Rafael, J.A. (1997) Knockdown efficiency of natural pyrethrum and survival rate of living arthropods obtained by canopy fogging in Central Amazonia. In *Canopy Arthropods* (ed. N. Stork, J. Adis, & R. Didham), pp. 67–81. Chapman & Hall, London.

Adis, J., Basset, Y., Floren, A., Hammond, P.M., & Linsenmair, K.E. (1998a) Canopy fogging of an overstorey tree — recommendations for standardization. *Ecotropica*, **4**, 93–97.

Adis, J., Harada, A.Y., da Fonseca, C.R.V., Paarmann, W., & Rafael, J.A. (1998b) Arthropods obtained from the Amazonian tree species "Cupiuba" (*Goupia glabra*) by repeated canopy fogging with natural pyrethrum. *Acta Amazonica*, **28**, 273–283.

Aizen, M.A. & Feinsinger, P. (1994) Habitat fragmentation, native insect pollinators and feral honey bees in Argentine "Chacos Serrano". *Ecological Applications*, **4**, 378–392.

Barker, M.G. (1997) An update on low-tech methods for forest canopy access and on sampling the forest canopy. *Selbyana*, **18**, 16–26.

Barker, M. & Standridge N. (2002) Ropes as a mechanism for canopy access. In *The Global Canopy Handbook* (ed. A. Mitchell, K. Secoy, & T. Jackson), pp 13–23. GCP, Oxford.

Basset, Y. & Charles, E. (2000) An annotated list of insect herbivores foraging on the seedlings of five forest trees in Guyana. *Anais da Sociedade Entomologica do Brasil*, **29**, 433–452.

Basset, Y. & Höft, R. (1994) Can apparent leaf damage in tropical trees be predicted by herbivore load or host-related variables? A case study in Papua New Guinea. *Selbyana*, **15**, 3–13.

Basset, Y., Aberlanc, H.P., & Delvare, G. (1992) Abundance and stratification of foliage arthropods in a lowland rain forest of Cameroon. *Ecological Entomology*, **17**, 310–318.

Basset, Y., Springate, N.D. Aberlanc, H.P., & Delvare, G. (1997) A review of methods for sampling arthropods in tree canopies. In *Canopy Arthropods* (ed. N. Stork, J. Adis, & R. Didham), pp. 27–52. Chapman & Hall, London.

Basset, Y., Novotny, V., Miller, S.E., & Kitching, R.L (2003a) Methodological advances and limitations in canopy entomology. In *Arthropods of Tropical Forests: Spatio-Temporal Dynamics and Resource Use in the Canopy* (ed. Y. Basset, V. Novotny, S.E. Miller, & R.L. Kitching), pp. 7–16. Cambridge University Press, Cambridge.

Basset, Y., Hammond, P., Barrios,H., Holloway, J.D. and Miller, S.E. (2003b) Vertical stratification of arthropod assemblages. In *Arthropods of Tropical Forests: Spatio-Temporal Dynamics and Resource Use in the Canopy* (ed. Y. Basset, V. Novotny, S.E. Miller, & R.L. Kitching), pp. 17–27. Cambridge University Press, Cambridge.

Bauerle, W.L., Hinckley, T.M., Cermak, J., Kucera, J., & Bible, K. (1999) The canopy water relations of old-growth Douglas-fir trees. *Trees (Berlin)*, **13** (4), 211–217.

Behan-Pelletier, V. & Winchester, N.N. (1998) Arboreal oribatid mite diversity: colonising the canopy. *Applied Soil Ecology*, **9**, 45–51.

Behan-Pelletier, V.M., Tomlin, A., Winchester, N., & Fox, C. (1996) Sampling protocols for microarthropods. In *A Workshop Report on Terrestrial Arthropod Sampling Protocols for Graminoid Ecosystems* (ed. A.T. Finnamore). http://www.eman-rese.ca/eman/reports/publications/sage [accessed May 8, 2004].

Blanton, C.M. (1990) Canopy arthropod sampling: a comparison of collapsible bag and fogging methods. *Journal of Agricultural Entomology*, **7**, 41–50.

Bowden, J. (1982) An analysis of the factors affecting catches of insects in light-traps. *Bulletin of Entomological Research*, **72**, 535–556.

Compton, S.G., Ellwood, M.D.F., Davis, A.J., & Welch, K. (2000) The flight heights of chalcid wasps (Hymenoptera, Chalcidoidea) in a lowland Bornean rain forest: fig wasps are the high fliers. *Biotropica*, **32**, 515–522.

Cooper, R.J. & Whitmore, R.C. (1990) Arthropod sampling methods in ornithology. *Studies in Avian Biology*, **13**, 29–37.

Costa, J.T. & Crossley, D.A. Jr. (1991) Diel patterns of canopy arthropods associated with three tree species. *Environmental Entomology*, **20**, 1542–1548.

Crossley, D.A. Jr., Callahan, J.T., Gist, C.S., Maudsley, J.R., & Waide, J.B. (1976) Compartmentalization of arthropod communities in forest canopies at Coweeta. *Journal of the Georgia Entomological Society*, **11**, 44–49.

Devy, M.S. (1998) Breeding systems in bee-pollinated canopy forests of Southwestern Ghats, India. *Selbyana*, **19**, 274.

Dial, R. & Tobin, S.C. (1994) Description of arborist methods for forest canopy access and movement. *Selbyana*, **15**, 24–37.

Didham, R.K. (1997) Dipteran tree-crown assemblages in a diverse southern temperate rainforest. In *Canopy Arthropods* (ed. N. Stork, J. Adis, & R. Didham), pp. 320–343. Chapman & Hall, London.

Erwin, T.L. (1982) Tropical forests: their richness in Coleoptera and other arthropod species. *Coleopterist's Bulletin*, **36**, 74–75.

Erwin, T.L. (1983) Tropical forest canopies: the last biotic frontier. *Bulletin of the Entomological Society of America*, **29**, 14–19.

FAO (1999) *State of the World's Forests*. FAO, Rome.

Floren, A. & Linsenmair, K.E. (1997) Diversity and recolonisation dynamics of selected arthropod groups on different tree species in a lowland rainforest in Sabah, Malaysia with special reference to Formicidae. In *Canopy Arthropods* (ed. N. Stork, J. Adis, & R. Didham), pp. 344–381. Chapman & Hall, London.

Floren, A. & Linsenmair, K.E. (2000) Do ant mosaics exist in pristine lowland rain forests? *Oecologia (Berlin)*, **123**, 129–137.

Foggo, A., Ozanne, C.M.P., Hambler, C., & Speight, M.R. (2001) Edge effects, tropical forests and invertebrates. *Plant Ecology*, **153**, 347–359.

Guilbert, E. (1998) Studying canopy arthropods in New Caledonia: how to obtain a representative sample. *Journal of Tropical Ecology*, **14**, 665–672.

Hallé, F. & Blanc, P. (eds.) (1990) *Biologie d'une Canopeé de Forêt equatoriale. Rapport de Mission. Radeau des Cimes Octobre — Novembre 1989, Guyane Française*. Montpellier II et CNRS-Paris VI, Montpellier / Paris.

Hammond, P.M. (1992) Species inventory. In *Global Biodiversity: Status of the Earth's Living Resources* (ed. B. Groombridge), pp. 17–39. Chapman & Hall, London.

Hammond, P.M., Stork, N.E., & Brendell, M.J.D. (1997) Tree-crown beetles in context: a comparison of canopy and other ecotone assemblages in a lowland tropical forest in Sulawesi. In *Canopy Arthropods* (ed. N. Stork, J. Adis, & R. Didham), pp. 184–223. Chapman & Hall, London.

Heatwole, H. & Higgins, W. (1993) Canopy research methods: a review. *Selbyana*, **14**, 23.

Hill, C.J. & Cermak, M. (1997) A new design and some preliminary results for a flight intercept trap to sample forest canopy arthropods. *Australian Journal of Entomology*, **36**, 51–55.

Hill, D. Roberts, P., & Stork, N. (1990) Densities and biomass of invertebrates in stands of rotationally managed coppice woodlands. *Biological Conservation*, **51**, 167–177.

Hollier, J.A. & Belshaw, R.D. (1993) Stratification and phenology of a woodland Neuroptera assemblage. *The Entomologist*, **112**, 169–175.

Hurlbert, S.H. (1984) Pseudoreplication and the design of ecological field experiments. *Ecological Monographs*, **54**, 187–211.

Inoue, T., Yumoto, T., Hamid, A.A., Seng, L.H., & Ogino, K. (1995) Construction of a canopy observation system in a tropical rainforest of Sarawak. *Selbyana*, **16**, 24–35.

Intachat, J. & Holloway, J.D. (2000) Is there stratification in diversity or preferred flight height of geometroid moths in Malaysian lowland tropical forest? *Biodiversity and Conservation*, **9**, 1417–1439.

Intachat, J. & Woiwood, I.P. (1999) Trap design for monitoring moth biodiversity in tropical rainforests. *Bulletin of Entomological Research*, **89**, 153–163.

Johnson, M.D. (2000) Evaluation of an arthropod sampling technique for measuring food availability for forest insectivorous birds. *Journal of Field Ornithology*, **71**, 88–109.

Kitching, R.L., Bergelson, J.M., Lowman, M.D., McIntyre, S., & Carruthers, G. (1993) The biodiversity of arthropods from Australian rainforest canopies: general introduction, methods, sites and ordinal results. *Australian Journal of Ecology*, **18**, 181–191.

Kitching, R.L., Orr, A.G., Thalib, L., Mitchell, H., Hopkins, M.S., & Graham, A.W. (2000a) Moth assemblages as indicators of environmental quality in remnants of upland Australian rain forest. *Journal of Applied Ecology*, **37**, 284–297.

Kitching, R.L., Vickerman, G., Laidlaw, M., & Hurley, K. (2000b) *The Comparative Assessment of Arthropod and Tree Biodiversity in Old-World Rainforests: the Rainforest CRC / Earthwatch Protocol Manual*. Rainforest CRC, Cairns.

Kitching, R.L., Basset, Y., Ozanne, C.M.P., & Winchester, N.N. (2002) Canopy knockdown techniques. In *The Global Canopy Handbook* (ed. A. Mitchell, K. Secoy, & T. Jackson), pp. 134–139. GCP, Oxford.

Lowman, M.D. & Wittman, P.K. (1995) The last biological frontier? Advancements in research on forest canopies. *Endeavour (Cambridge)*, **19**, 161–165.

Lowman, M.D., Kitching, R.L., & Carruthers, G. (1996) Arthropod sampling in Australian subtropical rain forests: how accurate are some of the more common techniques? *Selbyana*, **17**, 36–42.

Majer, J.D. & Recher, H.F. (1988) Invertebrate communities on Western Australian eucalypts: a comparison of branch clipping and chemical knockdown procedures. *Australian Journal of Ecology*, **13**, 269–278.

Majer, J., Recher, H.F., Perriman, W.S., & Achuthan, N. (1990) Spatial variation of invertebrate abundance within the canopies of two Australian eucalypt forests. *Studies in Avian Biology*, **13**, 65–72.

Majer, J., Recher, H.F., & Ganesh, S. (2000) Diversity patterns of eucalypt canopy arthropods in eastern and western Australia. *Ecological Entomology*, **25**, 295–306.

Marini-Filho, O.J. (1999) Distribution, composition, and dispersal of ant gardens and tending ants in three kinds of central Amazonian habitats. *Tropical Zoology*, **12**, 289–296.

Martin, J.L. (1966) The insect ecology of red pine plantations in central Ontario. IV. The crown fauna. *Canadian Entomologist*, **98**, 10–27.

May, R.M. (1990) How many species? *Philosophical Transactions of the Royal Society, Series B*, **330**, 293–304.

McGeoch, M.A. & Gaston, K.J. (2000) Edge effects on the prevalence and mortality factors of *Phytomyza ilicis* (Diptera, Agromyzidae) in a suburban woodland. *Ecology Letters*, **3**, 23–29.

Mitchell, A. (2001) Introduction—canopy science: time to shape up. *Plant Ecology*, **153**, 5–11.

Mitchell, A., Secoy, K., & Jackson, T. (eds.) (2002) *The Global Canopy Handbook*. GCP, Oxford.

Moffet, M. & Lowman, M.D. (1995) Canopy access techniques. In *Forest Canopies* (ed. M.D. Lowman & N. Nadkarni), pp. 3–26. Academic Press, San Diego.

Moran, V.C. & Southwood, T.R.E. (1982) The guild composition of arthropod communities in trees. *Journal of Animal Ecology*, **51**, 289–306.

Moran, V.C., Hoffmann, J.H., Impson, F.A.C., & Jenkins, J.F.G. (1994) Herbivorous insect species in the tree canopy of a relict South African forest. *Ecological Entomology*, **19**, 147–154.

Nadkarni, N.M. & Longino, J.T. (1990) Invertebrates in canopy and ground organic matter in a neotropical montane forest, Costa Rica. *Biotropica*, **22**, 286–289.

Nadkarni, N.M. & Parker, G.G. (1994) A profile of forest canopy science and scientists—who we are, what we want to know and obstacles we face: results of an international survey. *Selbyana*, **15**, 38–50.

Ohmart, C.P., Stewart, L.G., & Thomas J.R. (1983) Phytophagous insect communities in the canopies of three *Eucalyptus* forest types in south-eastern Australia. *Australian Journal of Ecology*, **8**, 395–403.

Oliver, I., & Beattie, A.J. (1996) Designing a cost-effective invertebrate survey: a test of methods for rapid assessment of biodiversity. *Ecological Applications*, **6**, 594–607.

Ozanne, C.M.P. (1991) The arthropod fauna of coniferous plantations. D.Phil thesis, Oxford University.

Ozanne, C.M.P. (1996) The arthropod communities of coniferous forest trees. *Selbyana*, **17**, 43–49.

Ozanne, C.M.P. (1999) A comparison of the canopy arthropods communities of coniferous and broad-leaved trees. *Selbyana*, **20**, 290–298.

Ozanne, C.M.P., Speight, M.R., & Evans, H.F. (1988) Spray deposition and retention in the canopies of five forest tree species. *Aspects of Applied Biology*, **17** (2), 245–246.

Ozanne, C.M.P., Foggo, A, Hambler, C., & Speight, M.R. (1997) The significance of edge-effects in the management of forests for invertebrate biodiversity. In *Canopy Arthropods* (ed. N. Stork, J. Adis, & R. Didham), pp. 534–550. Chapman & Hall, London.

Ozanne, C.M.P., Speight, M.R., Hambler, C., & Evans, H.F. (2000) Isolated trees and forest patches: patterns in canopy arthropod abundance and diversity in *Pinus sylvestris* (Scots Pine). *Forest Ecology and Management*, **137**, 53–63.

Ozanne, C.M.P., Anhuf, D., Boulter, S.L., et al. (2003) Biodiversity meets the atmosphere: a global view of forest canopies. *Science*, **301**, 183–186.

Paarmann, W. & Kerck, K. (1997) Advances in using the canopy fogging technique to collect living arthropods from tree-crowns. In *Canopy Arthropods* (ed. N. Stork, J. Adis, & R. Didham), pp. 53–66. Chapman & Hall, London.

Paarmann, W. & Stork, N.E. (1987) Canopy fogging, a method of collecting living insects for investigation of life history strategies. *Journal of Natural History*, **21**, 563–566.

Reynolds, B. & Crossley, D.A. Jr. (1995) Use of a canopy walkway for collecting arthropods and assessing leaf area removed. *Selbyana*, **16**, 21–23.

Ring, R.A. & Winchester, N.N. (1996) Coastal temperate rainforest canopy access systems in British Columbia, Canada. *Selbyana*, **17**, 22–26.

Roberts, H.R. (1973) Arboreal Orthoptera in the rain forests of Costa Rica collected with insecticide: a report on grasshoppers (Acrididae) including new species. *Proceedings of the Academy of Natural Sciences, Philadelphia*, **125**, 46–66.

Rodgers, D. & Kitching, R.L. (1998) Vertical stratification of rainforest collembolan (Collembola: Insecta) assemblages: description of ecological patterns and hypotheses concerning their generation. *Ecography*, **21**, 392–400.

Schowalter, T.D. & Ganio, L.M. (1998) Vertical and seasonal variation in canopy arthropod communities in an old-growth conifer forest in southwestern Washington, USA. *Bulletin of Entomological Research*, **88**, 633–640.

Schowalter, T.D., Webb, W.J., & Crossley, D.A. Jr. (1981) Community structure and nutrient content of canopy arthropods in clearcut and uncut forest ecosystems. *Ecology*, **62**, 1010–1019.

Schowalter, T.D., Stafford, S.G., & Slagle, R.L. (1988) Arboreal arthropod community structure in an early successional coniferous forest ecosystem in Western Oregon. *Great Basin Naturalist*, **48**, 327–333.

Shubert, H. and Ammer, U. (1998) Comparison of arthropod fauna in canopies of natural and managed forests of southern Germany. *Selbyana*, **19**, 298.

Southwood, T.R.E. & Henderson, P.A. (2000) *Ecological Methods*. 3rd edn. Blackwell Science, Oxford.

Southwood, T.R.E., Moran, V.C., & Kennedy, C.E.J. (1982) The assessment of arboreal insect fauna—comparisons of knockdown sampling and faunal lists. *Ecological Entomology*, **7**, 331–340.

Speight, M.R. & Wainhouse, D. (1989) *Ecology and Management of Forest Insects*. Oxford University Press, Oxford.

Springate, N.D. & Basset, Y. (1996) Diel activity of arboreal arthropods associated with Papua New Guinea trees. *Journal of Natural History*, **30**, 101–112.

Stork, N. (1991) The composition of the arthropod fauna of Bornean lowland rainforest trees. *Journal of Tropical Ecology*, **7**, 161–180.

Stork, N. (1993) How many species are there? *Biodiversity and Conservation*, **2**, 215–232.

Stork, N.E. & Best, V. (1994) European Science Foundation—results of a survey of European canopy research in the tropics. *Selbyana*, **15**, 51–62.

Stork, N.E. & Hammond, P. (1997) Sampling arthropods from tree-crowns by fogging with knockdown insecticides: lessons from studies of oak tree beetle assemblages in Richmond Park (UK). In *Canopy Arthropods* (ed. N. Stork, J. Adis, & R. Didham), pp. 3–26. Chapman & Hall, London.

Wagner, T. (2000) Influence of forest type and tree species on canopy-dwelling beetles in Budongo forest, Uganda. *Biotropica*, **32**, 502–514.

Walter, D.E. & Behan-Pelletier, V. (1999) Mites in forest canopies: filling the size distribution shortfall? *Annual Review of Entomology*, **44**, 1–19.

Walter, D.E., Seeman, O., Rodgers, D., & Kitching R.L. (1998) Mites in the mist: how unique is a rainforest canopy-knockdown fauna? *Australian Journal of Ecology*, **23**, 501–508.

Watanabe, H. (1997) Estimation of arboreal and terrestrial arthropod densities in the forest canopy as measured by insecticide smoking. In *Canopy Arthropods* (ed. N. Stork, J. Adis, & R. Didham), pp. 401–416. Chapman & Hall, London.

Winchester, N.N. (1997) Canopy arthropods of coastal Sitka spruce trees on Vancouver island, British Colombia, Canada. In *Canopy Arthropods* (ed. N. Stork, J. Adis, & R. Didham), pp. 151–168. Chapman & Hall, London.

Winchester, N.N. (2002) Canopy micro-arthropod diversity: suspended soil exploration. In *The Global Canopy Handbook* (ed. A. Mitchell, K. Secoy, & T. Jackson), pp. 140–144. GCP, Oxford.

Winchester, N.N. & Behan-Pelletier, V. (2003) Fauna of suspended soils in an *Ongkea gore* tree in Gabon. In *Arthropods of Tropical Forests: Spatio-Temporal Dynamics and Resource Use in the Canopy* (ed. Y. Basset, V. Novotny, S.E. Miller, & R.L. Kitching), pp. 102–109. Cambridge University Press, Cambridge.

Winchester, N.N. & Ring, R.A. (1996a) Northern temperate coastal Sitka spruce forests with special emphasis on canopies: studying arthropods in an unexplored frontier. *Northwest Science*, **70**, (special issue), 94–103.

Winchester, N.N. & Ring, R.A. (1996b) Centinelan extinctions: extirpation of Northern Temperate old-growth rainforest arthropod communities. *Selbyana*, **17**, 50–57.

Yamashita, Z. & Ishii, T. (1976) Basic structure of the arboreal arthropod fauna in the natural forest of Japan. Ecological studies of the arboreal arthropod fauna 1. *Report of the Environmental Science, Mie University*, **1**, 81–111.

Index of methods and approaches

Methodology	Topics addressed	Comments
Chemical knockdown	Investigation of within-canopy variation in density and species richness. Association of populations and communities with individual trees. Studies of the spatial distribution of organisms across habitats. Absolute estimates of population density and species richness. Assessment of community structure, e.g. guild. Collection of live specimens for subsequent experimental work on population dynamics and feeding strategies.	Collects insects in flight through the canopy and surface dwellers on the leaves, flowers, fruit, twigs, branches, and trunk of the tree. Homoptera, Psocoptera, Collembola, Coleoptera. All groups of insects collected; less effective for Lepidoptera. Does not reliably collect insects that spin leaves together, or that inhabit leaf domatia and epiphytes, or that bore into bark.
Branch clipping and bagging	Assessment of the vertical stratification of communities. Investigation of diel movement within the forest canopy. Specific questions about plant–herbivore relationships. Questions relating to presence and absence of species, absolute estimates of population density, guild structure, and heterogeneity of distribution. Studies requiring count of insect densities per unit of plant biomass and surface area.	Particularly effective for sedentary insects; collects wide range of groups. Large mobile insects are under-sampled, e.g. Odonata, midge clouds (Chironomidae).
Aerial and arboreal traps: Malaise, interception, emergence, and light	Questions about canopy community structure. Testing of specific hypotheses about the use of particular strata of the canopy. Relative estimates of population density and species richness.	Interception traps: dominant groups Coleoptera, Diptera and Hymenoptera. Combination traps: flightless arthropods, e.g. oribatid mites. Underestimate the contribution of sedentary and flightless arthropods to the community. Light traps: Lepidoptera, Coleoptera,

Continued

Methodology	Topics addressed	Comments
	Questions about movement of insects within the canopy space.	Hemiptera.
Moss cores, suspended soils and bark sprays	Questions about specific insect–plant relationships.	Acarina, Araneae, Collembola, Psocoptera.
	Resource partitioning within the forest canopy.	
	Absolute estimates of population density and species richness.	

CHAPTER 8

Sampling methods for water-filled tree holes and their artificial analogues

S.P. YANOVIAK AND O.M. FINCKE

Introduction

Insects of small aquatic habitats found in plants, called phytotelmata (plant-held waters; Varga 1928), have attracted the attention of naturalists for the greater part of a century (Fish 1983). For biological investigations, the relatively small accumulations of water occurring in bromeliads, pitcher plants, and tree holes offer several methodological advantages over lakes, streams, and other comparatively large systems (e.g. Maguire 1971). First, phytotelmata are discrete and can be treated as individual units for sampling and faunal surveys. Second, these habitats are often abundant where they occur, permitting sample sizes appropriate for statistical analyses. Finally, the macrofauna of phytotelmata is often specialized and of manageable diversity and abundance. This is especially true of the aquatic insect inhabitants (e.g. Kitching 2000). Water-filled tree holes are among the most tractable of small aquatic systems, in part because they are relatively persistent, and can be mimicked with plastic cups, bamboo sections, or other inexpensive materials. Despite these unique features of tree holes and their specialized inhabitants, the extent to which processes affecting their biodiversity and community structure can be generalized to larger systems remains to be seen.

Natural tree holes

Water-filled tree holes are formed by the collection of rainwater in natural cavities occurring in the above-ground woody portions of trees (e.g. Kitching 1971a). They exist in hardwood forests all over the world (Fish 1983, Kitching 2000), and are the most abundant standing water systems in some tropical forests. Tree holes occur in a variety of shapes and sizes. In the lowland moist forest of Panama, they may be superficially categorized as slit-shaped, bowl-shaped (Fig. 8.1), or pan-shaped (Fig. 8.2), based on the morphology of the hole aperture and the ratio of water volume to surface area (Fincke 1992a). Temperate tree holes have been classified according to the presence or absence of a continuous lining of tree bark on the hole interior (Kitching 1971a). Although tree holes occur in the crowns of trees and may exceed 50 liters in size (Fincke

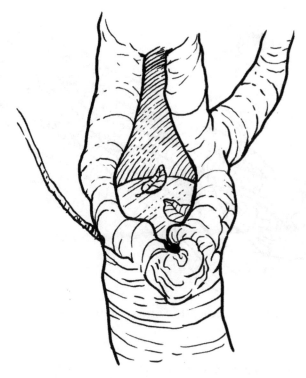

Fig. 8.1 Typical cup- or bowl-shaped tree hole in Panama.

1992a, Yanoviak 1999a, 1999b), most are much smaller, and many occur below 2 meters, where they are easily accessible. As such, they are excellent focal habitats for investigations of aquatic insect behavior, population biology, and community ecology.

A variety of macroorganisms use tree holes as breeding sites, and many species breed exclusively in this habitat. Aquatic insects dominate the assemblages of macrofauna in tree holes; larvae of true flies (Diptera) are generally the most common inhabitants (e.g. Snow 1949, Kitching 2000, Yanoviak 2001a). Tree holes are also the primary breeding sites for many disease vectors, including mosquitoes (Diptera: Culicidae; Galindo et al. 1955) and biting midges (Diptera: Ceratopogonidae; Vitale 1977). Tropical tree holes have the most diverse fauna and harbor an array of predators that are absent in temperate holes (e.g. odonates and tadpoles of dendrobatid frogs; Kitching 1990, Fincke 1992a, 1998, Orr 1994). Aquatic insect assemblages of tree holes are sufficiently diverse in terms of taxonomy and ecological function to permit theory-based studies, yet distinct and simple enough to be manageable for students with limited entomological background.

Here we present methods for non-destructive sampling of aquatic insects and

Fig. 8.2 Pan-shaped tree holes formed by the collection of rainwater in the trunk of a fallen tree.

other macroorganisms from water-filled tree holes based on our experience in Neotropical forests. Our goals are to describe a thorough approach to sampling tree holes, and to identify potential problems associated with data collection and interpretation, which also apply to other types of phytotelmata. All of the concerns we address may not be applicable to tree holes in all types of forests. For example, holes in temperate forests often lack predators, support lower insect diversity, and are subject to stronger seasonal effects, which may influence the frequency and timing of sampling required for a thorough inventory of tree hole occupants. We conclude with some caveats that should be considered before drawing general ecological or evolutionary inferences from tree holes systems. Belkin et al. (1965) and Service (1993) provide additional useful information and references regarding insect sampling from tree holes, with emphasis on mosquito larvae.

Sampling techniques for natural tree holes

Accurate estimates of aquatic insect abundance and diversity in most water-filled tree holes can be obtained with simple procedures and equipment (Fig. 8.3). The most common approach is removal of contents of the hole to a pan for counting. Researchers have devised a variety of techniques to accomplish this task, but reasonably complete samples are obtained by removing detritus and water from the hole, and sieving out the macroorganisms (e.g.

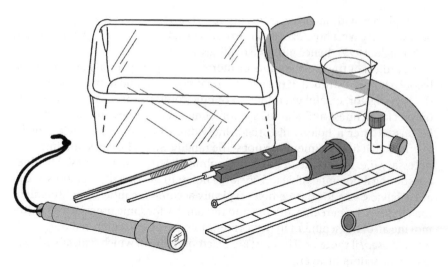

Fig. 8.3 Basic equipment used for sampling water-filled tree holes.

Jenkins & Carpenter 1946, Bradshaw & Holzapfel 1983, Walker & Merritt 1988, Copeland 1989, Barrera 1996).

Sub-sampling is one alternative approach to data collection from tree holes. Kitching (1971b) invented a core sampler that extracts a fraction of the hole volume with each use. This device can provide density data for population studies of some taxa (Kitching 1972a, 1972b), and it collects deep sediments, but the size and rigidity of the corer limit its use to a subset of holes with sufficiently large openings (Barrera 1988). Moreover, insects are often non-randomly distributed within and among tree holes (e.g. Barrera 1996), and it is unlikely that sub-samples collected with a corer would be useful in general surveys or community-level studies.

Although techniques will vary according to the nature of the investigation, thorough tree hole sampling can be summarized as a five-step process:

1 organisms in the undisturbed hole are noted with the aid of a flashlight, and water chemistry parameters are measured;
2 detritus and sediments are removed;
3 fluid contents are removed;
4 the hole is repeatedly flushed with clean water;
5 the interior walls of the empty hole are inspected with a flashlight.

These steps are useful for documenting the macrofauna of the most commonly encountered tree holes: those of small to medium volume (e.g. <5 liters) in which the water surface is exposed and accessible. Larger holes, and holes with narrow or slit openings, are more problematic and require improvised sampling techniques based on specific hole characteristics. Measuring water chemistry variables can be especially difficult in holes with narrow openings, and is best accomplished with small electronic probes. Tree hole water chemistry and temperature vary with hole size, and fluctuate considerably over a 24-hour period (e.g. Fincke 1999), hence multiple readings are preferable.

Regardless of tree hole volume or morphology, step 1 should be completed before a hole is disturbed. If the water is relatively clear and fauna are known to the investigator, careful examination of hole contents can yield accurate data on species richness and abundance for some taxa. Species are more likely to be missed after a hole is disturbed; individuals may hide in crevices or be overlooked if fine sediments do not settle rapidly. Steps 2–5 are sometimes unnecessary (i.e. in very small holes with minimal detritus), or excessively time-consuming if not impossible in very large holes. In aseasonally wet forests (e.g. at La Selva, Costa Rica), tree holes typically accumulate much more sediment than in forests where they dry out and remain dry for some time each year. Removing all of the sediment in the former cases can be extremely tedious, and is not necessary if the taxa of interest are macroorganisms, which typically remain above the sediment layer.

The type of equipment used for completion of steps 2–4 depends on the size and shape of the hole, but almost any hole can be sampled with common materials (Fig. 8.3). In small tree holes, water and soft sediments are removed to a large graduated cylinder for volume measurement using a large suction pipette (e.g. a turkey baster). The contents are then transferred to a white plastic pan for counting. Detritus is removed by hand or with long forceps, rinsed in the tree hole water, and set aside in another pan. A hole should always be probed with a stick or pencil before using bare hands to remove detritus. Tropical tree holes occasionally contain scorpions, land crabs, and ponerine ants, which, if unnoticed, can quickly ruin an otherwise productive field trip. Flushing by repeated filling (two or three times) with water collected from the hole tends to dislodge most organisms remaining in the hole (e.g. Lounibos 1981). If additional water is used for flushing, it should be held in a second pan to avoid dilution or contamination of chemicals and nutrients in the original hole water.

Larger holes can be emptied by using a flexible garden hose to siphon water into a pail. Detritus is removed by hand (or by using a wooden ruler or trowel to lift small packs of leaves), rinsed in the tree hole water and set aside. Rather than completely refilling the hole with a large quantity of water, rinsing the walls with a few liters of clean water is an effective way to dislodge remaining insects.

While the detritus and water collected from flushing are allowed to settle in their pans, the interior walls of the hole can be inspected with a flashlight to spot elusive organisms. Damselfly larvae commonly cling to the walls, and dragonfly larvae are often found covered with sediment at the bottom of the hole, where they can be quite cryptic (Fincke 1992a). With some experience, one can easily recognize elusive species and they can often be counted without removal.

Agitating leaves and other detritus in the collected water usually rids them of any clinging organisms; the composition of detritus can be noted, and litter can then be returned to the hole. After sediments settle in the pan, the clear water is decanted off. This concentrates macroorganisms such as odonates, syrphids, and tadpoles, making them easier to find and count. A small flashlight, which helps focus the investigator's attention to small areas, makes counting much easier, especially on overcast days or under dense forest canopy. A large grid (e.g. 4×4 cm) drawn on the bottom of the pan is also helpful when insects are very abundant. For large tree holes, very numerous insects such as mosquito larvae can be removed in batches to small cups, which permits one to count with greater accuracy. Sub-samples of taxa unfamiliar to the investigator can be collected live for rearing and identification in the laboratory. Small plastic bags (e.g. Nasco® Whirl-Paks) or vials provide the best means of transporting live specimens. Depending on the climate, it may be necessary to transport samples in a cooler with ice to prevent overheating. After subsamples are collected, the remaining organisms and original water can be returned to the hole, and the collection pans are rinsed before the next hole is sampled. Following this protocol, ten or more small to mid-size tree holes in the forest understory can be thoroughly sampled in a day.

Important considerations for natural tree hole experiments

Adequate sample size is a concern in the design of any field experiment (see Chapter 1), and can be problematic in long-term studies of tree holes, which are dynamic systems. Some of the largest holes form suddenly when a tree falls and depressions in the trunk fill with water, but most of these holes do not persist for more than a season or two (depending on the tree species). Even holes in living trees, which often hold water for decades, can vary in volume considerably from year to year, gradually filling completely with mud, or suddenly rotting through. On Barro Colorado Island (BCI), Panama, for example, of 44 water-filled holes in live trees checked in 1982, 6.8 percent had rotted through two years later, compared with 58 percent of those in fallen trees ($n = 12$) and 44 percent of those in dead, upright trees ($n = 9$). Of 23 water-filled holes checked in 1984, 28.6 percent of those in live trees ($n = 21$) had rotted through by the time they were again checked 10 years later. From these data, we estimate a turnover rate for water-filled tree holes in live trees between 2.8 and 3.4 percent per year. Thus, studies of longer than a year should always use more than the minimum number needed for sufficient statistical power in an experiment, with the percentage of additional holes depending on the proportion of study holes in living vs. dead trees.

Another problem associated with tree hole studies is the large number of variables that can affect community properties and interactions among resident aquatic insects. For example, diversity and abundance tend to increase with tree hole volume (Sota 1998, Fincke 1999, Yanoviak 1999b), and predator effects may be stronger in smaller holes (Fincke 1994). This problem is best overcome by surveying a large number of holes several weeks before the start of the experiment, then focusing manipulations on a subset of holes that fall within an acceptable range of variation. Some fauna are found only in very large holes (e.g. *Agalychnis callidryas* tadpoles), whereas others may be more common in shaded holes (e.g. Heteroptera: Veliidae), holes high in dissolved oxygen (e.g. *Physalaemus pustulosus* tadpoles), or holes with abundant, fruity detritus (e.g. Diptera: Syrphidae) (Fincke 1999, Yanoviak 1999c, 2001a). Therefore, biodiversity surveys should incorporate a broad range of hole types and, where possible, note the detritus composition. Because the fauna of tree holes is depauperate relative to that of streams or lakes, overlooking a few species can make a significant difference in conclusions drawn about biodiversity within or between forests.

Tree holes located within 2 m of the ground are best for replicated experiments due to the time and hazards associated with canopy work. However, tree hole height can affect community properties and distributions of some species (Galindo et al. 1951, 1955; Lounibos 1981). In Panama, for example, species richness in tree holes generally declines with increasing height above the ground (Yanoviak 1999b). Thus, diversity surveys and community-level studies should include tree holes from the ground to the canopy. Holes in the crowns of trees are easier to find than to sample. Overflow stains (Snow 1949) and drinking monkeys (Yanoviak 1999b) pinpoint the locations of canopy tree holes to the ground-based observer, but usually only a small percentage are accessible. Moffett and Lowman (1995) reviewed methods for canopy access; the single-line climbing technique (Perry 1978) is the most effective for canopy tree hole work. Once in a tree crown, the investigator can tie in to a fixed point, leave the main rope, and move laterally along branches to sample tree holes. This is a slow and often difficult process, with minimal data resulting from extensive time and energy expenditure. Despite the risk of pseudoreplication (Hurlbert 1984), the most efficient strategy for canopy tree hole work is to focus climbing efforts on tree species that typically possess many holes per crown, and repeatedly sample holes that are readily accessible.

For those who cannot or choose not to climb trees, cranes or canopy walkways (Moffett & Lowman 1995) provide alternative access to the canopy. However, both of these methods require the use of artificial holes that can be positioned in accessible areas. A rope and pulley system can also be used to raise artificial tree holes into the canopy (e.g. Loor & DeFoliart 1970), but the instability of the containers makes them prone to disturbance from wind and canopy mammals, and may result in lost data. Sampling the colonists of artificial holes secured or suspended in tree crowns or at midstory will at least provide a list of organisms that likely use natural tree holes at the same level (Yanoviak 1999b).

Artificial tree holes

Many of the problems associated with sampling natural phytotelmata for ecological experiments can be overcome by using artificial analogues. Simple containers can be used to mimic a variety of phytotelmata, such as *Heliconia* spp bracts (Naeem 1988) and bromeliads (Frank 1985, 1986; Haugen 2001). A major advantage of artificial plant containers is that water volume, nutrient input, and the initial presence or absence of some species can be standardized. Plastic analogues are generally inexpensive and can be censused completely in much less time than the same number of natural habitats of similar size. Most importantly, artificial containers generally attract the same fauna as the natural systems (e.g. Pimm & Kitching 1987, Fincke et al. 1997, Yanoviak 2001a) and will even be readily defended by territorial odonates and frogs (Fincke 1992b, 1998; Haugen 2001).

Almost any container filled with rainwater and a small amount of leaf litter will function as an artificial tree hole for short-term experiments. Tree hole analogues with varying degrees of realism can be constructed from bamboo sections (e.g. Lounibos 1981), automobile tires (e.g. Juliano 1998), stone vases (e.g. Sota et al. 1994), or plastic pots (e.g. Fincke 1992a). Galindo et al. (1951, 1955) described two bamboo trap designs (closed-top and open-top), and discussed differences in mosquito species composition between the types. Closed-top traps with small lateral openings mimic a specific tree hole morphology that is difficult to sample, thus they provide a useful addition for tree hole experiments or surveys. Some containers (e.g. tires and stone vases) are weak replicas of tree holes, but attract many tree hole mosquito species and are often used in vector control studies.

We prefer to use plastic containers for artificial tree holes because they are readily available, lightweight, and durable. Of the several sizes and shapes of containers we use to replicate water-filled tree holes in tropical forest studies, three types seem to give the best results.

Because most natural holes are less than 1.0 liter, the artificial hole we often use is a 0.65-liter plastic cup (Churchill Container Corp., Shawnee, KS; Fig. 8.4). A second type is constructed from a 1.5-liter plastic funnel (Detailed Designs/Injectron, Inc., NO. FN-01, USA) in which the spout is removed and the bottom hole is closed from the inside with a rubber stopper (Fig. 8.5). A funnel design that is flat on one side facilitates secure attachment to a tree. To mimic larger holes, we use either a 6.65-liter oil drain pan (Koller Enterprises, Inc., Fenton, MO) or larger (9.0 liter) brown plastic wash tub (Action Industries, Inc., Cheswick, PA; Fig. 8.6). The latter has convenient handles contiguous with the rim that make attachment easier, and its considerable depth results in proportionally less water loss during the inevitable tipping that occurs after attachment. These types of artificial tree holes will survive years of exposure and closely approximate the shape of similar-sized natural holes.

Artificial tree holes in the form of cups and funnels may be tied to small tree trunks or branches, whereas larger pan-type holes are either secured to forked branches in the canopy or to the trunks of fallen trees in the understory (Fig. 8.6). Polypropylene rope (6 mm diameter) is best for securing artificial holes, but a stronger material (i.e., wire) is required if ants or termites are

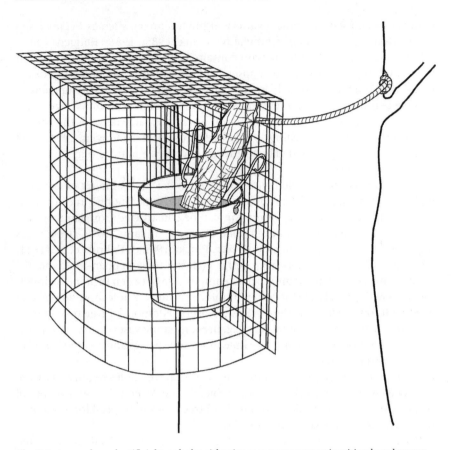

Fig. 8.4 A cup-shaped artificial tree hole with wire cage to prevent oviposition by odonates.

nesting in the tree (Yanoviak 1999b). Plastic-coated flexible wire hooks can be used to hang a cup or funnel from rope around the tree (Fig. 8.5), allowing rapid removal and replacement when frequent sampling is planned. Pans and small artificial holes sampled less often can be secured with rope passed once around the tree and through perforations or handles in the container rim (Fig. 8.4). These methods cause no obvious harm to the tree.

We fill artificial holes with rainwater and put a partially submerged piece of tree bark or balsa wood (Novak & Peloquin 1981) in them as a perch for ovipositing insects. Recently fallen leaf litter collected from the forest floor is added as a nutrient base for the aquatic community (Fish & Carpenter 1982). An initial volume of uncompressed litter within 25–50% of the total artificial hole volume is appropriate for general studies, but the quantity of litter used will depend on the nature of the experiment, the type of forest, and the season. Litter fall reflects seasonal and species-specific patterns of leaf fall and fruiting (e.g. Foster 1982), resulting in variation in nutrient input over space and time.

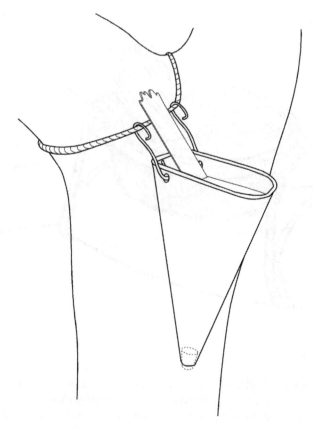

Fig. 8.5 A funnel-shaped artificial tree hole.

For example, 30-day litter accumulation in 0.65-liter cups (71 cm² opening) placed in the BCI forest ranged from 0.0 to 1.2 g dry mass ($x = 0.45 \pm 0.07$ g s.e.; Yanoviak 2001b), and a single fruit fall can result in a pulse of superabundant nutrients (Fincke et al. 1997). To keep nutrients above some minimum for experimental purposes, it may be necessary to periodically add small amounts of litter (e.g. 10% of hole volume) or a substitute nutrient (e.g. fish food or yeast) to some holes.

Apart from providing a standardized physical environment, artificial tree holes also allow some control over potentially important biological variables, such as nutrient input or colonization by key taxa. For example, modifying an artificial tree hole by covering it with a large-mesh wire screen cage (Fig. 8.4) prevents most natural nutrient input, but allows colonization by mosquitoes and most other macroorganisms (Fincke et al. 1997). This cage design also effectively excluded odonates from artificial tree holes in Panama, where they are the most common top predators in this system (Fincke 1998). Similar screening

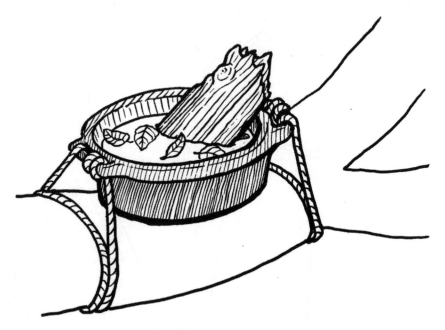

Fig. 8.6 A pan-shaped artificial tree hole.

was only moderately effective at excluding odonates from natural tree holes, in part because eggs laid in the bark prior to screening could not be detected and removed (Yanoviak 2001b). Because the screen cages exclude most falling detritus, additional leaf material must be added to experiments lasting more than a few weeks. Deciding on the quantity of additional nutrient input is not a trivial problem, particularly if growth rates or biodiversity are being measured. Two medium-sized leaves added bimonthly to our 0.65-liter cup-shaped holes kept the abundance of mosquito larvae similar to controls that were open to natural leaf fall, whereas adding 0.05 g of yeast bimonthly resulted in higher than normal levels of mosquito larvae (O.M. Fincke, unpublished). Litter that falls into an adjacent, uncovered, but otherwise identical container could be added to the experimental hole on a regular basis, making the nutrient input more closely reflect natural conditions.

Sampling techniques for artificial tree holes

Artificial tree holes offer a big advantage over natural tree holes because they can be easily emptied completely; time is the only limiting factor in getting accurate counts of the fauna. As for natural tree holes, chemical parameters should be measured and a preliminary census of the fauna made before the artificial hole is disturbed. A cup- or funnel-type hole is then untied (or unhooked) from the tree and its contents poured into a white pan. Using the methods described earlier for natural holes, one can census organisms in a 0.65-liter hole in under 30 minutes, and in a 1.5-liter hole in under 60 minutes. Large pan-type artificial holes can be left in place for sampling but may require several hours to census, depending on the focus of the study.

Important considerations for artificial tree hole experiments

Artificial tree holes provide an excellent means of controlling multiple variables and increasing sample sizes for experimentation. However, because artificial tree holes are not integral parts of trees, researchers using them should be aware of four differences that might affect their results:

1 artificial tree holes are typically younger than holes in living trees, and thus lack potentially relevant biological history (e.g. accumulations of feces, refractory detritus, and sediments);

2 they may receive less stemflow than natural holes in upright trees;

3 their contents have no direct contact with living wood;

4 their inner sides are much smoother than the creviced surface of natural tree holes, which may provide protection for some species or life history stages.

Stemflow inputs and contact with wood are potentially important because both can affect nutrient dynamics and insect productivity in tree holes (e.g. Carpenter 1982, Walker et al. 1991), and stemflow contributes to washout disturbance (Washburn & Anderson 1993). Contact between tree hole water and living wood allows the exchange of materials (e.g. tannins, sap, nitrogenous wastes) between the water and the tree, whereas this exchange and any potential tree species effects on community structure would not occur in artificial tree holes. Abiotic conditions in artificial holes can differ significantly from natural holes of similar size (Table 8.1). However, most tree hole inhabitants tolerate a wide range of pH and dissolved oxygen (Fincke 1999, Yanoviak 1999a), and

Table 8.1 Comparison of abiotic variables in 11 artificial holes and 25 natural tree holes at La Selva, Costa Rica. Means were calculated from measurements taken three times per day (see Fincke 1998 for methods). Ranges in parentheses. Significant differences between natural and artificial holes indicated by $*p < 0.05$, $**p < 0.01$ (t-tests).

Hole type	Volume (liters)	Temperature (°C)	pH	Dissolved oxygen (ppm)
Artificial	0.8 ± 0.2 (0.1–2.0)	27.2 ± 0.5** (24.0–29.0)	5.5 ± 0.2* (3.4–6.2)	2.7 ± 0.3** (0.7–3.9)
Natural	0.9 ± 0.1 (0.1–2.0)	25.0 ± 0.2 (24.7–29.7)	4.7 ± 0.1 (3.4–6.0)	1.0 ± 0.1 (0.3–2.2)

such differences should not affect colonization or survivorship of most macro-fauna (although this may not be true for microorganisms). Artificial holes are particularly good mimics of natural holes in fallen trees (Fig. 8.2), which typically receive limited stem flow, do not contact living tissue, and are relatively young.

Some simple procedures can be used to add realism to the artificial system if necessary. Inoculation of artificial holes with water from natural holes (e.g. during setup and occasionally thereafter) can quickly establish and maintain the microbial assemblage, which is a critical part of tree hole food webs (e.g. Fish & Carpenter 1982, Walker et al. 1991). The rope used to secure cups to trees often conducts stemflow to the cup interior (S.P. Yanoviak, personal observation), and additional stemflow can be directed into a hole by placing the emergent portion of bark or balsa wood against the tree trunk (Fig. 8.4) or by tacking a small piece of plastic onto the tree and allowing it to drain into the hole.

Detritus composition and container color are two additional considerations for those using artificial tree holes in field experiments. The type of litter added to a hole can affect insect productivity and aquatic community structure (Carpenter 1982, Fish & Carpenter 1982, Walker et al. 1997, Yanoviak 1999d), so the composition of litter in a hole (in terms of fragment size, species, age, etc.) should either be consistently haphazard or standardized. Habitat color influences insect colonization in artificial tree holes (Yanoviak 2001c) and other types of phytotelmata (Frank 1985, 1986). Although some workers use clear plastic pans to mimic tree holes in temperate forests (e.g. Srivastava & Lawton 1998), we recommend black or dark brown containers. In Panama, black containers attracted more species than blue or green containers (Yanoviak 2001c). Clear plastic pots can be painted black on the outside, and tubs of any color can be made more realistic by lining the inside with a piece of black plastic (garbage bags work well) that hangs down over the outside edge.

Statistical methods for water-filled tree holes

In most cases, data gathered from replicated tree hole experiments can be analyzed using standard statistical techniques (e.g. ANOVA). Repeated-measures ANOVAs are often used to compare treatment means when the same artificial or natural holes are sampled multiple times (e.g. Fincke et al. 1997, Yanoviak 1999b). Because a large number of ecological and physical variables can be measured in each tree hole, multivariate analyses may be appropriate for many research questions (e.g. Barrera 1988, 1996). It is common for one tree hole to contain zero individuals while hundreds of mosquitoes are present in another. The $\log(x + 1)$ transformation will usually normalize this extreme variation (Sokal & Rohlf 1981). Note that some holes are depauperate of both predators and prey species simply because of resource limitation or abiotic factors; it is important to differentiate between those factors and low diversity resulting from biological interactions (e.g. Fincke et al. 1997).

Problems in interpretation of comparative data

Consideration of spatial and temporal scale is critical when using tree holes as a system to test ecological or evolutionary theory. Tree holes, like bromeliad phytotelmata described by Picado (1913), are analogous to a "subdivided swamp" for most macrofauna using them. Because the resource is subdivided, colonization by certain taxa may be limited with respect to volume, height above the ground, or even morphology of the tree hole opening (see also Frank & Lounibos 1987; Fincke 1992a). Whereas individual tree holes are discrete, replicable units, the scale of the "swamp," which is ecologically comparable to a lake or stream, would be all the tree holes in a forest, which is neither discrete nor easily replicable. For example, top predators decrease diversity within water-filled tree holes on BCI (Yanoviak 2001b). But do forests (at similar latitude) lacking major tree hole predators have greater diversity of tree hole species than forests without those predators? Answering that question requires pooling diversity across replicate holes, with and without predators. Even then, unless the sample of tree holes is representative of the natural distribution with respect to volume, height, and age since the last filling, conclusions may vary.

Finally, most tree hole denizens represent only the larval stage of a species; adults typically are not limited to using a single hole over their reproductive life span, and may have species-specific dispersal distances. In evolutionary studies, for example, the scale of interest would not be simply the fitness of individuals using a given hole, but rather the fitness derived from all the tree holes used over an individual's reproductive life span (e.g. Fincke & Hadrys 2001). Hence, conclusions about community or population processes may be premature without knowledge of the seasonality, longevity, and dispersal ability of the adults in question.

Conclusions

Although there is a growing number of studies documenting the insect fauna of water filled tree holes around the world (Kitching 2000, Yanoviak 2001a), current knowledge remains overwhelmingly biased towards potential disease vectors. Despite considerable interest in the ecology of this system, few studies have addressed the importance of microbial diversity and ecology in tree holes (e.g. Walker & Merritt 1988; Walker et al. 1991). Decomposer microbes (bacteria and fungi) form a critical link between the nutrient base (e.g. leaf litter) and secondary consumers (e.g. mosquito larvae) in tree holes (Fish & Carpenter 1982). Various other microorganisms, such as microcrustaceans, rotifers, and protozoans, also occur in tree holes (Kitching 2000, Yanoviak 2001a), and may function as prey or competitors with the macrofauna. Microbial ecology has been largely overlooked in tropical tree holes, and several basic questions remain to be answered for this system in general. For example, what regulates microbial diversity and productivity in tree holes? How does the composition of detritus affect decomposer assemblages? Does microbial diversity influence macro-organism diversity or productivity? Are microbial assemblages more species-rich in tropical tree holes? The ecology of microorganisms has been examined in other phytotelmata (e.g. Addicott 1974, Cochran-Stafira & von Ende 1998, Carrias et al. 2001), and these studies exemplify the kinds of investigations that are needed in tree holes. Likewise, few studies have addressed the ecological importance of inorganic nutrients (e.g. nitrogen and phosphorus) in tree holes (e.g. Carpenter 1982; Walker et al. 1991). Microbial and nutrient dynamics have been described for many large freshwater systems, and some of the techniques commonly used by stream and lake ecologists to quantify these parameters could be transferred to tree holes.

In summary, water-filled tree holes are tractable habitats for ecological and behavioral studies; sampling their insect fauna is a relatively simple process, and the use of artificial holes is an inexpensive way to increase sample size and control multiple factors for experiments. The extent to which inferences from tree hole data have a more general application for freshwater systems remains to be seen. Nevertheless, given their important ecological role, these aquatic micro-habitats merit much more attention than they have received, especially in tropical forests.

Acknowledgements

We are grateful to Coral McAllister for the illustrations. Comments from S. Stuntz and C. Ozanne, and discussions with L. P. Lounibos, improved the manuscript.

References

Addicott, J.F. (1974) Predation and prey community structure: an experimental study of the effect of mosquito larvae on the protozoan communities of pitcher plants. *Ecology*, **55**, 475–492.

Barrera, R. (1988) Multiple factors and their interactions on structuring the community of aquatic insects of treeholes. PhD thesis, Pennsylvania State University.

Barrera, R. (1996) Species concurrence and the structure of a community of aquatic insects in tree holes. *Journal of Vector Ecology*, **21**, 66–80.

Belkin, J. N., Hogue C. L., Galindo, P., Aitken, T. H., Schick, R. X., & Powder, W. A. (1965) Mosquito studies (Diptera, Culicidae). II. Methods for the collection, rearing and preservation of mosquitoes. *Contributions of the American Entomological Institute* **1**, 19–78.

Bradshaw, W.E. & Holzapfel, C.M. (1983) Predator-mediated, non-equilibrium coexistence of tree-hole mosquitoes in southeastern North America. *Oecologia*, **57**, 239–256.

Carpenter, S.R. (1982) Stemflow chemistry: effects on population dynamics of detritivorous mosquitoes in tree-hole ecosystems. *Oecologia*, **53**, 1–6.

Carrias, J.-F., Cussac, M.-E., & Corbara, B. (2001) A preliminary study of freshwater protozoa in tank bromeliads. *Journal of Tropical Ecology*, **17**, 611–617.

Cochran-Stafira, D.L. & von Ende, C.N. (1998) Integrating bacteria into food webs: studies with *Sarracenia purpurea* inquilines. *Ecology*, **79**, 880–898.

Copeland, R.S. (1989) The insects of treeholes of northern Indiana with special reference to *Megaselia scalaris* (Diptera: Phoridae) and *Spilomyia longicornis* (Diptera: Syrphidae). *Great Lakes Entomologist*, **22**, 127–132.

Fincke, O.M. (1992a) Interspecific competition for tree holes: consequences for mating systems and coexistence in neotropical damselflies. *The American Naturalist*, **139**, 80–101.

Fincke, O.M. (1992b) Consequences of larval ecology for territoriality and reproductive success of a neotropical damselfly. *Ecology*, **73**, 449–462.

Fincke, O.M. (1994) Population regulation of a tropical damselfly in the larval stage by food limitation, cannibalism, intraguild predation and habitat drying. *Oecologia*, **100**, 118–127.

Fincke, O.M. (1998) The population ecology of *Megaloprepus coerulatus* and its effect on species assemblages in water-filled tree holes. In *Insect Populations: in Theory and in Practice* (ed. J.P. Dempster & I.F.G. McLean), pp. 391–416. Kluwer, Dordrecht.

Fincke, O.M. (1999) Organization of predator assemblages in Neotropical tree holes: effects of abiotic factors and priority. *Ecological Entomology*, **24**, 13–23.

Fincke, O.M. (unpublished ms.) Constraints on adaptive cannibalism, clutch size, and offspring sex ratios in a shared tree hole nursery.

Fincke, O.M. & Hadrys, H. (2001) Unpredictable offspring survivorship in the damselfly, *Megaloprepus coerulatus*, shapes parental behavior, constrains sexual selection, and challenges traditional fitness estimates. *Evolution*, **55**, 762–772.

Fincke, O.M., Yanoviak, S.P., & Hanschu, R.D. (1997) Predation by odonates depresses mosquito abundance in water-filled tree holes in Panama. *Oecologia*, **112**, 244–253.

Fish, D. (1983) Phytotelmata: flora and fauna. In *Phytotelmata: Terrestrial Plants as Hosts for Aquatic Insect Communities* (ed. J.H. Frank & L.P. Lounibos), pp. 1–27. Plexus, Medford, NJ.

Fish, D. & Carpenter, S.R. (1982) Leaf litter and larval mosquito dynamics in tree-hole ecosystems. *Ecology*, **63**, 283–288.

Foster, R.B. (1982) Seasonal rhythms of fruitfall on Barro Colorado Island. In *Ecology of a Tropical Forest: Seasonal Rhythms and Long-Term Changes* (ed. E.G. Leigh, A.S. Rand, & D.M. Windsor), pp. 151–172. Smithsonian Institution, Washington, DC.

Frank, J.H. (1985) Use of an artificial bromeliad to show the importance of color value in restricting the colonization of bromeliads by *Aedes aegypti* and *Culex quinquefasciatus*. *Journal of the American Mosquito Control Association*, **1**, 28–32.

Frank, J.H. (1986) Bromeliads as ovipositional sites for *Wyeomyia* mosquitoes: form and color influence behavior. *Florida Entomologist*, **69**, 728–742.

Frank, J.H. & Lounibos, L.P. (1987) Phytotelmata: swamps or islands? *Florida Entomologist*, **70**, 14–20.

Galindo, P., Carpenter, S.J. & Trapido, H. (1951) Ecological observations on forest mosquitoes of an endemic yellow fever area in Panama. *American Journal of Tropical Medicine*, **31**, 98–137.

Galindo, P., Carpenter, S.J., & Trapido, H. (1955) A contribution to the ecology and biology of tree hole breeding mosquitoes of Panama. *Annals of the Entomological Society of America*, **48**, 158–164.

Haugen, L. (2001) Privation and uncertainty in the small nursery of Peruvian tadpoles: larval ecology shapes the parental mating system. PhD thesis, University of Oklahoma.

Hurlbert, S.H. (1984) Pseudoreplication and the design of ecological field experiments. *Ecological Monographs*, **54**, 187–211.

Jenkins, D.W. & Carpenter, S.J. (1946) Ecology of the tree hole breeding mosquitoes of nearctic North America. *Ecological Monographs*, **16**, 31–47.

Juliano, S.A. (1998) Species introduction and replacement among mosquitoes: interspecific resource competition or apparent competition? *Ecology*, **79**, 255–268.

Kitching, R.L. (1971a) An ecological study of water-filled tree-holes and their position in the woodland ecosystem. *Journal of Animal Ecology*, **40**, 281–302.

Kitching, R.L. (1971b) A core sampler for semi-fluid substrates. *Hydrobiologia*, **37**, 205–209.

Kitching, R.L. (1972a) The immature stages of *Dasyhelea dufouri* Laboulbene (Diptera: Ceratopogonidae) in water-filled tree-holes. *Journal of Entomology (ser. A)*, **47**, 109–114.

Kitching, R. L. (1972b) Population studies of the immature stages of the tree-hole midge *Metriocnemus martinii* Thienemann (Diptera: Chironomidae). *Journal of Animal Ecology*, **41**, 53–62.

Kitching, R. L. (1990) Foodwebs from phytotelmata in Madang, Papua New Guinea. *The Entomologist*, **109**, 153–164.

Kitching, R. L. (2000) *Food Webs and Container Habitats: the Natural History and Ecology of Phytotelmata*. Cambridge University Press, Cambridge.

Loor, K.A. & DeFoliart, G.R. (1970) Field observations on the biology of *Aedes triseriatus*. *Mosquito News*, **30**, 60–64.

Lounibos, L.P. (1981) Habitat segregation among African treehole mosquitoes. *Ecological Entomology*, **6**, 129–154.

Maguire, B. Jr. (1971) Phytotelmata: biota and community structure determination in plant-held waters. *Annual Review of Ecology and Systematics*, **2**, 439–464.

Moffett, M.W. & Lowman, M.D. (1995) Canopy access techniques. In *Forest Canopies* (ed. M.D. Lowman & N.M. Nadkarni), pp. 3–26. Academic Press, San Diego.

Naeem, S. (1988) Predator–prey interactions and community structure: chironomids, mosquitoes and copepods in *Heliconia imbricata* (Musaceae). *Oecologia*, **77**, 202–209.

Novak, R.J. & Peloquin, J.J. (1981) A substrate modification for the oviposition trap used for detecting the presence of *Aedes triseriatus*. *Mosquito News*, **41**, 180–181.

Orr, A.G. (1994) Life histories and ecology of Odonata breeding in phytotelmata in Bornean rainforest. *Odonatologica*, **23**, 365–377.

Perry, D.R. (1978) A method of access into the crowns of emergent and canopy trees. *Biotropica*, **10**, 155–157.

Picado, C. (1913) Les broméliacées épiphytes comme milieu biologique. *Bulletin Scientifique de la France et de la Belgique*, **47**, 215–360.

Pimm, S.L. & Kitching, R.L. (1987) The determinants of food chain lengths. *Oikos*, **50**, 302–307.

Service, M.W. (1993) *Mosquito Ecology: Field Sampling Methods*. 2nd edn. Kluwer, Dordrecht.

Snow, W.E. (1949) The Arthropoda of wet tree holes. PhD thesis, University of Illinois, Urbana.

Sokal, R.R. & Rohlf, F.J. (1981) *Biometry*. W.H. Freeman, New York.

Sota, T. (1998) Microhabitat size distribution affects local difference in community structure: metazoan communities in treeholes. *Researches on Population Ecology*, **40**, 249–255.

Sota, T., Mogi, M., & Hayamizu, E. (1994) Habitat stability and the larval mosquito community in treeholes and other containers on a temperate island. *Researches on Population Ecology*, **36**, 93–104.

Srivastava, D.S. & Lawton, J.H. (1998) Why more productive sites have more species: experimental test of theory using tree-hole communities. *The American Naturalist*, **152**, 510–529.

Varga, L. (1928) Ein interessanter Biotop der Biocönose von Wasserorganismen. *Biologisches Zentralblatt*, **48**, 143–162.

Vitale, G. (1977) *Culicoides* breeding sites in Panama. *Mosquito News*, **37**, 282.

Walker, E.D. & Merritt, R.W. (1988) The significance of leaf detritus to mosquito (Diptera: Culicidae) productivity from treeholes. *Environmental Entomology*, **17**, 199–206.

Walker, E.D., Lawson, D.L., Merritt, R.W., Morgan, W.T., & Klug, M.J. (1991) Nutrient dynamics, bacterial populations, and mosquito productivity in tree hole ecosystems and microcosms. *Ecology*, **72**, 1529–1546.

Walker, E.D., Kaufman, M.G., Ayres, M.P., Riedel, M.H., & Merritt, R.W. (1997) Effects of variation in quality of leaf detritus on growth of the eastern tree-hole mosquito, *Aedes triseriatus* (Diptera: Culicidae). *Canadian Journal of Zoology*, **75**, 707–718.

Washburn, J.O. & Anderson, J.R. (1993) Habitat overflow, a source of larval mortality for *Aedes sierrensis* (Diptera: Culicidae). *Journal of Medical Entomology*, **30**, 802–804.

Yanoviak, S.P. (1999a) Community ecology of water-filled tree holes in Panama. PhD thesis, University of Oklahoma, Norman.

Yanoviak, S.P. (1999b) Community structure in water-filled tree holes of Panama: effects of hole height and size. *Selbyana*, **20**, 106–115.

Yanoviak, S.P. (1999c) Distribution and abundance of *Microvelia cavicola* Polhemus (Heteroptera: Veliidae) on Barro Colorado Island, Panama. *Journal of the New York Entomological Society*, **107**, 38–45.

Yanoviak, S.P. (1999d) Effects of leaf litter species on macroinvertebrate community properties and mosquito yield in Neotropical tree hole microcosms. *Oecologia*, **120**, 147–155.

Yanoviak, S.P. (2001a) The macrofauna of water-filled tree holes on Barro Colorado Island, Panama. *Biotropica*, **33**, 110–120.

Yanoviak, S.P. (2001b) Predation, resource availability, and community structure in Neotropical water-filled tree holes. *Oecologia*, **126**, 125–133.

Yanoviak, S.P. (2001c) Container color and location affect macroinvertebrate community structure in artificial treeholes in Panama. *Florida Entomologist*, **84**, 265–271.

Index of methods and approaches

Methodology	Topics addressed	Comments
General surveys	Descriptive data on community structure.	Collections are taken from a large number of holes over several seasons. Provides basic natural history data from which further questions and experiments are developed.
Quantitative sub-sampling	Distribution and abundance of a given species.	May be accomplished with a corer or similar tools. Often the only practical option for very large tree holes.
Species exclusion or addition	Predator effects on community structure; interspecific interactions.	Exclusion methods depend on organism size and behavior, and may not be 100% effective.
Manipulation of litter inputs	Effects of basal resources on community structure.	Qualitative and quantitative characteristics of litter are important considerations
Forest canopy access	Effects of environmental gradients on species distributions.	Ratio of effort and time expenditure to quantity of data recovered may be prohibitive. Artificial tree holes provide a viable option

CHAPTER 9

Sampling devices and sampling design for aquatic insects

LEON BLAUSTEIN AND MATTHEW SPENCER

Introduction

In this chapter, we consider two major problems associated with accurately estimating populations and community structure of aquatic insects: sampling devices appropriate for specific questions, and errors associated with estimates from sampling. Types of sampling devices for aquatic insects are numerous. This reflects the fact that among aquatic insects there is great diversity in mobility, behavior, and microhabitat use. Consequently, many sampling devices are only useful for a specific group of species in a specific type of habitat. In this chapter, we will briefly describe a small subset of these devices. Next, we will consider errors in estimating population size and species richness. In this "errors" section, we give prominence to the often overlooked problem of making comparisons across environmental conditions or experimental treatments when sampling efficiency may vary across these conditions. We also consider how many samples are needed to estimate various parameters such as population densities or species richness, given a defined level of accuracy and precision. Finally, we very briefly consider a few ethical considerations when sampling in aquatic environments. While we attempt to give some coverage to sampling the different habitat types, we do give particular emphasis to the habitat that we are most familiar with — small lentic habitats.

Before addressing these problems, because there is occasionally ambiguity in the literature, we begin by defining a few terms used in this chapter: absolute density, relative abundance, and sampling efficiency. *Absolute density* refers to the real density — number per unit area or volume. *Relative abundance* (or abundance index) refers to the number collected per sampling effort, which is quantitative and may be used for comparative purposes but is not an estimate of the real density. This can be the number caught in a one-meter sweep without knowing how many individuals escaped the net, or the number caught per light trap, which by itself tells us nothing about the real density but may allow us to make comparisons. *Sampling efficiency*, in the case of sampling devices that measure the number caught per unit area or volume, gives the proportion of individuals in that sampled area that are caught. Knowing sampling efficiency in such cases allows us to convert a relative abundance to an absolute density. In other cases, sampling efficiency is a relative term that is not linked to actual densities.

For example, we might determine that a light trap is twice as efficient at trapping one family of beetles as it is at trapping a second family, even though we might not be able to relate numbers caught to actual densities.

A survey of sampling devices

Sampling devices have been categorized according to: (i) type of habitat for which they are suitable; (ii) whether they yield absolute density estimates versus abundance indices; (iii) whether the insect actively enters the device (e.g. light trap) or is passive but is caught (e.g. sweep net); (iv) time and cost (Merritt et al. 1996, Turner & Trexler 1997). Our survey of sampling devices is organized largely according to the inverse of number (iii) — i.e. we categorize according to the activity of the collecting individual and not the activity of the insect. For active-operator devices, the operator moves the device to capture the insects. In a passive-operator device, the device is stationary and insects enter on their own or, in the case of lotic environments, are swept into the device by stream flow. In general, passive-operator devices have several advantages: they tend to have better precision and accuracy than active-operator devices, and environmental disturbance is minimized. For example, two passive-operator sampling devices, an aquatic light trap and minnow trap, cause little environmental disturbance, whereas actively sweeping with a D-net can uproot and tear vegetation in the sampled area. Disturbing the environment may not be permissible, or may be ethically questionable. Moreover, disturbing the environment may also be undesirable if repeated samples are necessary and earlier sampling affects densities and species composition in subsequent samples. A disadvantage of passive-operator devices is that the operator must come at least twice for one sample — first to set the device in place and again to collect the sample.

Our survey is very brief and provides minimal instructions for use. For a more comprehensive list and description of sampling devices, we suggest beginning with a very useful table by Merritt et al. (1996, Table 3) that classifies devices according to various factors and provides references for further information. A short written description for how to use a sampling device is a poor substitute for observing a highly experienced operator in the field, particularly for active-operator sampling devices. One excellent proxy is a video cassette prepared by Resh et al. (1990) in which sampling is demonstrated for approximately 30 devices in the field.

Sampling with nets and dippers

Nets are active-operator sampling devices, and are probably the most common group of devices for sampling aquatic insects. For most nets, the operator actively sweeps the net through the water. In the case of tow nets for sampling pelagic insect species such as *Chaoborus*, the net is attached to a line and is pulled across the water. Absolute density estimates of sweeps or tows can be made in theory if both the distance that the net is moved through the water and the sampling efficiency are known. However, since sampling efficiency is generally not known, sweeping often serves as a quantitative measure of "number per sweep" or "number *caught* per unit volume" rather than "number per unit volume". Nets vary in mesh sizes. There is a trade-off between mesh size and catch efficiency — smaller mesh size will catch a wider range of size classes but catch efficiency will be reduced, particularly for larger and more mobile species. A longer bag can partially remedy this problem. Sweep nets have frames of various shapes but those used for benthic organisms are generally D-shaped. The flat part of the frame is at the bottom to maximize the fit of substrate contour with the frame. Both precision and efficiency in sampling benthic insects should be lower in stony substrate than in fine substrate. Filamentous algae and macrophytes probably reduce both precision and efficiency for sweep nets for insects occupying all levels of the water column.

Typical "dippers" differ from sweep nets in that the collecting devices are solid containers rather than mesh and thus the volume of water sampled for a single sample is confined to the size of the container. Dippers are generally used to collect organisms at, or close to, the water–air interface. The most common of such samplers is the mosquito dipper. As implied by its name, it is used for sampling mosquito immatures though it can also be used to sample associated species in shallow aquatic habitats (e.g. Washino & Hokama 1968). An extensive literature review assessing dipping can be found in Service (1993, Chapter 2). Although there are many variations, the dipper usually consists of a one-pint (473 ml) or half-liter container attached to a one-meter pole. The container is generally white, making the dark larvae more detectable. Dippers also include soup ladles for sampling small habitats such as water-filled tires and small rock pools. The actual technique for making the dip sample varies greatly among researchers. Resh et al. (1990) illustrate dip sampling by dragging the dipper along the water surface as one might sweep a net. Others, by rotating the wrist, "carve" out a volume of water. Still others place the dipper in the water, allowing suction to fill in the volume of the dipper. Advantages of the dipper as a sampling device are that it is inexpensive and light. Generally, a number of dips collected randomly, or across a transect, are concentrated together through a net to constitute a single sample. While sampling programs using dippers are generally considered to give *relative* abundance estimates, there have been attempts to calibrate dipping in order to yield an *absolute* density estimate (e.g. Stewart & Schaefer 1983). Andis et al. (1983, reported in Service 1993) found surprisingly strong, positive correlations between numbers of mosquito larvae per dip and number of larvae collected in a unit area sampler.

Area or column samplers

Despite the fact that sweep netting and dipping provide a sample of known volume, they are rarely considered to provide absolute densities because there is often great difficulty in determining sampling efficiency. Probably for this reason, these sampling devices tend to be called "semi-quantitative" samplers in the literature (for example, see Merritt et al. 1996). There are other sampling devices in which the operator actively samples a known area or volume of habitat with less error. A popular device for sampling benthos in lotic habitats (riffles) is the Surber sampler. Contributing to its popularity is that it is easy to transport (it is foldable and light) and to use. This device consists of a quadrat and an attached net perpendicular to the quadrat. The quadrat is randomly placed on the substrata upstream from the operator. Each rock within the quadrat is slightly lifted and benthos are dislodged by rubbing the rock surfaces with one's hand. These dislodged individuals are then swept into the attached drift net by the stream flow. For higher efficiency, these rocks can also be placed into a plastic bag or bucket and brought to shore for additional inspection. After this has been done to all rocks within the quadrat, the substrate inside the quadrat can be vigorously rubbed with the operator's hand to dislodge any remaining fauna. The device is then lifted and brought to shore where the net is inverted and its contents placed in a white pan for species identification and enumeration on the spot, or preserved and processed later in the laboratory. This device is suitable for water depths of up to 30 cm and for velocities where the operator and the sampler can maintain positions. If done meticulously, both precision and accuracy are considered to be quite high and there is probably less variance in among-operator sampling efficiency than with many active-operator devices. However, rocks that lie only partially inside the quadrat must be dealt with in systematic matter. This becomes more and more problematic as rock size increases. Similarly, the depth of the sample into the substrate must also be standardized because macroinvertebrate vertical distributions vary with species (Rutherford & MacKay 1985).

A functionally similar sampling device that also yields absolute density estimates of stream benthos is the Hess sampler. This device consists of a cylinder that has a mesh screen on one side to allow flow and a long attached net on the other side. The cylinder is placed on the substrate. A foam lining at the cylinder bottom makes for a good seal with the substrate. One's hand is then placed inside the cylinder, dislodging organisms from the substrate, and stream flow through the screen front results in the deposition of the dislodged organisms inside the net.

In soft substrate, core samplers can quantitatively determine abundance of hyporheic (substrate dwelling) species. These core samples also allow for determination of vertical distributions, though it should be noted that some preservation techniques may warp the core sample, thus distorting the real vertical distributions (Rutledge & Fleeger 1988). Another method for quantitatively estimating absolute density of hyporheic species in soft substrate is the use of grab samplers. These devices contain a jaw-like apparatus with a rectangular opening at the bottom. The jaws plunge into the substrate when dropped from above, encompassing a known area and volume of fine substrate. They either close upon contact with the substrate (Petersen grab) or a weight is sent down the attached line afterwards to trigger the closure of the grab (Ekman dredge). If a rock or debris prevents the complete closure of the trap, then part of the sample is lost and the sample is unusable.

Some devices simultaneously sample the entire water column. Column samplers may be cylindrical or rectangular in cross section. They are forced down through the water into the substrate, thus trapping species inside the volume of the sampler. Such column samplers work best where there is soft substrate because the bottom must be sealed to prevent escape of organisms. The length of the device must of course exceed the water depth. The insects in the column sample can then be collected by pumping the water out through a sieve, or by continually sweeping with a small net inside the column sampler until additional sweeps capture no additional organisms. The sample can then estimate densities of neustonic, pelagic, and benthic species. However, as is the case with many active-operator devices, mobile insects are likely to be sampled with lower efficiency as they are more likely to escape.

While most column samplers are designed for plunging them from the air down to the bottom, Resh et al. (1990) demonstrate a "bottom-up" water column sampler—i.e. it is pulled up from the bottom. This bottom-up or "pull-up sampler" can also be used in flexible vegetation. It consists of a pole with a sharp point and a net attached perpendicular near the base of the pole. The pole is forced into the substrate such that the net frame lies parallel to and on the substrate. The pole can then be rotated 180 degrees so that the net lies below a relatively undisturbed water column. After some re-equilibration time, the pole is lifted up catching species within the water column. If there is vegetation, the pole can be first lifted just above the water. The vegetation that extends outside the frame is then snipped along the net frame. In this way, only the vegetation inside the column is collected, and this can give a density estimate of macrophyte biomass in addition to abundance of the insects.

Sampling with natural and artificial substrates

Some passive-operator sampling devices use artificial habitats that serve as substrates for colonization or oviposition, and substrate samplers have become increasingly popular for estimating density of stream benthos. Some samplers use natural substrate. For example, rocks from streams can be collected, washed, and possibly sterilized, then placed in wire mesh baskets entrenched into the same kind of substrate. If left in long enough, the substrates within the samplers approach the densities and species compositions outside the samplers. Other substrate samplers use an artificial substrate that simulates natural substrate. For example, clay tiles can be placed in the stream to simulate the rocks (Lamberti & Resh 1985). Rutherford (1995) used artificial grass substrates anchored to natural substrate in streams to assess insect colonization and dispersion. Immediately after removing the artificial substrates, she sprayed them with an aerosol anaesthetic (Cytocool®) to freeze invertebrates, in order to facilitate measuring the spatial distribution. Still other substrate samplers do not simulate real substrate but instead simply provide a standardized substrate for colonization that can be compared across time or space. A commonly used one, the Hester–Dendy sampler (Hester & Dendy 1962) consists of a set of discs or plates connected and separated by a central pole, with plates spaced wide enough to allow colonization by macroinvertebrates (e.g. Caquet et al. 1996, Turner & Trexler 1997). As these artificial substrates are standardized in terms of architecture and surface area, abundances derived from such samples in different places left in the habitat for the same period could be compared. For an excellent and extensive critique of artificial substrates for sampling freshwater benthic macroinvertebrates, see Rosenberg and Resh (1982).

Sampling with mesocosms

Mesocosms such as outdoor artificial pools, streams, or enclosures that are open at the top, though generally thought of as bodies of water to be sampled, can serve as sampling devices in their own right (Blaustein & Schwartz 2001). For example, artificial pools have been used as a sampling device to examine how risk of predation (e.g. Chesson 1984, Resetarits 2001), food level (e.g. Blaustein & Kotler 1993), and many other factors influence oviposition site selection by aquatic insects. The number of eggs, egg strings, or egg rafts in a particular mesocosm can represent a single sample.

One should keep in mind that if one is interested in comparing the relative abundances of different species colonizing artificial pools, the physical attributes of the artificial pools can greatly influence the answer. For example, size of experimental aquatic mesocosms varies greatly among experiments (Petersen et al. 1999), and predator species may be much more likely to colonize larger artificial pools than smaller ones (Pearman 1995, Wilcox 2001).

In terms of how many samples are necessary to address ecological questions, a general rule of sampling is that the more samples we take, the more likely we are to statistically detect small treatment differences (we deal in depth with this question later). This generality may not be the case in mesocosm experiments when the experiment depends on oviposition by an insect of limited population size (i.e. the total number of individuals colonizing mesocosms does not increase proportionally with the number of mesocosms). Suppose, as was the case in the study by Morin et al. (1988), we wished to understand how colonizing insect herbivores might compete with tadpoles. Morin et al. set up a number of artificial pools containing known numbers of tadpoles. Some pools were left open to allow colonization by herbivorous insects while others were covered with screening to prevent oviposition. Tadpoles growing in the pools with open colonization (by herbivorous insects) were adversely affected. Had they used considerably more replicated pools, the density of colonizing herbivorous insects per pool would likely have been lower. Had this been the case (fewer herbivorous insects per pool if more total pools), the competitive effect of herbivorous insects on tadpoles would have been lower. Not only would the magnitude of the effect have been lower, but as a consequence the probability of a type II error would probably have been higher. So, in such cases, the generality of "the more sampling units the better" is not necessarily true, depending on the experimental design of mesocosm experiments.

Sampling by traps

Traps are sampling devices in which the operator is passive and the insects actively enter the traps. In the case of drift nets used in streams (Matthaei et al. 1998), the active agent is not the organism but the flowing water. If water velocity and trap efficiency are known, a quantitative estimate of the drift density can be calculated. However, for traps in general, it is the insect that is the active agent and trap catch for interception devices depends on swimming speeds and direction. If the trap actually *attracts* the insects, then it does not measure densities but instead some number that is a relative count to the number captured in another trap. Gee minnow traps appear to *initially* act as interception devices if no bait is added. The trap is made of metal mesh and contains inverted funnels on both ends. Active insects crawl or swim through the funnel into the trap and generally cannot find their way back out. After the first individuals enter the minnow trap, the trap's contents may then begin to attract or repel other individuals and species. Also, because the sample is live, predation may occur inside the traps at high rates. Minnow traps are only effective traps for insects if mesh size is small. Turner and Trexler (1997), who used a large mesh size (6 mm), caught very few insect individuals but Blaustein (1988), who used a smaller mesh size (2 mm) captured coleopterans and some hemipteran species in abundance. Minnow traps can be used to capture insects live, but if the trap is totally submerged, insects that require atmospheric oxygen will eventually drown.

While drift nets and minnow traps serve largely as interception traps for horizontally moving individuals, emergence traps intercept and capture individuals emerging from the water, and can estimate absolute density of emerging insects. Emergence traps are generally pyramidal or conical in shape. The open base is often set just below the water surface. They can be attached to floats so that they are in the correct vertical position even if the water depth changes. Emerging insects climb the funnel-shaped trap into a collecting container at the top.

Some traps have attractants and thus cannot provide absolute density estimates (unless a mark–recapture or removal program is used), but instead provide relative indices of abundance. These attractants can be light (e.g. Washino & Hokama 1968), food (e.g. Vance et al. 1995), or some type of oviposition attractant (e.g. Trexler et al. 1998).

Visual observation and photography

Insect populations can be sampled, or in some cases complete counts can be made, by visual counts or photography. This is most often done for surface-dwelling organisms. Quadrats can be set up for such counts but in the case of small habitats such as rock pools, the total number can be counted. This can often be done easily for stages found at the water surface. For example, we have used total counts of mosquito egg rafts laid in artificial or natural rock pools (e.g. Blaustein et al. 1995) and the number of chironomid pupal exuviae on the water surface (Blaustein et al. 1996). Counts can be made for odonate emergence by counting the exuviae left by emerging individuals on surfaces above the water. Pelagic or benthic species may also be counted sometimes in clear, shallow water. We have done so to estimate densities of chironomid larval cases (S.S. Schwartz et al. unpublished data). Surprisingly few studies have used photography as a sampling method to measure density and spatial distribution in aquatic insects. Resh et al. (1990) demonstrate use of a container with a transparent plate (similar to a diving mask) placed onto the water surface that allows photography of benthic organisms in clear, shallow water. Digital cameras and image analyzers not only can facilitate counts, but reduce errors in counting and measuring.

Sampling errors

The problem of measurement error is the central concept in the design of any sampling program. All measurements are subject to errors, of which there are many kinds (Rowe 1994). In particular, ecologists need to be aware of systematic errors, random errors, and measurement interactions. *Systematic errors* are consistent biases in the estimate of some variable. Examples include: when size fractions of a population are small enough to escape through the mesh of a net; when some individuals escape a moving net (Fleminger & Clutter 1965); when some individuals that colonized an artificial substrate sampler are lost upon making the collection (Rosenberg & Resh 1982); the overestimation of body mass from an incorrectly calibrated balance. *Random errors* are differences between measured and true values arising from chance factors. Examples of random error are the number of animals from a population of given density that happen to be caught on a given sweep of a net, or variation between estimates of the mass of a single individual due to the variable amount of water adhering to its surface. We discuss both systematic and random errors at length below. *Measurement interactions* occur when the true value of a quantity is changed while attempting to measure it. For example, we may underestimate the density of a mobile insect such as a gerrid by taking a regular grid of net sweeps because the insects may flee from the sampled area in response to the sampling activities. We will not have much more to say about measurement interactions, but it is worth remembering that they often trade off against random errors: the more intensive the sampling, the lower the random errors but the greater the risk that the system is altered.

Are systematic errors important?

Classical statistics takes no account of systematic errors, because systematic errors are rarely amenable to statistical analysis. Physics textbooks typically make the assumption that systematic errors are small enough to be ignored (Taylor 1982), and biostatistics texts (e.g. Zar 1984, Sokal & Rohlf 1995) usually confine their discussion of systematic errors to the desirability of ensuring that there are none. On the contrary, analyses of historical trends in estimates of basic physical constants (which are probably the most reliable measurements of any natural quantities) suggest that systematic errors are large, important, and difficult to eliminate (Shlyakhter & Kammen 1992).

There are cases, usually experimental, in which the presence of systematic errors may be unimportant (e.g. how does the number of Ephemeroptera change with nutrient concentrations?). A constant additive systematic error cancels out in an additive model (such as an analysis of variance on untransformed data), and a constant proportional systematic error cancels out in a multiplicative model (such as an analysis of variance on log-transformed data).

Factors affecting sampling efficiency

Sampling devices that yield the number of individuals collected per unit area or volume sampled do not give accurate (and generally give under-) estimates of absolute densities prior to correcting for sampling efficiency. For example, individuals located in the area to be sampled may escape a moving net (Fleminger & Clutter 1965) or contents of artificial substrate samplers may be lost when lifting the samplers (Rosenberg & Resh 1982). If absolute densities are necessary, it may be possible to calibrate the sampling device—i.e. to determine the sampling efficiency of the device by sampling under conditions where exact densities are known. We deal with specifics of calibration in the next section. Calibration may not be necessary if relative densities (e.g. number per 1 m sweep), and not absolute densities are sufficient. However, in both cases, sampling efficiency for a particular sampling device may vary with environmental conditions or across species or size classes. This becomes particularly problematic when the researcher is comparing densities in different environments and sampling efficiency varies between environments. Similarly, it is particularly problematic when the researcher compares densities of different species or size classes within a species and it is assumed that sampling efficiency is constant across these categories when in fact it is not. In such cases, part or all of a difference that might be found may not be a true difference but instead the result of differential sampling efficiency. We believe these problems to be common ones that are often ignored. We present some examples below, drawing largely from our own experiences.

Sampling efficiency can vary among species or among size classes within a species

If the goal of sampling is to acquire accurate estimates of age (or size) class structure of a population, it would be ill-advised to assume that the sampling device samples different age classes with equal efficiency. Two age classes within a species that display different behavior or different swimming/crawling speeds are likely to be sampled at different efficiencies. For example, the minnow trap samples different age classes of fish differentially (Blaustein 1989) and this is very likely the case for aquatic insects as well. A second example is dipping, which does not sample different instars of mosquito larvae with the same efficiency (reviewed in Service 1993). In general, later instars are more likely to escape active-operator sampling devices than early instars.

Similarly, without knowing the sampling efficiency of a specific sampling device for each species of interest, a very different picture of community structure might emerge depending on the sampling device used. For example, without taking into consideration possible differences in sampling efficiency among species, dipterans would be considered the dominant aquatic insect species in rice fields based on dipping, but coleopterans and hemipterans would be the dominant species based on aquatic light traps (Washino & Hokama 1968).

Sampling efficiency can vary as a function of density

Sampling efficiency with dipping may drop with increasing mosquito larval density because an alarm reaction by mosquito larvae, which causes them to descend, may increase with increasing larval density (Thomas 1950, reported in Service 1993). If this is the case, and if there are large true differences in the densities of two treatments, then the difference observed by dipping between two treatments may be underestimated. The opposite—i.e. an overestimate of a treatment effect—is also possible. Imagine that some treatment effect, e.g. a pesticide, results in truly lowering the density of odonate naiads. At low densities (the pesticide-treated ponds), most individuals may occupy the underside of rocks, a specific habitat that is sampled at a low efficiency. At high densities (the non-treated ponds), perhaps many individuals are then relegated to a different microhabitat, such as the upper side of rocks, that is sampled at a higher efficiency. In this case, without considering the differential efficiency as a function of density, the treatment effect would be overestimated.

Sampling efficiency can vary with respect to vegetation type and density

Sampling devices, particularly those in which the collector actively samples, have differential sampling efficiencies in open versus vegetated habitats, in different densities of vegetation, or in different types of vegetation (Turner & Trexler 1997). For example, sweep-netting benthos with a D-net may work quite well, yielding high sampling efficiency, in flexible submergent vegetation such as *Chara* or *Najas* species, but this device is pretty much useless in rigid emergent vegetation such as rice or cattail. Efficiency in passive-operator devices across different vegetation types likely varies less than with active-operator sampling devices, but may still exist. For example, we would expect that aquatic light traps should attract a smaller proportion of individuals as density of vegetation increases. Turner and Trexler (1997) assessed invertebrate species richness in different vegetation types using many types of sampling devices. Had they used only a stovepipe (column) sampler, they could have concluded that invertebrate species richness was quite similar in sawgrass and spikerush habitats. Had they used only a Hester–Dendy sampler, they would have concluded that species richness was higher in sawgrass.

Similarly, this problem is very likely to occur when comparing forested (i.e. shaded) habitat versus habitat open to direct solar radiation. Open habitats will likely contain high densities of filamentous algae, which in turn will affect sampling efficiency of active-operator devices such as sweep nets.

Sampling efficiency can vary across substrate type

Active-operator sampling devices of benthos along a coarse (stony) substrate are likely to be less efficient than sampling in fine substrate. Based on our sampling salamander larvae with D-frame sweep nets in pools with large stones versus pools with fine mud substrate, we might have concluded that there were many more larvae in the fine substrate pools than in the stony-bottom pools. However, some of these pools dried shortly after sampling them with D-nets, giving us the opportunity to get a rough idea of our sampling efficiency since we could then get a rather accurate count of the total number of larvae in these pools. We found that larval densities were not lower in the stony pools, as our sampling indicated, but that sampling efficiency was considerably lower in this habitat. This lesson should apply for many benthic insects as well.

Another pilot study of ours illustrates the potential problem of differentiating between real differences in size class distributions of a species across habitats and differential sampling or counting efficiency across habitats. We attempted to compare the size structure of libellulid dragonfly nymphs in two locations in a small pond: shallow water close to the shore, and deep water far from the shore. We used sweeps with an aquatic D-net in both locations and measured the head widths of each nymph we caught in the field. The preliminary data suggested that larvae were, on average, larger near the shore (mean head width 2.9 mm, range 1.3–6.4 mm, $n = 42$) than far from shore (mean head width 1.5 mm, range 0.6–6.0 mm, $n = 122$). It is quite possible that there is a real difference in size class structure at the two locations (if, for example, most oviposition occurs on vegetation near the center of the pond, young larvae tend to be found close to the oviposition site, and older larvae disperse). However, the near-shore sample was filled with mud while the far-shore sample was much cleaner. Efficiency to catch different size classes with a sweep net may differ under these different habitats. Even more likely, the probability of detecting the smaller individuals in the muddy sample may have been lower when processing in the field.

Sampling efficiency may be influenced by indirect effects of an interacting species

A species that is manipulated and alters the environmental conditions can affect sampling efficiency. Suppose that we wish to assess the effect of a predator on larval chironomid densities in rock pools. Our experimental design consists of control (no predator) pools and pools containing the predator. Perhaps we choose to use a sweep net to estimate abundances of various invertebrate taxa. Predators, either by reducing herbivores or by nutrient recycling, can cause increased amounts of filamentous algae in small pools (e.g. Blaustein et al. 1996). Sweep-netting through waters containing heavy mats of filamentous algae (predator pools) is likely to be differentially efficient compared with waters without such mats. Now suppose that we measure 50 percent fewer chironomid larvae in predator plots than in non-predator plots based on sweep samples. Is all or part of this reduction due to the direct consumptive effects of the predator on chironomid larvae? Alternatively, is it possible that the predator has little or no effect on the chironomid densities but that sampling efficiency for chironomids is simply much higher in control (low filamentous algae) pools?

Sampling efficiency can be influenced by differential behavioral responses across treatments

Commonly manipulated factors in aquatic studies such as sublethal effects of chemicals and risk of predation can influence behavior. Over the past two decades, considerable information has accumulated that many aquatic insects, in response to risk of predation, will reduce their activity (Lima 1998). The predator may also affect the behavior of the prey species, which may in turn affect the proportion of prey caught. For example, chironomid larvae swim in the water column. Thus, we might be able to assess relative densities of chironomid larvae by sweep-netting the water column or, depending on water visibility, counting swimming chironomid larvae. We did this in shallow artificial pools where predaceous fire salamander larvae were manipulated and before any build-up of filamentous algae (S.S. Schwartz et al. unpublished data). When we compared the number of chironomid larvae per sweep sample in the presence and absence of salamanders, the predator caused nearly a 100 percent reduction of chironomid larvae counted. With these data alone, we might conclude that this predator reduces densities of chironomid larvae by nearly 100 percent.

However, when we counted the number of chironomid larvae in their cases or the number of chironomid pupal exuviae on the water surface, we found little or no difference between control and predator pools. Thus, it seems that the predators largely influence the behavior (reduced swimming) but actually have little if any effect on the abundance of chironomids. Here, the treatment factor (the predator) altered the proportion of chironomid larvae caught in our net.

Dealing with systematic errors

In many cases, the aim of a sampling program is to estimate the true value of a quantity (e.g. "how many dragonfly larvae are there in the pond?"). Ecologists may want to compare their estimates, not only with estimates from another experimental treatment sampled in the same way, but with published estimates for other species or other habitats, or with predicted values obtained from theory. In the previous section, we emphasized that knowledge of systematic sampling error is important even for estimates of relative abundances when sampling efficiency varies across environmental conditions.

In order of preference, here are some ways in which one might attempt to deal with systematic errors when sampling.

1 Directly estimate and correct for systematic errors

It may be possible to set up situations in which the true answer is known, and calculate a calibration curve. For example, Stewart and Schaefer (1983) wanted to calibrate estimates of the density of larval mosquitoes in rice fields obtained by sampling 1 m² enclosures with dippers. They set up enclosures into which known numbers of larvae were introduced, and estimated the mean number of larvae per dip. The calibration problem is then simply to find a good description of the relationship between the true number and the measured value: in this case, a linear regression was used, but other kinds of relationships (logarithmic, quadratic, etc.) might better fit the data. To be of practical use, the calibration curve must fit the data well because the goodness of fit determines the precision with which true values can be estimated. The true value should be the predictor (because it is assumed to be without sampling error) and the measured value should be the response. When subsequently applying the calibration curve, one needs to estimate the true value given the measured value, by rearranging the equation. This is known as inverse prediction (Sokal & Rohlf 1995). For example, if the calibration curve is

$$\hat{Y} = a + bX \tag{9.1}$$

where $\hat{Y}$ is the predicted measurement, a is the intercept, b is the slope and X is the true value, then one should estimate true values from measurements using

$$\hat{X} = \frac{Y - a}{b} \tag{9.2}$$

where $\hat{X}$ is the inverse-predicted true value, Y is the observed measurement and a and b are the intercept and slope from Equation 9.1. It would not be correct to estimate a calibration curve by regression using the measured values as predictors and the true values as responses, because this is contrary to the assumption that the predictor is without error. One last point about such calibration curves is that the values of the parameter estimates are meaningful in themselves. In a simple linear calibration, a non-zero intercept indicates an additive component of systematic error and a slope that is different from one indicates a proportional component of systematic error.

There are cases—e.g. a large, spatially heterogeneous lake—in which it would be very difficult to construct a calibration curve. In these situations, the following methods should be more practical.

2 Take test samples with different methods likely to have different kinds of systematic biases

If the difference between the mean estimates obtained by different methods is much lower than the standard error of any of those estimates, one can be reasonably confident that the systematic errors are small enough to ignore, and use whichever method is most convenient. If time and money allow, one might continue to use several methods. Southwood (1978, p. 4) suggests weighting the estimate from each method by the inverse of its variance.

It is important that the methods are sufficiently different from each other so that they are not likely to have the same kind of systematic error. For example, kick samples taken with three different sizes of net and three different durations of sampling probably all suffer from the same kinds of bias. On the other hand, kick samples, grab samples, and artificial substrate samples likely suffer from quite different biases, so showing that all three gave similar results would be a strong argument that the biases are small enough to ignore. How small a difference is "small enough to ignore"? This is a matter of judgment, but one needs to think about the absolute difference between the results from different methods, the smallest difference between two measurements which one would think of as "important," and the standard errors of estimates obtained by each method. A difference between methods that is large relative to the smallest "important" difference and to the standard errors of both methods is cause for concern. Southwood (1978, p. 4) suggests formal statistical tests, but common sense also helps.

3 Take samples by several sufficiently different methods (as above), estimate the size of the systematic errors, and carry these systematic errors through subsequent calculations

We illustrate this with a simple case study. To estimate the ratio of mosquito pupae to larvae in a small rock pool (0.6 m long × 0.3 m wide × 0.09 m deep), we tried two sampling techniques. The water was clear, so we first counted all the pupae and larvae we could see in one minute (which we felt was long enough to count all those visible at any time). We repeated this count ten times. Then we swept an aquarium net twice along the length of the pool, counted the numbers of pupae and larvae caught in the net and returned them to the pool. We also took ten samples in this way. For each replicate sample in each method, we calculated the ratio of observed pupae to larvae. We estimated the expected mean and standard error of the ratio with each number of sampling units from one to ten, using a non-parametric bootstrap (Hilborn & Mangel 1997).

As we would expect, the mean ratio from 1000 bootstrap replicates does not change with the number of sampling units for each method, and the standard error decreases as the number of sampling units increases (Fig. 9.1). However, visual counts give a consistently lower and less variable estimate of the ratio of pupae to larvae (10 sampling units: bootstrap mean 0.69, standard error 0.05) than net sweeps (10 sampling units: bootstrap mean 0.95, standard error 0.10). The difference between the bootstrap means from 10 sampling units is 0.26, which is clearly not negligible (5.77 standard errors of the visual counts, or 2.73 standard errors from the sweeps).

Which, if either, is the better estimate? Our first guess might be visual counts because of the smaller standard error. However, it is quite possible that visual

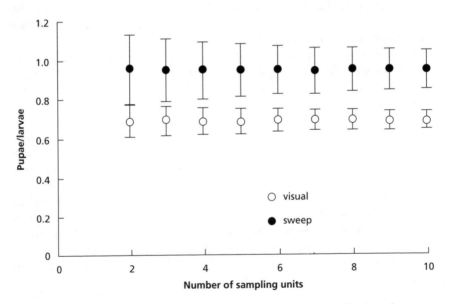

Fig. 9.1 Mean estimates of the ratio of mosquito pupae to larvae in a small rock pool obtained by visual counts (open circles) and net sweeps (filled circles), with standard errors. Means and standard errors were estimated by a non-parametric bootstrap with 1000 replicates for each number of sampling units.

counts are highly repeatable (precise) but biased. For example, it might have been harder to see pupae than larvae because of differences in behavior. The standard error for the sweeps might be larger because each sweep captures a small and variable fraction of all the individuals in the pool. On the other hand, we might be overestimating the ratio of pupae to larvae by net sweeps because larvae are better than pupae at avoiding the net. We have no reason to believe that there is any particular relationship between the sizes of the random and systematic components of error. For a small rock pool, with enough patience, we could probably get much closer to the true ratio of pupae to larvae by emptying the pool, filtering all the water through a net, and searching the material remaining in the pool. Even so, there would still be biases. The smallest larvae are much smaller than pupae, so we might be more likely to miss larvae than pupae. We could also set up artificial pools with known numbers of pupae and larvae and estimate a calibration curve for either sampling method (as described above). However, there are many cases in which no obviously better method is available, and a calibration curve cannot be constructed. The best way of expressing our ignorance is to use both methods, and carry out all subsequent analyses using the values from each method separately. Formally, we could use the interval [0.69, 0.95] as a way of expressing the range of possible values of the pupae : larvae ratio, and we would obtain other intervals representing the results of any subsequent calculations.

Figure 9.2a shows the separate proportion histograms for the bootstrap estimates of the ratio of pupae to larvae from each sampling method (with ten sampling units in each case, and 1000 bootstrap replicates). We might be tempted to average them, but this would almost certainly be wrong. The result of averaging the two distributions in Fig. 9.2a has a mean of 0.82, with 95 percent of values lying between 0.70 and 0.92 (Fig. 9.2b). This seems to suggest that the true ratio of pupae to larvae is exactly halfway between the means obtained by each of the two methods, and that a true value as extreme as the mean of either of the methods alone is quite unlikely. This will only be correct if the systematic errors in the two methods are equal and opposite, for which we have no evidence. Possibility theory (Dubois & Prade 1988) provides an alternative way to deal with these uncertainties. We might reasonably decide that the median estimate from each method (0.69 for visual counts or 0.95 for sweeps), or any value between these medians, was an "entirely possible" value, given the information currently available. We could give the interval [0.69, 0.95] the possibility level 1. The lowest bootstrap estimate of the ratio we ever obtained in 1000 replicates was 0.53, and the highest was 1.20 (from visual and sweep data respectively). We could treat this interval as the range of values that are "just possible," with a possibility level just above zero. Between possibility levels zero and one, there are infinitely many other intervals, each with a different possibility level. In particular, we might be interested in the interval between the lowest observed lower 95 percent confidence limit and the highest observed upper 95 percent confidence limit (in this case, 0.60 to 1.13). This is not itself a confidence

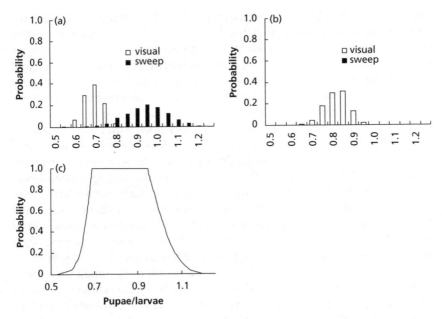

Fig. 9.2 (a) Proportion histograms for bootstrap estimates of the ratio of mosquito pupae to larvae in a small rock pool (open bars are visual counts, solid bars are sweeps, 10 sampling units and 1000 bootstrap replicates in each case). The labels on the abscissa are the midpoints of every second bin. (b) Proportion histogram for the mean ratio of mosquito pupae to larvae, averaging bootstrap estimates from visual counts and sweeps as in (a) and assuming errors are independent. The abscissa is labeled as in (a). (c) A fuzzy number representing the possible values for the ratio of mosquito pupae to larvae, derived from the bootstrap estimates in (a).

interval, but it is an interval that encloses the 95 percent confidence intervals for the ratio of pupae to larvae obtained by both sampling methods. Thus, we should give this interval possibility level 0.05 (although the exact possibility level is not particularly important so long as the ordering of possibility levels is right). Figure 9.2c shows the set of intervals for a range of possibility levels. This is known as a fuzzy number. Fuzzy numbers are often more satisfactory than probability theory for dealing with uncertainty (Ferson & Kuhn 1992), and can be manipulated by a consistent set of arithmetical operations (Dubois & Prade 1988).

The fuzzy number we have defined here would change if we used different sampling methods. If we used only one sampling method, we would obtain a single best estimate. By using more sampling methods, we apparently become less certain about the best estimate of the ratio of pupae to larvae. If we really do not know how each method is likely to be biased, this is a reasonable reflection of our subjective uncertainty. Of course, we have no guarantee that the true value lies within the range of values that we consider to be possible. Even if we used many different sampling methods, they might all be biased in the same direction. We can make this unlikely by using several methods that are likely to be subject to completely different kinds of bias. In our case study we can conclude only that, based on our current knowledge, the ratio of pupae to larvae in the rock pool we studied is possibly between 0.69 and 0.95, and that the 95 percent confidence interval is 0.60 to 1.13.

Random errors: how many sample units?

Random errors, combined with the desired precision to estimate a value, determine the number of sample units. Asking a well-defined question that includes the desired precision will help to ensure that the sampling program is designed appropriately. We will discuss some design considerations for seven such questions.

1 What is the average density of a species in a pond (with a standard error no greater than 5 percent of the mean)?

To answer this question, one needs to know variability among sampling units. Taking a few pilot samples can yield this information. A general formula for the number of samples required is

$$n = \left(\frac{s}{E\bar{x}}\right)^2 \tag{9.3}$$

where n is the number of samples required, s is the estimated standard deviation, E is the required standard error/mean ratio and $\bar{x}$ is the estimated mean (Southwood 1978). If the precision needed can be expressed in terms of the standard error, this formula can be applied no matter what the distribution of the population.

It is often useful to express precision as a function of the width of a confidence interval rather than a standard error. Calculating the number of samples needed to achieve a given width of confidence interval is more complicated, because the confidence interval depends on the population distribution as well as the standard error. At the planning stage, approximate confidence intervals based on the t distribution (assuming a normal distribution of the sampled variable) may be good enough, especially if working with large means:

$$n = \left(\frac{t_{\alpha,n-1}s}{D\bar{x}}\right)^2 \approx \left(\frac{2s}{D\bar{x}}\right) \quad \text{if } n \text{ large and } \alpha = 0.05 \tag{9.4}$$

where D is the desired ratio of half the width of the $100(1 - \alpha)$ percent confidence interval to the mean and $t_{\alpha,n-1}$ is the critical value of the t distribution for a given α and n (Southwood 1978). How large is "n large"? From a set of statistical tables (e.g. Rohlf & Sokal 1995), it can be seen that t converges to a stable value as n increases, and that for α of 0.05, t is approximately 2 if n is more than about 30 (to the level of accuracy needed in planning a sampling program).

Equation 9.4 strictly applies only to normally distributed data. However, most populations are not normally distributed. For example, counts can only be zero or positive integers. Although transformations can be used to make the data approximately normal, one is more likely to be interested in properties of the untransformed data (for example, the arithmetic mean) than of trans-formed data (for example, the geometric mean that results from back-trans-forming a log-transformed dataset). If working with count data, a formula based on the Poisson or negative binomial distributions may be appropriate. The number of samples needed in these cases is

$$n = \frac{\left(t_{\alpha,n-1}s\right)^2}{D^2}\left(\frac{1}{\bar{x}}+\frac{1}{k}\right) \approx \frac{4}{D^2}\left(\frac{1}{\bar{x}}+\frac{1}{k}\right) \quad \text{if } n \text{ large and } \alpha = 0.05 \tag{9.5}$$

where k is an estimate of the negative binomial exponent (Krebs 1989). For a Poisson distribution, k is ∞. With smaller values of k (corresponding to a more aggregated population), the number of samples needed to achieve a given width of confidence interval will be larger. Several methods can be used to ob-tain a preliminary estimate of k. Iterative solution of the following equation is the best method if the number of pilot samples is fairly large:

$$N\ln\left(1+\frac{\bar{x}}{k}\right) = \sum_{x=0}^{\infty}\frac{A(x)}{k+x} \tag{9.6}$$

where $A(x)$ is the sum of frequencies of sampling units having more than x indi-viduals, and N is the number of sampling units in the pilot study (Crawley 1993). To solve for k, a first guess can be made (the estimate of k from Equation 9.7 below is reasonable) and left- and right-hand sides of Equation 9.6 are cal-culated. The guess is too large if the left-hand side is greater than the right-hand side, and vice versa, so the value should be adjusted such that the equation bal-ances (minimizing the difference between the two sides using a spreadsheet or mathematical software is the quickest way). However, if N is small, it may not be possible to obtain a solution to Equation 9.6. In these cases, a rough estimate is

$$k = \frac{\bar{x}^2}{s^2 - \bar{x}} \tag{9.7}$$

After trying the best estimate of k in Equation 9.5, it may be a good idea to also try a slightly lower value (which will give a slightly higher estimate of the num-ber of samples needed). If a small change in k makes a large difference in the number of samples needed, one should be conservative and take more samples than is thought necessary.

What is a reasonable level of precision to aim for? The natural variability of the population sets a limit to the precision that can be achieved in practice. For benthic invertebrates, a 95 percent confidence interval around ±30–55 percent of the mean is often considered to be "moderate" precision, while a 95 percent confidence interval around ±10–25 percent of the mean is "high" precision (Norris et al. 1992). It is also important to think about the size of the sampling units. When the mean density of the organism is high, smaller sample units can be more cost-effective, but at low densities larger sampling units avoid the problem of zero counts, which are often difficult to analyze (Norris et al. 1992). However, ecological considerations (it is not sensible to use sampling units larger than the scale at which one is interested in estimating density) and practical constraints (many sampling devices are only available in a few different sizes) may constrain choices.

Where should samples be taken? Deciding to estimate average density implies that microhabitat variations (such as shallow areas vs. deep areas) are not of particular interest. Nevertheless, to avoid any bias (and to allow the opportunity of examining small-scale patterns in density if one later decides this is necessary), the aim should be to sample each microhabitat in proportion to the fraction of the total habitat size that it contributes. Systematic or stratified random sampling is a good way to achieve this, although pure random sampling will be reasonable if many sampling units are taken. One pitfall is that the appropriate arrangement of sampling locations depends on whether one is sampling organisms that use the whole water column, or organisms that use only the surface or the benthos. If sampling organisms that use the whole water column, one needs to arrange sampling units so that each kind of microhabitat is represented in proportion to the fraction of habitat volume that it contributes. If sampling organisms that use only the surface or the benthos, one needs to arrange sampling units so that each kind of microhabitat is represented in proportion to the fraction of habitat area that it contributes. These two alternatives are only the same if depth is constant throughout the habitat.

2 How many individuals of a species are there in a sediment sample (to within 5 percent of the estimated asymptotic number)?

This objective seems similar to the previous one (estimating the average density in a defined region), yet it is sometimes more efficiently approached with a very different sampling design. In the previous case, one would get a biased estimate of average density if sampling effort were concentrated where one expects to find organisms. In this case, sampling where organisms are expected is the most efficient way time-wise. Of course it is worth checking a few unlikely places as well, to be sure that one's ideas about where to look were correct. Also, it should be remembered that if sampling is conducted in this way, one will not be able to convert estimates of abundance into a valid estimate of average density across the whole sample. Failing to recognize this distinction is one of the likely causes of the negative relationship between sampling area and estimated density that is often reported across studies (Blackburn & Gaston 1999, Gaston et al. 1999, Johnson 1999).

We expressed our desired precision as "to within 5 percent of the estimated asymptotic number." This is a sensible approach if the sample can be searched fairly completely, but there is some constraint on total searching effort. For example, suppose live counts of chironomid larvae in sediment cores are desired, but there is concern that if trying to maintain the cores in the laboratory for more than a day, the number of larvae may change (perhaps there are predators in the sample). One could search the sample under a dissecting microscope, removing each larva as it is found and recording the time. At the end of the day, one could plot the cumulative number of larvae against the time at which each larva was found. The cumulative number of larvae should flatten out at high sampling effort. If an asymptotic function to this curve is fitted and the estimated asymptote is within 5 percent (or whatever value one decides is satisfactory) of the final total, the sampling method is satisfactory.

3 What is the ratio of densities of two species (with a standard error no larger than 20 percent of the mean)?

First, one should ask whether it is really necessary to estimate a ratio. Ratios and other derived variables often have much higher standard errors than directly measurable variables (Taylor 1982, Jasienski & Bazzaz 1999). The sampling distributions of ratios do not lend themselves to standard statistical methods (Atchley et al. 1976). The relationships between variables with common components (for example, between X/Y and Y) are mathematically constrained. This can lead to two kinds of problems. First, if the shared measurement error is large (for example, if most of the measurement error in X/Y results from measurement error in Y), any relationship between the variables is determined mainly by this shared error, and is unlikely to be informative (Prairie & Bird 1989). Second, variables with very strong mathematical constraints may not contain much biological information. For example, the ratio of the number of predatory to non-predatory species has a more or less constant value close to 1, across many food webs (Cohen 1978). However, this is simply a mathematical necessity, once one realizes that most species in most food webs are both predators and prey by these definitions (Closs et al. 1993). Ratios should be used carefully when they measure quantities of genuine interest and with due regard for their statistical peculiarities.

Ratios of dependent variables are the source of much confusion in ecological literature. For example, Krebs (1989) suggests the following estimate of the mean ratio $\hat{R}$ of two variables x and y:

$$\hat{R} = \frac{\bar{x}}{\bar{y}} \tag{9.8}$$

This is often misleading, as is the associated standard error suggested by Krebs (following Cochran 1977). Unless x and y are independent, the ratio of the means (Equation 9.8) is not the same as the mean ratio. The general formula for the mean ratio is

$$\hat{R} = \frac{\bar{x}}{\bar{y}} - \text{cov}\left(\frac{x}{y}, y\right) \cdot \frac{1}{\bar{y}}$$

(9.9)

where $\text{cov}\left(\frac{x}{y}, y\right)$ is the covariance between x/y and y (Welsh et al. 1988). Welsh et al. (1988) and Kirchner (1998) give formulae for means and variances of several useful functions, but for many cases the formulae only apply to certain special distributions (e.g. normal or lognormal), or to independent variables. Of course, one can calculate the mean ratio and its standard error directly from the ratio in each sampling unit, which is much easier. Simple formulae for confidence intervals are rarely available. We suggest the following:

a For a rough idea of the number of samples needed, calculate the mean ratio and its standard error, and use normal approximate confidence intervals to estimate the number of samples needed for a given precision (Equation 9.4). Because the normal approximation is unlikely to be very accurate, it may be a good idea to take more samples than the formula suggests.

b In cases where the level of sampling effort must be determined accurately (for example, if the cost per unit effort is high), use simulation to estimate the expected width of confidence intervals for a given level of sampling effort. After taking some pilot samples, find a parametric distribution that describes each component of the ratio reasonably well, and estimate the parameters (e.g. the mean for a Poisson distribution, the mean and k for a negative binomial distribution, the mean and variance for a normal distribution). Estimate the correlation between the variables. For a range of proposed numbers of sampling units, generate many (perhaps 1000) random datasets with the appropriate number of sampling units, distribution of each component variable, and correlation between them. Special software is available, or one could code one of the simple algorithms for generating correlated random variables (e.g. Nelsen 1986, 1987). Estimate the confidence limits as the $100(\alpha/2)$ and $100(1 - \alpha/2)$ percentiles of the distribution of mean ratios over all random datasets of a given number of sampling units. Choose the lowest number of sampling units for which the confidence interval is narrow enough.

4 *What is the difference in density of a species between two ponds or habitats (with a 95 percent confidence interval of the difference no wider than 5 individuals per m^2)?*

To keep things simple, we will assume that some transformation of the distribution of sample mean density estimates in each pond can make the estimates more or less normally distributed with similar variances. Our illustration is based on pages 223–5 in Sokal and Rohlf (1995). Given two sample means $\bar{Y}_1$ and $\bar{Y}_2$, our best estimate of the difference Δ between them is simply $\bar{Y}_1 - \bar{Y}_2$. The standard error of the difference (s_Δ) is

$$s_\Delta = \sqrt{\left[\frac{(n_1 - 1)1s_1^2 + (n_2 - 1)s_2^2}{n_1 + n_2 - 2}\right]\left(\frac{n_1 + n_2}{n_1 n_2}\right)} \qquad (9.10)$$

where s_i is the sample standard deviation and n_i is the number of sampling units taken from population i. The $100(1 - \alpha)$ percent confidence interval for the difference between the means can then be calculated from the t distribution

$$(\bar{Y}_1 - \bar{Y}_2) \pm t_{\alpha[v]}s_\Delta \qquad (9.11)$$

where the degrees of freedom (v) are $n_1 + n_2 - 2$. The best approach is to take pilot samples in each pond and estimate $\bar{Y}_1 - \bar{Y}_2$, s_1 and s_2. Then we calculate how the width of the 95 percent confidence interval changes as we substitute different values for n_1 and n_2, and choose values that are sufficient to achieve our aim. The width of 95 percent confidence interval chosen as acceptable is a way of indicating the range of estimates that we would be prepared to think of as more or less the same.

For example, we ran a pilot study to assess the use of minnow traps to compare density estimates of libellulid naiads in shallow and deep water in a small pond. We set up five traps in each habitat type and counted the number of larvae they contained after three hours. The raw data are shown in Table 9.1. Given the size of these means, we think we would like to know the difference between them $(9.2 - 5.4 = 3.8)$ with a 95 percent confidence interval no wider than 2 individuals per trap (this is approximately ± 26 percent of the difference,

Table 9.1 Numbers of libellulid larvae sampled by minnow traps in three hours in two different habitat types (shallow water near shore and deep water far from shore) in a small pond.

	Shallow	Deep
	9	7
	6	10
	12	6
	12	3
	7	1
Mean	9.2	5.4
Variance	7.7	12.3

or "high" precision for benthic invertebrates (Norris et al. 1992). As the number of samples is small, we cannot be sure what kind of distribution would best describe the data, so we use the untransformed values. This will only give us an approximate estimate of the sample size required. Figure 9.3 shows the estimated width of the 95 percent confidence interval on the difference between the means, for a range of sample sizes. To achieve a confidence interval no wider than 2 individuals per trap, we would need at least 80 traps per habitat. This is not practical; given the size of the pond, we could barely fit so many traps into each habitat at the same time. We could either settle for a less precise estimate of the difference with fewer traps or find a less variable sampling method.

We have chosen to emphasize the estimation of a confidence interval for the difference between two means rather than a *p* value for two reasons. First, calculating the size of a difference and its confidence interval is much more informative than simply stating a *p* value (Harlow et al. 1997). *P* values combine the size of a difference and the precision with which this estimate is known into a single number. The same *p* value could come from a small difference with high precision or a large difference with low precision, yet we would interpret these two results quite differently. Second, meta-analyses are increasingly important in ecology (Osenberg et al. 1999), and are based on estimates of effect size rather than *p* values. Routinely thinking about effect size rather than *p* values will thus improve our understanding of our data, and serve as the foundation for understanding modern statistical tools. Estimating confidence intervals on measures of effect size is closely related to power analysis. Thinking about power at the design stage is important because if we don't design our sampling program so as to have sufficient power, we will not only waste our time, but may be tempted to draw misleading inferences. For a readable introduction to power analysis, see Murphy and Myors (1998). Formulae for calculating the sample size needed to

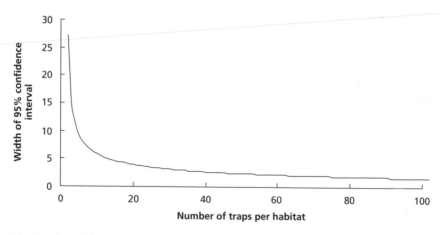

Fig. 9.3 The width of the 95% confidence interval for estimating the abundance of libellulids per trap as a function of the number of minnow traps set.

achieve a specified power for a specified difference between means are given by Zar (1984, pp. 133 & 193) and Sokal and Rohlf (1995, p. 263), and are discussed by Norris et al. (1992). What level of power should we aim for? Power below 0.5 is worse than useless. Assuming that there is really a difference between two means (and the probability of there being absolutely no difference is infinitely small), any failure to detect such a difference is a type II error. With power less than 0.5, we would therefore make fewer errors by flipping a coin than by doing a survey or experiment (Murphy & Myors 1998). The minimum level of power for which one should aim is often suggested to be 0.8 (Murphy & Myors 1998). Even with a power of 0.8, the chance of two experiments on identical systems yielding consistent results (either both statistically significant or both not statistically significant) is only 0.68 (Gurevitch & Hedges 1999).

5 How does the age structure of the species change through the season, on a time scale short enough to estimate the survival probability of age classes of 2 weeks each, with a standard error less than 0.2 in each estimate?

Such a question can be addressed by mark–recapture techniques. Marking soft-bodied aquatic insects is more problematic (though radioactive isotopes may be used; e.g. Croset et al. 1976) than marking hard-bodied organisms such as adult beetles (Nürnberger 1996, Svensson 1999). Krebs (1989) suggests a number of methods for estimating survival from marked individuals, each with different requirements and different ways of obtaining standard errors. If one is not able to mark individuals, survival probabilities can be estimated from time series of samples. This is difficult (Caswell & Twombly 1989, Manly 1990, 1997, Wood 1997). Equal sampling intervals, the same as the width of the age classes, will make things a bit easier. If using stages rather than age classes, the sampling interval should be no longer than the minimum duration of the shortest stage, and again using equal sampling intervals is desirable. To avoid wasting sampling effort, one should first make up some plausible data, using known survival parameters with added random error based on fairly pessimistic estimates of the amount of sampling variability one expects to encounter. Then the data should be run through the estimation process one intends to use. If unable to recover the known parameters, the sampling program will unlikely be worthwhile.

6 How many samples are needed to obtain a good estimate of the species composition of a community?

Most species have low relative abundances (May 1975), and it would take a tremendous amount of effort to enumerate them all. If we intend to study the "real" distribution of relative abundances in the community, we should be prepared for an exhaustive sampling program (e.g. Siemann et al. 1999). On the other hand, we might only be interested in those species that are abundant enough to be important in the community. A sensible approach is to decide on (and explicitly state!) a working definition of "important" in terms of relative abundance, and make a rough estimate of the sampling effort needed to detect the rarest important species. To do this, we need to know the abundance–frequency distribution in the community, and the relationship between abundance and detection probability. To solve this problem before carrying out a sampling program, we will have to make some guesses. We might also be interested in evaluating the effectiveness of a sampling program that has already been conducted, in which case the necessary data will already exist.

Abundance–frequency curves provide a way to think about the consequences of a given decision about the lowest level of abundance we would like to detect. For example, suppose that we want to detect the most common y percent of the species. How rare is the rarest of these y percent? Or suppose we decide that we are not interested in detecting a species with an abundance lower than x, what proportion of species in the community will we be ignoring? Figure 9.4 shows the abundance–frequency distribution for invertebrates (mainly insect larvae) in snag habitats of a subtropical blackwater river (data from Benke et al. 1984, upper snag site). For many communities, the abundance–frequency distribution is approximately lognormal (May 1975). Fitting a lognormal

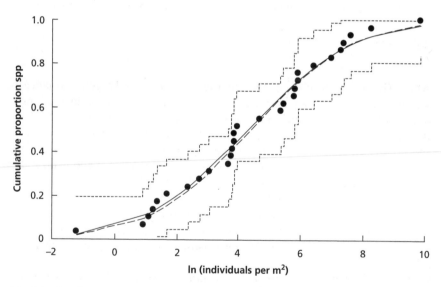

Fig. 9.4 Abundance–frequency distribution for invertebrates in snag habitats of a subtropical blackwater river (data from Benke et al. [1984], upper snag site). The circles are the observed data, with short dashed lines indicating the 95% Kolmogorov–Smirnov confidence interval for an intrinsic hypothesis (one in which we estimate the mean and variance of a fitted distribution from the data themselves; Sokal 1995). The solid line is a fitted lognormal distribution, assuming that the lowest observed abundance is the lowest observable abundance. The long dashed line is a fitted lognormal distribution, estimating the lowest observable abundance from the sampling effort.

distribution to observed data usually requires a correction for the lowest abundance that could have been observed (described by Magurran 1988 and by Krebs 1989). In count data, the correction is usually the logarithm of 0.5 (the lower boundary of the log abundance class in which only one individual was observed). However, the data of Benke et al. (1984) are measured as numbers per m^2, so we have to divide the (natural) logarithm of 0.5 by the total area sampled. A total of 96 samples were taken, with areas usually between 0.04 and 0.01 m^2, so the lowest detectable abundance might possibly have been as low as $\ln(0.5 \times 96/0.04) = -2.04$, or as high as $\ln(0.5 \times 96/0.01) = -0.65$. The lowest natural log abundance actually observed was -1.20, so we fitted lognormal distributions with lowest detectable natural log abundances of -2.04 and -1.20. These alternatives make little difference to the predicted distribution (Fig. 9.4). Moreover, both fitted distributions lie well within the 95% Kolmogorov–Smirnov confidence interval (Sokal & Rohlf 1995) for the observed distribution. In a retrospective analysis, we could conclude that the lognormal distribution was a reasonable description of our data.

For a prospective analysis, we are unlikely to have data like those in Fig. 9.4, but we might still be able to make a plausible guess at the abundance–frequency distribution. The "canonical lognormal" distribution predicts a relationship between the number of species and the variance in log abundance

$$S = \sigma\sqrt{\pi/2}\exp\left(\frac{(\sigma\ln 2)^2}{2}\right) \qquad (9.12)$$

where S is the true number of species in the community and σ^2 is the variance in log abundance. It has been forcefully argued that this relationship is a consequence of the way in which niche space is partitioned (Sugihara 1980). If we find Sugihara's argument convincing, and we can guess the true number of species and either the mean or the total log abundance, we can also make a guess at the abundance–frequency distribution. Hypotheses about how the world works are very often used as the justification for selecting a particular distribution to describe data. For example, normal or lognormal distributions are widely used even when data are too scarce to discriminate among the set of distributions that might reasonably be used, because the central limit theorem provides a theoretical basis for the normal and lognormal (Hattis & Burmaster 1994).

The sample actually contained 29 distinguishable taxa at the upper snag site, with a total of 33,300 individuals per m^2. The estimated variance in natural log abundance from the fitted lognormal distributions was 7.05 to 7.51 (assuming that the lowest detectable natural log abundance was either -2.04 or -1.20, as above). Suppose that, based on past experience or literature surveys, we had guessed that we might find somewhere between 20 and 40 species, and somewhere between 20,000 and 40,000 individuals per m^2 (even if we don't know very much about the habitat, we will probably be able to make guesses like these). The mean abundance might then be as low as 500 or as high as 2000 individuals per m². Using Equation 9.12, we would predict a variance of 7.3 to 9.7 in natural log abundance.

Once we have fitted or guessed a distribution, we can answer questions like "how rare is the rarest of the most abundant y percent of species?" and "what proportion of species will be missed if we ignore those with abundance lower than x?" For example, how rare is the rarest of the most abundant 95 percent of species in the data of Benke et al. (1984)? This is equivalent to finding the 5th percentile of abundance, for which we need the inverse normal distribution function, and the mean and variance of natural log abundance (the inverse normal distribution function is available in most spreadsheets and statistical programs). Given the means and variances we guessed, the 5th percentile of abundance might be as low as 2.98 or as high as 23.49 individuals per m². The curves we fitted to the observed distribution have a 5th percentile of 0.84 to 1.05 individuals per m², so the abundance at the 5th percentile is actually considerably lower than our guess. As another example, suppose we decide that we will not attempt to detect species with a density lower than 1 individual per m². What proportion of species will we be ignoring? This requires the cumulative normal distribution function, which is also readily available. From our guessed means and variances, we would expect to be ignoring somewhere between 0.2 percent and 2 percent of the species. From the fitted curves, we would expect to be ignoring somewhere between 5 percent and 6 percent of the species. Our rough guesses were clearly a little overoptimistic, but they are close enough to be useful at the planning stage. For a retrospective analysis, we could use the abundance curve fitted to the data we actually observed.

Assuming we have decided on the rarest species of interest, how much effort will we need to expend to be reasonably sure of detecting these species? We would need to have some idea of the probability of detecting an individual animal in a sampling unit, and the distribution of true abundances in sampling units for a given mean abundance. In a prospective analysis, we might try to estimate the probability of detecting an individual animal from a few preliminary trials (in which we could make artificial samples containing known numbers of individuals). In a retrospective analysis, we could estimate the probability that some individuals were missed by repeat counts of samples. The distribution of true abundances in sampling units depends on how the mean and the variance of abundance are related. Again, this can be established either from preliminary trials or retrospectively. We can estimate the number of sampling units needed to detect the species with a given probability as

$$N = \frac{\ln(1-\alpha)}{\ln(1-p)} \tag{9.13}$$

where N is the number of sampling units needed, α is the desired probability of detecting the species, and p is the probability of the species being present in a

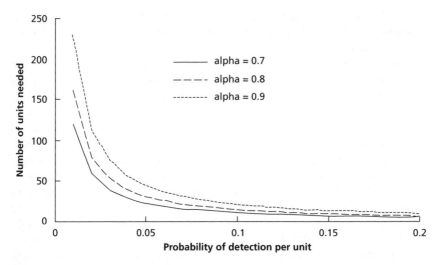

Fig. 9.5 Estimating the number of sampling units needed to detect a species (Equation 9.13). The abscissa is the probability that a species will be detected in a single sampling unit (p), and the ordinate is the number of sampling units needed (N). Curves are plotted for three different values of the desired probability of detecting the species in the entire sample (α). Solid line, $\alpha = 0.7$; dashed line, $\alpha = 0.8$; dotted line, $\alpha = 0.9$.

single sampling unit (McArdle 1990). Figure 9.5 shows the relationship between p and N for three different values of α. The number of sampling units needed depends very strongly on the probability that the species will be detected in a single sampling unit, so we will need to work much harder to achieve a desired probability of detecting a rare than a common species.

7 On which prey species does a given predator feed (including all species that make up more than 5 percent of prey individuals)?

Properties of food webs such as the average number of feeding interactions per species are sensitive to sampling effort (Martinez et al. 1999). Yield–effort curves like the species accumulation curves we discussed in the previous section can be used to determine how many feeding interactions are likely to have been missed across the whole food web. In general, a very large amount of effort is needed to thoroughly document a food web. The number of possible feeding interactions is N^2, where N is the number of species, and many of these possible interactions may occur very rarely. For example, Polis (1991) illustrated a yield–effort curve for the number of prey species of a single species of scorpion. After more than 2000 person-hours of field work on the whole food web, spread over five years, more than 100 different prey species had been recorded for the scorpion, with no sign of an asymptote.

If one is interested in the overall properties of food webs, there may be no alternative but to attempt a very large sampling program. On the other hand, one might only want to know about those prey species that make up a substantial part of the diet of a predator. One could plot the frequency with which each prey species is recorded in the diet (over all sampling units) against the cumulative sampling effort at which the prey species was first recorded. On average, prey species that make up a large proportion of the diet will tend to be first recorded earlier in the sampling program than prey species that make up a small proportion of the diet. One could stop sampling when the relationship between frequency and effort at first recording falls below some threshold (which defines the frequency below which one thinks a prey species is not important).

Ethical considerations

Ethical considerations should be a part of any sampling program and may include avoidance of: (i) killing more individuals than needed even when the population is not under threat; (ii) increasing the probability of extinction of a species; and (iii) damaging or altering an environment. We suggest a few guidelines for ethical sampling.

Excessive killing, even when the species is not endangered, receives more attention when the animals sampled are mammals and other higher animals (e.g. Puttman 1995) than for insects. To minimize the number of individuals killed, it might be possible to collect individuals live in the field, count them and return them alive. Regardless of whether one feels that "excessive" killing of a common insect species is an ethical consideration, collecting too many individuals may actually significantly reduce the population and thus alter results of subsequent samples. Sampling, counting, and returning live specimens may also allow for more samples without the researcher actually significantly reducing populations through sampling.

Overall, damage to the environment should be a serious consideration for all field workers. Caution should be taken not to spill preservatives such as formaldehyde in the field. Active-operator sampling devices such as benthic nets or grabs can potentially perturb the environment while passive-operator sampling devices such as traps and colonizing substrates cause little damage. Researchers are also probably responsible for unintentional introductions of species due to propagules on sampling equipment or on boots. This may give false impressions of the degree of isolation of water bodies and the degree of separation of populations in later genetic studies. Unintentional introductions may also include disease agents. Herpetologists have expressed particular concern about the unintentional spreading of pathogens to anuran species which has prompted the DAPTF (Declining Amphibian Populations Task Force) to adopt a "fieldwork code of practice" which calls for sterilizing boots and sampling equipment with alcohol.

Acknowledgments

We thank Vince Resh for useful discussion. The writing of this chapter was supported by US–Israel Binational Science Foundation grants 95-0035 awarded to LB and Joel E. Cohen and 98-0390 awarded to LB and Marc Mangel.

References

Andis, M.D., Meek, C.L., & Wright, V.L. (1983) Bionomics of Louisiana riceland mosquito larvae I. A comparison of sampling techniques. *Mosquito News*, **43**, 195–203.

Atchley, W.R., Gaskins, C.T., & Anderson, D. (1976) Statistical properties of ratios. I. Empirical results. *Systematic Zoology*, **25**, 137–148.

Benke, A.C., Van Arsdall, T.C. Jr., & Gillespie, D.M. (1984) Invertebrate productivity in a subtropical blackwater river: the importance of habitat and life history. *Ecological Monographs*, **54**, 25–63.

Blackburn, T.M. & Gaston, K.J. (1999) Density, survey area, and the perfection (or otherwise) of ecologists. *Oikos*, **85**, 570–573.

Blaustein, L. (1988) Biotic interactions in rice fields: an ecological approach to mosquito control. PhD thesis, University of California, Davis.

Blaustein, L. (1989) Factors affecting the efficiency of minnow traps to sample populations of mosquitofish (*Gambusia affinis*) and green sunfish (*Lepomis cyanellus*). *Journal of the American Mosquito Control Association*, **5**, 29–35.

Blaustein, L. & Kotler, B.P. (1993) Oviposition habitat selection by *Culiseta longiareolata*: effects of immature conspecifics, tadpoles and food levels. *Ecological Entomology*, **18**, 104–108.

Blaustein, L. & Schwartz, S.S. (2001) Why study ecology in temporary pools? *Israel Journal of Zoology*, **47**, 303–312.

Blaustein, L., Kotler, B.P., & Ward, D. (1995) Direct and indirect effects of a predatory backswimmer (*Notonecta maculata*) on community structure of desert temporary pools. *Ecological Entomology*, **20**, 311–318.

Blaustein, L., Friedman, J., & Fahima, T. (1996) Larval *Salamandra* drive temporary pool community dynamics: evidence from an artificial pool experiment. *Oikos*, **76**, 392–402.

Caquet, Th., Lagadic, L., Jonot, O., et al. (1996) Outdoor experimental ponds (mesocosms) designed for long-term ecotoxicological studies in aquatic environment. *Ecotoxicology and Environmental Safety*, **34**, 125–133.

Caswell, H. & Twombly, S. (1989) Estimation of stage-specific demographic parameters for zooplankton populations: methods based on stage-classified matrix projection models. In *Estimation and Analysis of Insect Populations* (ed. L. McDonald, B. Manly, J. Lockwood, & J. Logan), pp. 93–107. Springer, New York.

Chesson, J. (1984) Effects of notonectids (Hemiptera: Notonectidae) on mosquitoes (Diptera: Culicidae): predation or selective oviposition? *Environmental Entomology*, **13**, 531–538.

Closs, G., Watterson, G.A., & Donnelly, P.J. (1993) Constant predator–prey ratios: an arithmetical artifact? *Ecology*, **74**, 238–243.

Cochran, W.G. (1977) *Sampling Techniques*. Wiley, New York.

Cohen, J.E. (1978) *Food Webs and Niche Space*. Princeton University Press, Princeton.

Crawley, M.J. (1993) *GLIM for Ecologists*. Blackwell, Oxford.

Croset, H., Papierok, B., Rioux, J.A., Gabinaud, A., Cousserans, J., & Arnaud, D. (1976) Absolute estimates of larval populations of culicid mosquitoes: comparisons of capture–recapture, removal, and dipping methods. *Ecological Entomology*, **1**, 251–256.

Dubois, D. & Prade, H. (1988) *Possibility Theory: an Approach to Computerized Processing of Uncertainty.* Plenum Press, New York.

Ferson, S. & Kuhn, R. (1992) Propagating uncertainty in ecological risk analysis using interval and fuzzy arithmetic. In *Computer Techniques in Environmental Studies IV* (ed. P. Zannetti), pp. 387–401. Elsevier, London.

Fleminger, A. & Clutter, R.I. (1965) Avoidance of towed nets by zooplankton. *Oceanography and Limnology,* **10,** 96–104.

Gaston, K.J., Blackburn, T.M., & Gregory, R.D. (1999) Does variation in census area confound density comparisons? *Journal of Applied Ecology,* **36,** 191–204.

Gurevitch, J. & Hedges, L.V. (1999) Statistical issues in ecological meta-analysis. *Ecology,* **80,** 1142–1149.

Harlow, L.L., Muliak, S.A., & Steiger, J.H. (eds.) (1997) *What If There Were No Significance Tests?* Lawrence Erlbaum. Mahwah, NJ.

Hattis, D. & Burmaster, D.E. (1994) Assessment of variability and uncertainty distributions for practical risk analyses. *Risk Analysis,* **14,** 713–730.

Hester, F.E. & Dendy, J.S. (1962) A multiple sampler for aquatic macroinvertebrates. *Transactions of the American Fisheries Society,* **91,** 420–421.

Hilborn, R. & Mangel, M. (1997) *The Ecological Detective.* Princeton University Press, Princeton.

Jasienski, M. & Bazzaz, F.A. (1999) The fallacy of ratios and the testability of models in biology. *Oikos,* **84,** 321–326.

Johnson, C.N. (1999) Relationships between body size and population density of animals: the problem of the scaling of study area in relation to body size. *Oikos,* **85,** 565–569.

Kirchner, T.B. (1998) Time, space, variability, and uncertainty. In *Risk Assessment: Logic and Measurement* (ed. M.C. Newman and C.L. Strojan), pp. 303–323. Ann Arbor Press, Chelsea, MI.

Krebs, C.J. (1989) *Ecological Methodology.* HarperCollins, New York.

Lamberti, G.A. & Resh, V.H. (1985) Comparability of introduced tiles and natural substrates for sampling lotic bacteria, algae and macroinvertebrates. *Freshwater Biology,* **15,** 21–30.

Lima, S.L. (1998) Nonlethal effects in the ecology of predator:prey interactions. *Bioscience,* **48,** 25–34.

Magurran, A.E. (1988) *Ecological Diversity and its Measurement.* Croom Helm, London.

Manly, B.F.J. (1990) *Stage-Structured Populations: Sampling, Analysis and Simulation.* Chapman & Hall, London.

Manly, B.F.J. (1997) A method for the estimation of parameters for natural stage-structured populations. *Researches on Population Ecology,* **39,** 101–111.

Martinez, N.D., Hawkins, B.A., Dawah, H.A., & Feifarek, B.P. (1999) Effects of sampling effort on characterization of food-web structure. *Ecology,* **80,** 1044–1055.

Matthaei, C.D., Werthmuller, D., & Frutiger, A. (1998). An update on the quantification of stream drift. *Archives of Hydrobiology,* **143,** 1–19.

May, R.M. (1975) Patterns of species abundance and diversity. In *Ecology and Evolution of Communities* (ed. M.L. Cody & J.L. Diamond), pp. 81–120. Belknap, Harvard.

McArdle, B.H. (1990) When are rare species not there? *Oikos,* **57,** 276–277.

Merritt, R.W., Resh, V.H., & Cummins, K.W. (1996) Design of aquatic insect studies: collecting, sampling, and rearing procedures. In *An Introduction to the Aquatic Insects of North America* (ed. R.W. Merritt & K.W. Cummins), pp. 12–28. Kendall/Hunt, Dubuque, IA.

Morin, P.J., Lawler, S.P., & Johnson, E.A. (1988) Competition between aquatic insects and vertebrates: interaction strength and higher-order interactions. *Ecology,* **69,** 1401–1409.

Murphy, K.R. & Myors, B. (1998) *Statistical Power Analysis: a Simple and General Model for Traditional and Modern Hypothesis Tests.* Lawrence Erlbaum, Mahwah, NJ.

Muscha, M.J., Zimmer, K.D., Butler, M.G., & Hanson, M.A. (2001) A comparison of horizontally and vertically deployed aquatic invertebrate activity traps. *Wetlands,* **21,** 301–307.

Nelsen, R.B. (1986) Properties of a one-parameter family of bivariate distributions with specified marginals. *Communications in Statistics: Theory and Methods*, **A15**, 3277–3285.

Nelsen, R.B. (1987) Discrete bivariate distributions with given marginals and correlation. *Communications in Statistics: Simulation and Computation*, **B16**, 199–208.

New, T.R. (1998) *Invertebrate Surveys for Conservation*. Oxford University Press, Oxford.

Norris, R.H., McElravy, E.P., & Resh, V.H. (1992) The sampling problem. In *Rivers Handbook* (ed. P. Calow & G.E. Petts), pp. 282–306. Blackwell, Oxford.

Nürnberger, B. (1996) Local dynamics and dispersal in a structured population of the whirligig beetle *Dineutus assimilis*. *Oecologia*, **106**, 325–336.

Osenberg, C.W., Sarnelle, O., & Goldberg, D.E. (1999) Meta-analysis in ecology: concepts, statistics, and applications. *Ecology*, **80**, 1103–1104.

Pearman, P.B. (1995) Effects of pond size and consequent predator density on two species of tadpoles. *Oecologia*, **102**, 1–8.

Petersen, J.E., Cornwell, J.C., & Kemp, W.M. (1999) Implicit scaling in the design of experimental aquatic ecosystems. *Oikos*, **85**, 3–18.

Polis, G.A. (1991) Complex trophic interactions in deserts: an empirical critique of food web theory. *American Naturalist*, **138**, 123–155.

Prairie, Y.T. & Bird, D.F. (1989) Some misconceptions about the spurious correlation problem in the ecological literature. *Oecologia*, **81**, 285–288.

Puttman, R.J. (1995) Ethical considerations and animal welfare in ecological field studies. *Biodiversity and Conservation*, **4**, 903–915.

Resetarits, W.J. (2001) Colonization under threat of predation: avoidance of fish by an aquatic beetle, *Tropisternus lateralis* (Coleoptera: Hydrophilidae). *Oecologia*, **129**, 155–160.

Resh, V.H., Feminella, J.W., & McElravy, E.P. (1990) *Sampling Aquatic Insects*. Videotape. Office of Media Services. University of California, Berkeley, CA.

Rohlf, F.J. & Sokal, R.R. (1995) *Statistical Tables*. W.H. Freeman, New York.

Rosenberg, D.M. & Resh, V.H. (1982) The use of artificial substrates in the study of freshwater benthic macroinvertebrates. In *Artificial Substrates* (ed. J. Cairns Jr.), pp. 175–235. Ann Arbor Science Publishers/Butterworth Group, Ann Arbor, MI.

Rowe, W.D. (1994) Understanding uncertainty. *Risk Analysis*, **14**, 743–750.

Rutherford, J.E. (1995) Patterns of dispersion of aquatic insects colonizing artificial substrates in a southern Ontario stream. *Canadian Journal of Zoology*, **73**, 458–468.

Rutherford, J.E. & MacKay, R.J. (1985) The vertical distribution of hydropsychid larvae and pupae (Trichoptera : Hydropsychidae) in stream substrates. *Canadian Journal of Zoology*, **63**, 1306–1355.

Rutledge, P.A. & Fleeger, J.W. (1988) Laboratory studies on core sampling with application to subtidal meiobenthos collection. *Limnology and Oceanography*, **33**, 274–280.

Service, M.W. (1993) *Mosquito Ecology: Field Sampling Methods*. Elsevier, London.

Shlyakhter, A.I. & Kammen, D.M. (1992) Sea-level rise or fall? *Nature*, **357**, 25.

Siemann, E., Tilman, D., & Haarstad, J. (1999) Abundance, diversity and body size: patterns from a grassland arthropod community. *Journal of Animal Ecology*, **68**, 824–835.

Sokal, R.R. & Rohlf, F.J. (1995) *Biometry*. W.H. Freeman, New York.

Southwood, T.R.E. (1978) *Ecological Methods*. Chapman & Hall, London.

Stewart, R.J. & Schaefer, C.H. (1983) The relationship between dipper counts and the absolute density of *Culex tarsalis* larvae and pupae in rice fields. *Mosquito News*, **43**, 129–135.

Sugihara, G. (1980). Minimal community structure: an explanation of species abundance patterns. *The American Naturalist*, **116**, 770–787.

Svensson, B.W. (1999) Environmental heterogeneity in space and time: patch use, recruitment and dynamics of a rock pool population of a gyrinid beetle. *Oikos*, **84**, 227–238.

Taylor, J.R. (1982) *An Introduction to Error Analysis*. University Science Books, Sausalito, CA.

Trexler, J.D., Apperson, C.S., & Schal, C. (1998) Laboratory and field evaluations of oviposition responses of *Aedes albopictus* and *Aedes triseriatus* (Diptera: Culicidae) to oak leaf infusion. *Journal of Medical Entomology*, **35**, 967–976.

Turner, A.M. & Trexler, J.C. (1997) Sampling aquatic invertebrates from marshes: evaluating the options. *Journal of the North American Benthological Society*, **16**, 694–709.

Vance, G.M., VanDyk, J.K., & Rowley, W.A. 1995. A device for sampling aquatic insects associated with carrion in water. *Journal of Forensic Sciences*, **40**, 479–482.

Washino, R.K. & Hokama, Y. (1968) Quantitative sampling of aquatic insects in a shallow-water habitat. *Annals of the Entomological Society of America*, **61**, 785–786.

Welsh, A.H., Peterson, A.T., & Altmann, S.A. (1988) The fallacy of averages. *American Naturalist*, **132**, 277–288.

Wilcox, C. 2001. Habitat size and isolation affect colonization of seasonal wetlands by predatory aquatic insects. *Israel Journal of Zoology*, **47**, 459–475.

Wood, S.N. (1997) Inverse problems and structured-population dynamics. In *Structured-Population Models in Marine, Terrestrial and Freshwater Systems* (ed. S. Tuljapurkar & H. Caswell), pp. 555–586. Chapman & Hall, New York.

Zar, J.H. (1984) *Biostatistical Analysis*. Prentice-Hall, Englewood Cliffs, NJ.

Index of methods and approaches

Topic	Examples
Survey of sampling devices	Active vs. passive operator, absolute vs. relative estimates.
Kinds of sampling errors	Systematic bias, random errors, measurement interactions.
Sampling efficiency	Absolute and relative estimates, abiotic and biotic factors affecting sampling efficiency, calibration curves, number of samples needed.
Ethical issues	Avoid unnecessary killing of individuals and damage to habitats. Sampling can spread diseases.

Methodology	Uses
Active operator	
Sweep nets	Open water and flexible vegetation.
Dippers	Organisms close to surface in shallow water, e.g. immature mosquitoes.
	Small habitats e.g. water-filled tires, rock pools.
Surber/Hess	Benthos in lotic habitats.
Core samples	Soft benthos.
Grabs	Soft benthos.
Column samples	Water column above soft benthos.

Continued

Methodology	Uses
Passive operator	
Artificial substrates, e.g. rocks, tiles, artificial vegetation, Hester–Dendy	Rocky benthos, vegetated water bodies.
Mesocosms	Oviposition, dispersion.
Drift nets	Flowing water.
Minnow traps	Active organisms.
Emergence traps	Insects emerging from and leaving water.
Light traps	Night.
Food traps	Predators.
Oviposition attractant traps	Females.
Visual counts and photography	Small habitats, surface-dwelling organisms, benthic species in clear shallow water.

CHAPTER 10

Methods for sampling termites

DAVID T. JONES, ROBERT H.J. VERKERK, AND PAUL EGGLETON

Introduction

Termites (order Isoptera) are predominantly tropical in distribution. Their species richness is highest in lowland equatorial rain forests, and generally declines with increasing latitude (Collins 1983, Eggleton et al. 1994) and altitude (Gathorne-Hardy et al. 2001). Termite survival is limited by low temperatures and high aridity, and very few species occur beyond 45° latitude (Collins 1989). The forests of West Africa have the highest termite species richness, closely followed by South America, whereas the forests of Southeast Asia and Madagascar are considerably less diverse (Eggleton 2000, Davies et al. 2003). These regional diversity anomalies are also associated with significant differences in clade and functional diversity (Davies et al. 2003).

Termites are at the ecological center of many tropical ecosystems (Wilson 1992), and can achieve very high population densities. For example, in the forests of southern Cameroon, termites are one of the most numerous of all arthropod groups (Watt et al. 1997) with abundances of up to 10,000 per m², and live biomass densities up to 100 g per m² (Eggleton et al. 1996). Termites have a wide range of dietary, foraging, and nesting habits, with many species showing a high degree of resource specialization (Wood 1978, Collins 1989, Sleaford et al. 1996). The vast majority of species feed on dead plant material, while relatively few species feed on living plant tissue. On a humification gradient, from undecomposed dead wood and leaf-litter to humus in the soil, most detritiverous species consume material that occupies a relatively narrow range of the gradient (Donovan et al. 2001a). As the dominant arthropod detritivores, termites are important in decomposition processes (Wood & Sands 1978, Matsumoto & Abe 1979, Collins 1983) and play a central role as mediators of nutrient and carbon fluxes (Jones 1990, Lawton et al. 1996, Bignell et al. 1997, Tayasu et al. 1997, Sugimoto et al. 2000). Termite activity, such as tunneling, soil-feeding, and mound building, helps to maintain macropore structure, redistributes organic matter, and improves soil stability and quality (Lee & Wood 1971, Lobry de Bruyn & Conacher 1990, Black & Okwakol 1997, Holt & Lepage 2000, Donovan et al. 2001b). However, termites' influence on ecosystem processes at any site is likely to depend on the species composition and abundance of the local termite assemblage.

221

Approximately 2650 species of termites have been described to date (Kambhampati & Eggleton 2000), and less than 3 percent of these cause significant economic damage to buildings or related manmade structures (Pearce 1997). A similar proportion are serious pests of crops (Wood 1996). The termite fauna of urban environments is usually highly depauperate and characterized by wood-feeding species, unlike natural habitats that often support much greater species and functional diversity. For example, 136 species have been recorded in a single forest site in Cameroon, of which 73 percent are soil-feeders (Jones & Eggleton 2000).

Termite sampling methodologies have been discussed by Lee and Wood (1971), Baroni-Urbani et al. (1978), Nutting and Jones (1990), and Eggleton and Bignell (1995). Those reviews provide detailed results from numerous sampling studies, and are recommended as a rich source of referenced information on the subject. Our intention is not to duplicate those valuable reviews but to provide an overview of sampling methods and a framework for their applications, and to draw together recent developments in sampling technologies and strategies. In this chapter, we:

1 outline the difficulties encountered when sampling termites;
2 review all major sampling methods;
3 describe two sampling regimes that have been designed for use in tropical forests: one estimates the population density of local termite assemblages, the other is a rapid protocol for assessing species composition;
4 review sampling and monitoring methods for subterranean termite pests of buildings;
5 present a case study describing the methods used to monitor populations of an infestation of a pest species in England.

The difficulties of sampling termites

Being eusocial insects, termite colonies have a fixed location, and the sterile castes (workers and soldiers) are usually present throughout the year. Therefore, termites can be sampled directly, unlike many solitary and more mobile insects. However, effective sampling of termites presents considerable theoretical and practical problems. These problems stem from the very patchy spatial distribution of colonies and individuals within habitats, and the cryptic nature of most species. Sampling difficulties are at their most severe in tropical forests, where the structural complexity of the habitat combines with high levels of termite species richness, thus making many species difficult to find. Figure 10.1 shows schematically the complex distribution of termites in a West African forest, based on studies in southern Cameroon (Eggleton et al. 1995, 1996, Dibog 1998). The spatial distribution of termites in the forests of South America and Southeast Asia is shown schematically in Collins (1989).

Termites occupy a wide array of microhabitats, distributed vertically from

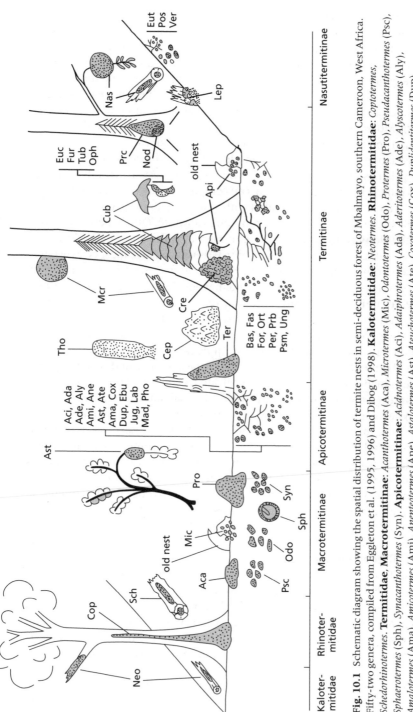

Fig. 10.1 Schematic diagram showing the spatial distribution of termite nests in semi-deciduous forest of Mbalmayo, southern Cameroon, West Africa. Fifty-two genera, compiled from Eggleton et al. (1995, 1996) and Dibog (1998). **Kalotermitidae**: *Neotermes*. **Rhinotermitidae**: *Coptotermes*, *Schedorhinotermes*. **Termitidae**. **Macrotermitinae**: *Acanthotermes* (Aca), *Microtermes* (Mic), *Odontotermes* (Odo), *Protermes* (Pro), *Pseudacanthotermes* (Psc), *Sphaerotermes* (Sph), *Synacanthotermes* (Syn). **Apicotermitinae**: *Acidnotermes* (Aci), *Adaiphrotermes* (Ada), *Aderitotermes* (Ade), *Alyscotermes* (Aly), *Amalotermes* (Ama), *Amicotermes* (Ami), *Anenteotermes* (Ane), *Astalotermes* (Ast), *Ateuchotermes* (Ate), *Coxotermes* (Cox), *Duplidentitermes* (Dup), *Eburnitermes* (Ebu), *Jugositermes* (Jug), *Labiotermes* (Lab), *Machadotermes* (Mad), *Phoxotermes* (Pho). **Termitinae**: *Apilitermes* (Api), *Basidentitermes* (Bas), *Cephalotermes* (Cep), *Crenetermes* (Cre), *Cubitermes* (Cub), *Euchilotermes* (Euc), *Fastigitermes* (Fas), *Foraminitermes* (For), *Furculitermes* (Fur), *Microcerotermes* (Mcr), *Noditermes* (Nod), *Ophiotermes* (Oph), *Orthotermes* (Ort), *Pericapritermes* (Per), *Proboscitermes* (Prb), *Procubitermes* (Prc), *Pseudomicrotermes* (Psm), *Termes* (Ter), *Thoracotermes* (Tho), *Tuberculitermes* (Tub), *Unguitermes* (Ung). **Nasutitermitinae**: *Eutermellus* (Eut), *Leptomyxotermes* (Lep), *Nasutitermes* (Nas), *Postsubulitermes* (Pos), *Verrucositermes* (Ver).

deep in the soil to the crowns of emergent trees. These numerous microhabitats represent real biological entities, but their exact limits can be difficult to define for the purposes of designing rigorous surveys. In practice, when sampling in the field, researchers have usually collected termites from one or more of four broad categories: mounds (epigeal nests that protrude above the surface of the soil), soil, dead wood at ground level, and arboreal habitats. Each category has distinct sampling problems and requires specific techniques (see *Sampling methods*). Moreover, the effort needed to gather statistically meaningful data using many of these methods can be very labor-intensive.

The variation in termite nest design (described by Noirot 1970), from simple diffuse galleries excavated in soil or wood, to the most structurally complex edifices built in the animal kingdom, can complicate sampling. In many species the nest is a single unit (called a calie) with a clearly delineated boundary, and is easily distinguishable from the surrounding substrate. Other species are polycalic: the colony is distributed among numerous calies but remains interconnected via subterranean tunnels or arboreal runways. Polycalic species occur in the lower termites, for example *Hodotermes* (Coaton & Sheasby 1975), *Schedorhinotermes* (Husseneder et al. 1998), and in wood-feeding species of Macrotermitinae, Termitinae, and Nasutitermitinae (e.g. Sands 1961, Holt & Easey 1985, Roisin & Pasteels 1986, Atkinson & Adams 1997). For practical reasons it may be impossible to locate all parts of a polycalic nest. The physical resilience of the nest fabric can also hinder sampling. Mounds can be very hard due to a high content of cemented soil in the matrix. In contrast, many wood-feeding species build nests made of carton material (masticated wood) that are more fragile.

Nests are not always restricted to a single microhabitat. The hypogeal (subterranean) nest of some species, for example *Macrotermes malaccensis* or *Prohamitermes mirabilis*, may sometimes protrude above the soil surface. Most incipient colonies begin life within dead wood or in the soil but some species may eventually develop a large and obvious epigeal nest. In a few cases, a single species can produce several nest types. For example, *Microcerotermes crassus* can build nests entirely within wood, exterior arboreal nests, hypogeal nests, or epigeal mounds, all within the same forest (Takematsu et al. 2003). Many species of Nasutitermitinae that produce exterior arboreal carton nests may often have a large proportion of the nest within the tree trunk or branch to which it is attached. All wood-nesting termites, however, except a few genera of Kalotermitidae and Termopsidae, maintain some association with the soil.

It can be difficult to verify the territorial limits of a colony because some species extend their foraging range far beyond the nest. Some *Macrotermes* maintain a complex network of subterranean tunnels (Darlington 1982), and others forage in exposed columns on the forest floor (Sugio 1995). *Hospitalitermes* is an extreme case in which the colony sends out soldiers and workers in processional columns that extend for up to 65 m across the forest floor, before ascending living trees to graze microepiphytes from the trunk and branches (Jones & Gathorne-Hardy 1995). *Longipeditermes longipes* is frequently observed feeding on leaf-litter on the forest floor (Matsumoto & Abe 1979, Collins 1984) but it was recently recorded in insecticidal fogging samples from tree canopies (Hoare & Jones 1998). The extent to which species that are assumed to be restricted to foraging at ground level may venture into arboreal habitats is unknown.

In some habitats, the size, number, and apparent dominance of mounds can give the impression that they are the most abundant and ecologically important part of the local termite assemblage. As a result, a majority of ecological studies in earlier decades were largely confined to mound-building species (Wood & Sands 1978). However, in the Mbalmayo Forest Reserve, southern Cameroon, only 12 percent of termite species in the local assemblage build epigeal or arboreal nests (Eggleton et al. 1996), implying that surveys limited to conspicuous mounds and nests will lead to serious underestimates of species richness. Nonetheless, mounds can contribute to the termite species richness of an area because they often harbor secondary occupants (called inquilines) as well as the mound-building species (Dejean & Ruelle 1995, Eggleton & Bignell 1997). Therefore, mounds should be checked carefully during diversity studies. Assemblage-level population studies based on mound sampling can also be inherently biased because in many habitats the mounds may not hold a significant proportion of termite abundance or biomass (Sands 1972). In Mbalmayo, less than 10 percent of the overall abundance was in mounds (Eggleton et al. 1995). As so many termites nest and forage in the soil, any assemblage-level study must adequately sample the soil.

Approaches to sampling

Much of the published data on termite species richness and population density in local assemblages are not strictly comparable because previous studies have used a variety of sampling methods and experimental designs, and different levels of collecting effort (Eggleton & Bignell 1995). As a consequence, there were limitations in the generalities and differences that could be inferred among study sites. However, with the systematic use of standardized sampling methods that more accurately characterize the structure of local termite assemblages (see *Two standardized methods for use in tropical forests*, below), the detailed structure of spatial patterns within and between regions is now being elucidated.

Any sampling regime will be a compromise between the specific questions the research is trying to address, and the available resources (time, money, labor, equipment, and taxonomic expertise). It may be relatively simple to study the biology of a single species in the field, but as the research is widened to include more parts of the local termite assemblage the sampling regime will become increasingly complicated. Sampling regimes can be designed to answer one of three distinct types of research question.

1 Population density. Studies aimed at measuring termite density may seek to estimate either the population within a single colony or the total abundance of

all species encountered in a unit area of habitat. Included in this category are studies aimed at estimating colony density by counting the number of mounds or nests per unit area.

2 Species composition. Studies aimed at investigating the species composition of an assemblage in a local area but without estimating population density.

3 Termite activity. Studies aimed at investigating activity such as foraging range, food preference and rates of consumption, or alate swarming.

These three categories require different approaches in sampling methods. However, some of the methods can be modified to answer questions in more than one category. Figure 10.2 offers a sequence of questions via which an appropriate sampling method can be chosen. This "decision tree" is meant only as a guide since local conditions and logistical considerations may impose other practical and statistical limitations when designing a sampling regime. All major sampling methods are discussed in the following section (summarized in the *Index of methods*, page 250).

Sampling methods

Sampling mounds

Many studies have focused on mounds (e.g. see Lee & Wood 1971, Pomeroy 1977) because they are relatively easy to locate, and because they can be a dominant feature of the landscape, particularly in savannas. Baroni-Urbani et al. (1978) reviewed the use of aerial photographs, line transects, and various quadrat methods to measure the density of mounds. Very large intraspecific variations in mound density across seemingly homogenous savanna systems are often observed, as for example with *Cubitermes sankurensis* in central Africa (Mathot 1967). Spatial dispersion can be examined by mapping the location of mounds and then employing a nearest-neighbor technique (e.g. see Wood & Lee 1971, Schuurman & Dangerfield 1997, Meyer et al. 1999).

In the time it takes to dig into a large and strongly built mound, the disturbance can cause much of the population to evacuate the hive. To prevent this, the population within a mound can be killed *in situ* by fumigation with methyl bromide. The entire nest contents can then be excavated and the termites removed from the nest debris by flotation in water. The whole sampling process (described by Darlington 1984) is labor-intensive, and depending on the size of the mound it can take five laborers up to three weeks to complete. However, Darlington (1984) showed that sampling large mounds of *Macrotermes* without fumigation caused the population and biomass to be underestimated by up to an order of magnitude. *Macrotermes* mound parameters (both internal and external dimensions) and nest population are approximately linearly related. Therefore, survey data on mound size and density can be used to estimate abundance per unit area (Darlington & Dransfield 1987, Darlington 1990). Young

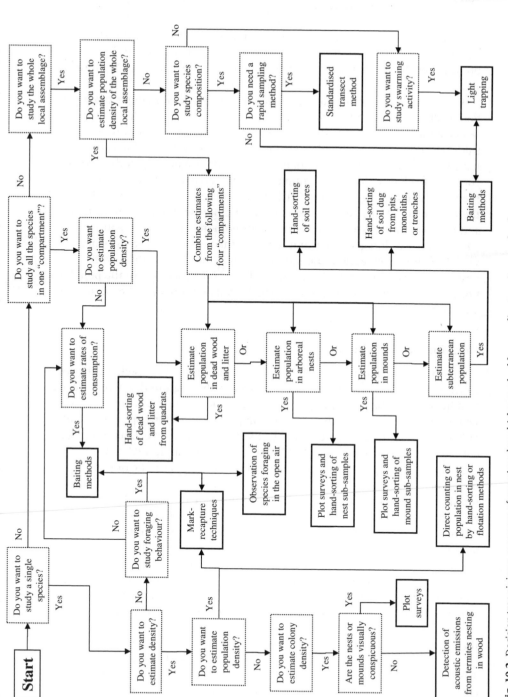

Fig. 10.2 Decision tree giving a sequence of questions by which a method for sampling termites may be chosen.

colonies will require separate number-to-biomass conversion factors because they can have different caste ratios, a higher proportion of larvae, and will often produce smaller individuals compared with mature colonies (Darlington 1991).

A quicker method of estimating population size is to take a sample of known volume from the mound, extract and count the termites, and multiply up to the total volume of the mound. Most mounds have irregular shapes, and therefore the total volume can be difficult to estimate. To overcome this problem, mounds have been measured as simple shapes, such as hemispheres (Sands 1961) or cones (Holt & Easey 1993). The sample of nest material must be taken quickly to prevent the termites retreating further into the mound. After breaking up the nest material, the termites are extracted either by flotation or by hand-sorting. Inaccuracies can arise because mound populations vary within each 24-hour cycle as foragers leave and return to the nest, and the density of individuals in any given area of the mound changes due to migration within the nest (Sands 1965). Ohiagu (1979) showed that more than half of the populations of four *Trinervitermes* species were in the soil rather than in the mounds at any one time.

Sampling termites in soil

Many species are restricted to the soil, with both nest and foraging galleries concealed underground without any indication above ground. Subterranean termites can only be sampled by removing units of soil and extracting the individuals by some method. Therefore, the important questions are what depth and volume of sampling unit should be used, and how many units should be collected? Baroni-Urbani et al. (1978) discuss these questions in detail.

In forests, subterranean termites characteristically occur in the organic layer of the soil profile (Collins 1989). A study in Malaysian rain forest reported that termites were mainly found in the top 15 cm and were rare below 25 cm (Abe & Matsumoto 1979). In Nigerian riparian forest, most termites were in the top 25 cm of the soil profile and showed no significant difference in abundance across seasons (Wood et al. 1982). However, in drier or more seasonal habitats the issue of sampling depth is complicated by vertical migration. Vertical distribution varies with species, soil type, and season, and no general correction factors can be applied (Wood & Sands 1978). In cultivated systems derived from woodland and savanna in west Africa, *Microtermes* is an abundant pest. In the wet season, *Microtermes* were usually concentrated in the upper 25 cm, whereas the proportion of the population below 50 cm greatly increased in the dry season as they moved deeper in the soil (Wood & Johnson 1978, Black & Wood 1989). This effect is probably less pronounced in tropical forests where the canopy limits fluctuations in soil temperature, but movement of termites may also be linked to rainfall events. In a seasonal humid forest in Cameroon, Dibog et al. (1998) found that both species richness and abundance in 10 cm deep soil samples were generally higher in dry periods compared with wet periods. However, no significant changes in overall species composition were observed.

Studies have used numerous sampling volumes, ranging from excavating very large pits or long, narrow trenches, to small soil cores. For example, Abe and Matsumoto (1979) dug one pit of area 1 m × 2 m to a depth of 25 cm, carefully removing and sorting the soil in smaller sub-units, whereas Collins (1979a) and Wood et al. (1982) took hundreds of soil cores of 10 cm diameter. Eggleton and Bignell (1995) outline the trade-offs involved in using different sampling sizes. Due to the highly heterogenous spatial distribution of termites, most small to medium sized soil samples will contain relatively low numbers of individuals or none at all, while a very few samples may have extremely high numbers if a nest or foraging party is encountered. Density estimates can therefore have high variance, making it difficult to demonstrate statistically significant differences among sites.

Hand-sorting is often recommended for extracting termites from soil samples. Wood et al. (1977) reported that 78–92 percent of all termites in soil samples were collected by hand-sorting, and applied a 12 percent loss factor in subsequent field studies (Wood et al. 1982). The technique is simple, requiring only trays on which to examine the soil and forceps for removing the termites. Termites can also be extracted by flotation methods (Strickland 1944, Salt 1952, Madge 1969) but Wood et al. (1977) considered these too time-consuming to be practical. Automatic extraction devices such as Berlese–Tullgren funnels are less suitable because the termites often die *in situ* as the soil dries out. In comparison, the Kempson extractor is better at removing soft-bodied invertebrates from soil samples because it is equipped with a thermostat that allows subtle temperature and moisture gradients to be maintained through the sample (Adis 1987). After testing the Kempson extractor, Silva and Martius (2000) suggested that it was as effective at removing termites as hand-sorting. However, their results were not statistically conclusive because the sample sizes were very small. The use of the Kempson extractor is impractical at some study sites because it needs about 15 days of continuous electricity supply.

Sampling termites in wood

Species that feed and nest within dead logs and branches can have huge populations. Individuals can be extremely difficult to dislodge from narrow galleries. For quantitative studies the only effective extraction method is to split the wood lengthways and remove the termites manually. Failure to sample larger items of dead wood may severely underestimate termite abundance (Collins 1983, Eggleton & Bignell 1995, Eggleton et al. 1996). The population in larger items of dead wood can be estimated by sampling sub-units by volume or weight and assuming a uniform population density throughout. However, few researchers have attempted to sample populations quantitatively in dead wood (see the standardized population sampling regime, below, for an example). No non-destructive techniques exist for estimating the size of colonies inhabiting living trees. Greaves (1967) describes methods for felling living trees and estimating the inhabiting populations of *Coptotermes*.

Collins (1983) and Jones (1996) used a semi-quantitative counting method to estimate the population in dead wood at forest sites in Borneo. The method involved splitting open dead wood with a machete and visually estimating the number of individuals by counting in units of 10s, 100s, or 1000s. Although this method has not been calibrated against direct counts, the Collins and Jones estimates appear reasonable when compared with densities recorded at a similar forest in Borneo using more rigorous methods (Eggleton et al. 1999).

Sampling termites in arboreal habitats

Arboreal nests on the trunks of trees and attached to understory vegetation can be easily sampled up to a height of about 2 or 3 m above ground level. However, because it is difficult to gain access to nests above this height, no satisfactory quantitative methods have been devised for sampling termites in forest canopies. Present knowledge of arboreal termite diversity is limited, and based mainly on casual samples of dead wood and nest material removed from tree crowns while using rope climbing techniques, canopy walkways, or after trees have been felled. One study (Ellwood et al. 2002) has revealed that a high proportion of large, epiphytic birds' nest ferns (*Asplenium nidus* complex) in the canopy of a forest in Borneo contain nests of *Hospitalitermes*. Insecticidal fogging is not suitable for dislodging termites, because those affected by the insecticide usually remain inside their nests and foraging tunnels.

"Dry-wood" termites (Kalotermitidae) such as *Neotermes*, *Cryptotermes*, and *Glyptotermes* usually nest wholly within dead branches, and their colonies number from a few hundred individuals (Harris 1950) up to about ten thousand (Maki & Abe 1986). The frequency with which kalotermitid alates are caught in light traps (Rebello & Martius 1994, Medeiros et al. 1999) suggests that they may be a more significant component of forest assemblages than previously thought. Several species of *Coptotermes* can pipe the inside of living trees and leave no external evidence of their presence. Many arboreal termites that build external carton nests on trees also produce covered runways down the trunk. This allows researchers to identify the termites by scraping away the sheeting and collecting the foragers. Researchers can also find arboreal species that nest in dead wood if dead branches attached to trees at ground level are removed and examined, or if sufficient fallen dry dead branches are collected from under tree crowns.

Sampling termites using baits

Cellulose baits simulate natural items of food such as fallen dead wood. The two most commonly used baits are wooden stakes and rolls of unscented toilet paper. Other materials have been used, such as translocated dungpats (de Souza 1993) and sawdust (Abensperg-Traun 1993). Baits are often set in grid formation (so-called "graveyard" trials), and in some cases the area is first cleared of naturally occurring fallen wood (Haverty et al. 1975). Wooden baits are usually cut from timber known to be susceptible to the local wood-feeding termites. However, field experiments have been ruined by neglecting to check whether purchased timber has been treated with insecticide. Stakes are driven into the ground and the top is left protruding above the soil to facilitate monitoring. Alternatively, baits may be laid out on the soil surface or buried (Sands 1972, Lenz et al. 1992, Dawes-Gromadzki 2003). Two stakes can be installed in contact with one another, as the interface between the stakes tends to encourage rapid colonization by subterranean termites. Researchers should not remove or disturb baits too frequently, as this will discourage termites from foraging on the bait. Usher and Ocloo (1974) tested the effect of stake size, shape, and position on the amount of damage caused by Macrotermitinae. They found that weight loss of wood increased as surface area increased, and that significantly more damage was recorded when stakes were completely rather than partially buried. For details of how baiting can be used to monitor and control pests species, see *Methods for sampling subterranean termite pests of buildings*, below.

Baits attract foraging termites, and therefore give estimates of relative intensity of foraging activity rather than relative population density. Baiting has been useful in studying inter- and intraspecific foraging activity (Sands 1972, Buxton 1981, Ferrar 1982, Pearce 1990, Pearce et al. 1990, Dawes-Gromadzki 2003), size of foraging territory (Haverty et al. 1975), and rates of food consumption (Haverty & Nutting 1974, Abe 1980). Baiting has also been used to estimate local species richness (de Souza 1993, Dangerfield & Mosugelo 1997, Taylor et al. 1998). However, this can be problematic because not all species are attracted to baits. Arboreal species that do not forage on the ground, and subterranean species that do not forage near the soil surface, may be excluded. Also, food preference trials have shown that not all termite species are attracted to the same bait materials (Haverty et al. 1976, Abensperg-Traun 1993, Dawes-Gromadzki 2003), implying that using a single bait type will under-sample the local species richness. While the degree of acceptance of cellulose baits depends on a variety of factors, it is notable that de Souza (1993) attracted 41 species (including soil-feeding species) using rolls of toilet paper when studying termite community structure in Brazilian cerrado.

Sampling using mark–recapture protocols

Population size can be estimated by mark–recapture protocols using radioisotopes (Spragg & Paton 1977, Easey & Holt 1989) or insoluble colored stains and dyes (Su et al. 1988, 1991, Evans 1997). However, when tested on two mound-building species the mark–recapture estimates varied widely within and among colonies, and could be 100 times larger than direct population counts (Evans et al. 1998). These errors occurred because several of the assumptions inherent in the protocol were violated: the fat-stain markers faded quickly and were transferred to unmarked individuals; marked individuals did not mix uniformly with unmarked individuals; foragers displayed feeding site fidelity; and the likelihood of recapture differed between castes and instars (Evans et al. 1998). Similar problems were encountered when the method was applied to subterranean nesting termites (Su et al. 1993, Forschler & Townsend 1996, Thorne et al. 1996), suggesting that mark–recapture protocols are unable to provide accurate population estimates.

Markers may be useful, however, in studies attempting to delineate colony boundaries or foraging distances. Fluorescent dyes can either be incorporated into baits or applied to workers as a dust. Particles of the dust have been detected after 48 h in the guts of workers of the highly destructive Australian giant termite, *Mastotermes darwiniensis*, 95 m from the initial site of application (Miller 1993).

Sampling alates using traps

Alates (or imagoes: the winged reproductive forms) must leave the nest to mate, and out-crossing can only be achieved if colonies of the same species synchronize the release of alates. Swarming is often associated with annual weather patterns and ambient climatic conditions (Nutting 1969). However, the precise physical and physiological factors that trigger alate release are still uncertain (Medeiros et al. 1999). Passive trapping devices such as flight interception and Malaise traps are not suitable for sampling termites because they collect very few alates (Rebello & Martius 1994). Light traps can be used to sample alates but the results must be interpreted with caution because the technique has several problems. The position of the trap strongly influences the number of species and the abundance of alates caught, because of the poor dispersal range of termites (Martius et al. 1996). Not all species show a clear preference for nocturnal swarming (Mill 1983), and alates released during the day may not be caught in traps run overnight. Some species produce relatively few alates, and others may not produce alates every year.

Depending on the degree of seasonality at the site, light traps may catch alates throughout the year (Martius et al. 1996) or only during a limited number of months (Medeiros et al. 1999). Because there is little interspecific synchronicity of swarming, traps operating over short periods will fail to capture many local species. Medeiros et al. (1999) found that continuous trapping over one year in Atlantic rain forest in northeastern Brazil captured only 55 percent of the species previously recorded at the same site when collecting by hand. In contrast, light traps may be more useful in urban habitats for species-specific studies. For example, alates of the Formosan termite *Coptotermes formosanus* have been monitored in light traps in New Orleans (USA) over a seven-year period (1989–95). Mean data showed a consistent increase over this time, suggesting that the species can adapt to that specific urban environment (Henderson 1996).

Recording the movement of termites

Several genera (including *Hospitalitermes, Lacessititermes, Longipeditermes, Constrictotermes,* and *Macrotermes*) form processional columns of soldiers and workers that march in the open to feeding sites. In such cases, close-up photographs of the column taken at regular intervals can be used to estimate the number of termites involved in the foraging activity (Collins 1979b, Miura & Matsumoto 1998). However, this photographic technique can give large discrepancies between the number of termites leaving the nest and the number returning (Collins 1979b).

Hinze and Leuthold (1999) used two new techniques for detecting and recording the movement of workers inside a laboratory colony of *Macrotermes bellicosus*. A metal detector monitored the movement of workers marked with small pieces of metal wire, while a photo detector counted both marked and unmarked termites entering and leaving the nest and the queen cell.

Detection of acoustic emissions

Recently, handheld acoustic emission devices have been developed to detect the feeding of hidden termite infestations in wood. This non-destructive technique can differentiate between the acoustic emissions of termites and other wood-boring insects, and has been used successfully to detect pest species of Rhinotermitidae and Kalotermitidae in buildings (Weissling & Thoms 1999, Thoms 2000) and urban trees (Mankin et al. 2002). However, the accuracy with which acoustic emissions can be used to predict population density still has to be demonstrated. Furthermore, the efficacy of this technology at detecting a range of termite species in natural environments has not been tested.

Two standardized methods for use in tropical forests

A sampling regime for estimating termite assemblage population density

Few researchers have tried to document the population density of an entire local termite assemblage in a diverse tropical habitat because of the considerable effort involved. The following plot-based sampling regime is designed for estimating the population density of the local termite assemblage, excluding arboreal termites at more than 2 m above ground level. Eggleton et al. (1999) used this regime in Borneo, adapting a similar regime first used in Cameroon (Eggleton et al. 1996). The basic sampling area is a 0.25 ha plot (50 m × 50 m), with an internal grid (10 m separation) marked with string to facilitate quadrat placement and mapping. Within each plot, three sampling methods are employed:

1 Twenty quadrats (each 2 m × 2 m) are placed using random coordinates. Quadrats falling on standing trees or other large obstacles are reassigned to new random coordinates. All dead wood and litter is removed from each quadrat and hand-sorted on site just outside the plot by a team of trained assistants. Litter is searched and woody material is split open, and all termites removed. Larger items of dead wood are sub-sampled by volume.

2 After removing the wood and litter, a soil pit of 30 cm × 30 cm × 25 cm depth is dug in the center of each quadrat and hand-sorted on site.

3 A systematic survey of visible mounds and arboreal nests is carried out over the entire area of the plot (searching up to a height of 2 m), making use of the internal grid. Nests are mapped and destructively sampled. Nest populations are estimated by sub-sampling by weight (see Eggleton et al. 1996).

The transect protocol

This transect-based protocol rapidly assesses the species composition of the local termite assemblage. The protocol, described by Jones and Eggleton (2000), was adapted from a similar method developed by Eggleton et al. (1996). The protocol has been used in many tropical forests around the world (Gathorne-Hardy et al. 2002, Davies et al. 2003).

The transect is 100 m long and 2 m wide, and divided into 20 contiguous sections (each 5 m × 2 m) and numbered sequentially. Two trained people sample each section for 30 minutes (a total of one hour of collecting per section). To standardize sampling effort, the collectors work steadily and continuously during each 30-minute period. In each section the following microhabitats are searched for termites: 12 samples of surface soil (each 12 cm × 12 cm, to 10 cm depth); accumulations of litter and humus at the base of trees and between buttress roots; the inside of dead tree stumps, logs, branches, and twigs; the soil within and beneath very rotten logs; all mounds and subterranean nests encountered (checking for inquiline species); arboreal nests, carton runways, and sheeting on vegetation up to a height of 2 m above ground level. The protocol allows the collectors to use their experience and judgment to search for and sample as many species in each section as time permits.

Jones and Eggleton (2000) tested this protocol in three forest sites where the local termite fauna was already comprehensively documented. Two transects were run at Danum Valley (Sabah, Borneo), one at Pasoh Forest Reserve (Peninsular Malaysia), and one at Mbalmayo Forest Reserve (Cameroon). At the three sites the transect samples contained 31 to 36 percent of the known local termite species pool (Table 10.1), giving a reasonably high degree of sampling consistency among sites. The taxonomic group composition (the proportion of species in each family, or subfamily in the case of the Termitidae) of the transect samples did not differ significantly from that of the known local fauna

Table 10.1 The number of termite species collected from transects in three forest sites (Jones & Eggleton 2000). The total number of known species recorded from each site is based on all available records, from labor-intensive sampling programs to casual collecting. These totals represent the best estimates of the species richness of each assemblage. Reproduced with permission from Blackwell Publishing Ltd.

Site	Species sampled in transect	Total known species	Proportion of total fauna in transect
Danum Valley, Sabah, Borneo (transect 1)	29	93	31.2%
Danum Valley, Sabah, Borneo (transect 2)	33	93	35.5%
Pasoh, Malaysia	29	80	36.3%
Mbalmayo, Cameroon	47	136	34.6%

at each site. Similarly, the functional group composition (the proportion of species in each feeding group) of the transect samples did not differ significantly from that of the known local fauna. In addition, the two transects run at Danum Valley gave very similar patterns, suggesting that the protocol produces consistent within-site results. One supervised training transect was shown to be sufficient experience to ensure that collectors were sampling to the level of efficiency that the protocol required.

Comparison of the two standardized methods

Although the two methods were designed to address different questions, it is useful to compare their relative merits. In tropical forests, the population sampling regime underestimates local species richness. This is because its strictly defined and prescriptive method only samples dead wood, termite nests, and a limited number of soil pits. In comparison, the transect protocol utilizes the expertise of the collector to search a wider array of suitable microhabitats within each section, thus increasing the likelihood of finding additional species. As a consequence, the transect accumulates species much more rapidly than population sampling (Fig. 10.3). Both methods avoid microhabitats above 2 m, but the transect protocol often collects arboreal nesting species that forage at ground level.

The population sampling regime is labor-intensive, and estimates of sampling efficiency (Table 10.2) suggest that it takes about four to five times more effort to obtain and identify roughly the same number of species as one transect. It should be noted that population sampling generates more specimens than the transect method and thus requires far greater taxonomic processing time for the

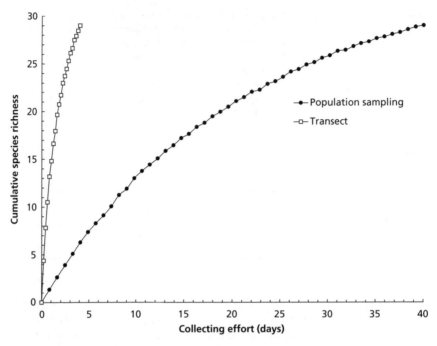

Fig. 10.3 Species accumulation curves showing the cumulative richness produced by sampling one plot using the population sampling method, and one transect in primary forest at Danum Valley (Sabah, Borneo). Cumulative richness is plotted against the collecting effort measured in person days (transect = 4 days; population sampling = 40 days). The cumulative totals were based on the smallest sampling unit for which species-level data were available. For the transect each sampling unit was one section, while for the population sampling this represents 20 soil pits, 20 dead wood quadrats, and 9 mounds (see text for description of methods). The curves are the mean of 500 random sequences of these units. After Jones & Eggleton 2000; reproduced with permission from Blackwell Publishing Ltd.

same number of species. Therefore, the transect protocol provides a much more rapid and cost-effective method for studying termite assemblage structure than population sampling regimes. The population sampling regime does, however, produce reliable estimates of termite population density and biomass (Eggleton et al. 1996) that can be used to quantify the impact of termites on ecosystem processes such as carbon fluxes (Bignell et al. 1997).

Methods for sampling subterranean termite pests of buildings

In regions of the world where termites cause major economic losses to buildings and associated structures, knowledge of the species present is an essential

Table 10.2 A comparison of the approximate effort required to conduct transects and plot-based population sampling regimes, the cumulative number of termite species collected, and the sampling efficiency of both methods, in forest at Danum Valley (Sabah, Borneo) and Mbalmayo (Cameroon) (Jones & Eggleton 2000). Sampling efficiency is defined as the number of species collected per unit effort, where effort is measured as the number of person-days required to collect and process the samples. Taxonomic processing is the time taken for one expert to sort and identify specimens, and in the case of the population sampling, to count specimens. See text for description of the population sampling methods. Reproduced with permission from Blackwell Publishing Ltd.

Site	Sampling method	Collecting time (days)	Taxonomic processing (days)	Total effort (days)	Cumulative number of species	Sampling efficiency (number of species collected per day)
Danum Valley	1 transect	4	8	12	29	2.42
	2 transects	8	16	24	40	1.67
	1 plot	40	20	60	29	0.48
	2 plots	80	40	120	38	0.32
	3 plots	120	60	180	47	0.26
Mbalmayo	1 transect	4	12	16	47	2.94
	1 plot	20	15	35	28	0.80
	2 plots	40	30	70	48	0.69

prerequisite to any management or colony elimination program. The vast majority of damage to buildings caused by termites worldwide is attributed to wood-feeding Rhinotermitidae from only two genera, *Coptotermes* and *Reticulitermes*. Moreover, all Rhinotermitidae have a subterranean habit, meaning that individual colonies require ground contact or a more or less continuous moisture supply.

Sampling of subterranean termites in localities with perceived or known threats from subterranean pest species can be undertaken for several reasons, including:

1 to assess the presence or absence of termites and, if present, to make collections to allow taxonomic identifications;

2 to undertake qualitative or quantitative assessments of termite assemblages, including ecological studies;

3 to assess the extent and severity of a known infestation;

4 to monitor the fate of a population following the implementation of a management or colony elimination program;

5 to estimate the actual or relative population size of a given species within a prescribed area.

Basic methodologies

A wide variety of methods have been used to sample termites in the built environment. Baiting, mark–recapture, and light-trapping are reviewed above, while the methods outlined below are frequently used to sample subterranean pests of buildings.

Sampling of potential food sources and colony nest sites

Investigation for wood-feeding termites in potential food sources (e.g. in-ground timber and other cellulose sources) and colony nesting sites is often the starting point of most studies or management programs (Verkerk 1990, Verkerk & Bravery 2001). A standard range of equipment is required for such surveys, including: a bright torch; mirrors (e.g. dental type) for viewing into confined spaces; a ladder to gain access to roof voids and arboreal nesting sites; a large screwdriver (the handle can be used for "sounding" timbers and the tip for probing); a sharp knife for cutting plasterboard, carpets, etc.; levers for lifting carpets, architraves, etc.; hammer and nails for butting up trap doors in timber floors; vials and 70 percent ethyl alcohol, labels, a fine paintbrush, and forceps, for collecting and preserving specimens. More sophisticated devices such as acoustic monitors (Potter et al 2001, Mankin et al 2002) and endoscopes (Fuchs et al 2004) have been used effectively to facilitate detection of termites or nesting sites in concealed areas or within trees.

Timber stakes or dowels

One of the most common methods for assessment of termite activity in and around buildings is the insertion of timber stakes or dowels into the ground or into potential nesting sites. Timbers should be of a species and condition known to be susceptible to "pest" termite species in the given locality. Stakes are typically 500 mm in length, and 50 × 25 mm in section, and are usually cut to a point at their base to ease installation in hard ground (Verkerk & Bravery 2001). Dowels may be punched directly into nests and are particularly useful to determine if colony elimination has been successful (Peters & Fitzgerald 2003).

Corrugated cardboard traps

Various methods of sampling termites have employed corrugated cardboard (untreated with fungicide) as a feeding medium within monitoring systems in subterranean management programs. Three such methods are described.
1 Reservoirs of corrugated cardboard (within timber boxes with slits in the base, or within plastic or aluminum foil containers) set in the soil, can be used as termite traps. Dampened cardboard is placed within these reservoirs in layers so

that termites may be detected or collected during periodic inspections. Termites may be encouraged into the reservoirs by linking strips of corrugated paper to each reservoir (Kirton et al. 1998).

2 Lengths of ABS pipe (perforated or unperforated) can be filled with rolled, dampened corrugated cardboard and can be installed vertically in the soil, with a section remaining above ground for access (e.g. Myles 1996, Haverty et al. 1999). The cardboard may be removed for the purpose of collecting termites without disturbing the pipe/soil interface. The tops of the pipes should be covered adequately to stabilize the environment within each pipe.

3 Lengths of PVC electrical conduit (e.g. 25 or 32 mm diameter) with holes (5–8 mm diameter) at 100–150 mm intervals can be packed with dampened, rolled corrugated cardboard prior to being set in trenches (c.100 mm beneath the soil surface), which are then backfilled with soil. The pipe system can be made continuous by way of angled connectors, with provision for access in cardboard reservoir traps at prescribed intervals (e.g. 25 m) to allow inspection (Verkerk 1990, Verkerk & Bravery 2001). These traps may take the form of cardboard-filled perforated buckets (with lids) or other suitable containers, partially set into the ground, which should be linked directly to the pipe system. Polystyrene and black polyethylene sheeting can be fixed over the traps to help stabilize environmental conditions within the containers. The system is highly flexible and can be used in a wide range of circumstances, but is particularly useful around the perimeter of buildings or other structures. It can also be adapted for installation directly into known colony nests so that changes in activity patterns can be monitored.

Commercially available monitoring/baiting systems

Since the mid-1990s, various combined termite baiting and monitoring systems have been marketed in many industrialized countries with subterranean termite problems (e.g. USA, Japan, Australia, Spain, Italy, France). Examples of such systems include Sentricon Colony Elimination System®, Sentri Tech®, Exterra® and Termigard®. These devices rely on individual in-ground stations which contain a termite food source. Following detection of activity in individual stations during periodic inspections (e.g. at monthly intervals), a termiticidal bait (usually a relatively slow-acting chitin synthesis or metabolic inhibitor) is added. The bait is transferred though the colony by trophallaxis, and cases of successful elimination of field colonies have been reported (e.g. Su & Scheffrahn 1996, Haagsma & Bean 1998, Peters & Fitzgerald 1999, Verkerk & Bravery 2001, Peters & Fitzgerald 2003). The devices can be used in a variety of configurations either as monitoring devices or as combined monitoring and baiting devices. The design of the monitoring station is important in influencing both termite attack and the sustainability of activity (Lewis et al. 1998).

Remote monitoring

A remote, electronic, monitoring device which transmits a signal to a datalogger when termites break a silver foil circuit painted onto polyethylene sheeting, in turn fixed to timber stakes, has shown considerable promise (Su 2002). Such remote systems are likely to have greatest applicability in high-value heritage sites such as in ancient or historic buildings.

CASE STUDY: INTENSIVE TERMITE MONITORING AND BAITING PROGRAM: DEVON, UNITED KINGDOM
Adapted from Verkerk & Bravery 2001

In May 1998, an established infestation of a southern European subspecies of subterranean termite (*Reticulitermes lucifugus grassei* Clement) was found approximately 1000 km north of its indigenous distribution (northern Spain/southwestern France), in a semi-rural, coastal setting in Saunton, Devon, UK. Infestations of such termites generally are based on expansive, diffuse, and interconnected (frequently "open") colonies arising from large numbers of neotenics (Clement et al. 2001), so infestations are therefore difficult to eliminate completely. There is some evidence that the termites were imported accidentally, possibly more than 30 years previously (Jenkins et al. 2001). Surveys and monitoring revealed a discrete, highly localized infestation extending over some 2400 m². Within this zone, two timber-framed houses surrounded by paving, outbuildings, mixed woodland, bracken, lawns, and gardens were affected. This case study briefly describes the monitoring systems which were implemented following the launch in June 1998 of a government-funded, consortium-based, 12-year program, the goal of which was to eliminate the infestation. At the time of writing, intensive monitoring has revealed the site to be free from termite activity for three and a half years, suggesting the program's goal may have been achieved. Monitoring as part of the government program will continue for 10 years from cessation of known activity.

Phase 1: initiation
Three key activities were undertaken during the first two months of the program:

General surveys and establishment of treatment, intensive monitoring and buffer zones
The mid-point of the north–south boundary between the two properties known to be infested was used to define the central point of the designated "eradication zone." This 1000 m diameter zone (Fig. 10.4) covers 29 independently owned properties concentrated on either side of the Saunton Road (traversing east–west through the center of the eradication zone). All 29 properties were surveyed using torches, probes, and timber "sounding" techniques for evidence of termites, with particular attention being paid to ground-floor (and sub-floor) areas, exteriors of buildings, outbuildings, trees and stumps, and other areas where evidence of subterranean termites was most likely to be detected if present (see *Sampling of potential food sources and colony nest sites*, above).

Detailed inspections of properties within the treatment zone
A 75 m radius "treatment" zone was designated, and properties and their grounds within this zone were subjected to detailed examination of all timbers susceptible to sub-

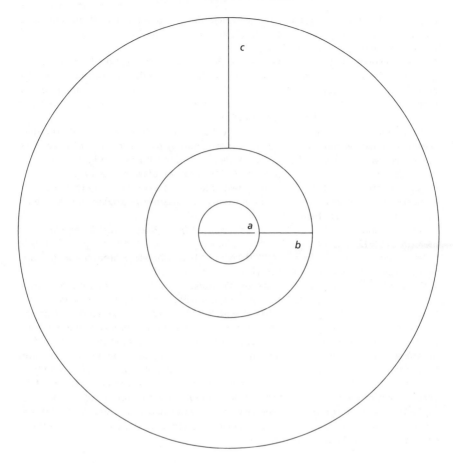

Fig. 10.4 Zoning system adopted for the UK Termite Eradication Programme (years 1–3). (a) = treatment zone (75 m radius from center point of known infestation); (b) = intensive monitoring zone (75–200 m radius from center point); (c) = buffer zone (200–500 m radius from center point). Any termite activity detected outside the treatment zone would have resulted in appropriate enlargement of all zones. Re-evaluation of the zoning system occurs annually.

terranean termite attack. The termite infestation as revealed by these surveys was found to extend over an approximately rectangular area of c.70 m length (east–west) and c.30 m width (north–south). Samples of termites were collected and the subspecies was identified and confirmed following cuticular hydrocarbon analyses by two independent laboratories in France. Samples were subsequently subjected to mitochondrial DNA and phylogenetic analyses (Jenkins et al 2001).

Installation of monitoring devices
The "intensive monitoring" zone was established between 75 m and 200 m radius from the notional center of the eradication zone (Fig. 10.4). A "buffer" zone was designated

from 200–500 m radius. In addition to the installation of 160 commercial monitoring/baiting devices (Sentricon Colony Elimination System® and Sentri Tech®), four other types of monitoring device were installed:

1 Wooden stakes. These were prepared from Scots pine *Pinus sylvestris* sapwood and measured 50 × 25 mm section × 500 mm length. Approximately 1000 were installed 3 m apart in more-or-less concentric rows with 10 m between rows throughout the intensive monitoring zone. In both the intensive monitoring and buffer zones, stakes were installed (at 2–3 m intervals) around all buildings and outbuildings.

2 An underground, perimeter "pipe and bucket" system. This comprised corrugated-cardboard-filled, perforated uPVC pipe (32 mm diameter) installed c.100 mm beneath the soil surface with corrugated-cardboard-filled, perforated bucket stations at approximately 25 m intervals around the perimeter of the grounds of both infested properties to provide a continuous detection system to help determine if termites reached these boundaries. The attractiveness of the cardboard was tested and confirmed by separate tests with termites feeding within the treatment zone.

3 Timber monitoring slates. These were prepared from *Pinus sylvestris* sapwood and measured 25 × 12 mm section × 180 mm length. They were installed into approximately 60 holes (25 mm diameter) drilled to reach the ground beneath specific areas of concrete and paving around the two main buildings known to be infested.

4 Conduit-based, grid monitoring system. On apparent collapse of the termite infestation (mid-1999) following installation of a bespoke treatment system based on impregnation of *Pinus sylvestris* sapwood "wafers" with hexaflumuron in an aqueous extract of a fungal attractant (*Gloeophyllum trabeum* Pers ex Fr), a supplementary grid monitoring system (at 3 m centers) was installed. This vertically orientated monitoring system (1 m depth) comprised two pairs of 500 mm length *Pinus sylvestris* sapwood timber slates (upper and lower pairs, each linked with cable ties), housed in heavily perforated (5 mm diameter) conduits (32 mm diameter). The devices were set at 3 m centers and covered the entire extent of the known infestation area. The upper (open) end of each conduit was capped flush with the ground or paving levels. The system allows for treatment of infestations by substitution of the upper pair of devices with timber-based baits, should activity be detected in individual monitoring devices.

Phase 2: monthly monitoring and baiting

Monthly monitoring began in the third month of the program, for the purpose of checking on activity in all devices and attaching commercial baits containing hexaflumuron (chitin synthesis inhibitor) (Recruit® or Sentry®) as necessary to active bait stations or constructional timbers. Termite abundance was assessed in relative terms according to the number of stations showing activity, and by using an arbitrary index of activity at each device where termites were present. Termite recruitment into bait stations was found to be very satisfactory, with 17 percent and 22 percent recruitment into the Sentri Tech® stations (92 in total) in July and August 1998 (one and two months respectively following installation of the stations). However, evidence of apparent avoidance behavior by the termites was noted by September 1998 in nearly all cases where commercial baits had been deployed, with no subsequent bait consumption. In addition, by October 1998, the area of termite activity detected by the available monitoring devices had extended to approximately 85 m along the east–west axis.

Phase 3: bait refinement phase

The indications of bait avoidance prompted an intensive laboratory-based investigation using cultures of *R. lucifugus grassei* (Devon strain collected from the field) as well as a less sensitive laboratory strain of *R. santonensis* for comparative purposes. The work culminated in the development of novel timber-based baits which were attractive to the target strain. Based on this work, a prophylactic baiting program was launched in early 1999.

Phase 4: prophylactic baiting program

By August 1999, 153 timber-based bait devices and 256 untreated monitoring devices had been installed in the treatment zone. Analysis of the extent of activity at common points in August 1998 and August 1999 suggested that the termite population at Saunton appeared to have been suppressed by at least 90 percent following the installation of the prophylactic system. However, to counter the possibility that apparent suppression had been caused by further bait avoidance, the supplementary grid monitoring system was installed within the treatment zone.

For two years monitoring visits continued monthly, with the exception of the "closed-season" months of December and January. By April 2000 a total of 719 devices had been installed in the treatment zone alone, 535 of these being untreated monitoring devices and 177 being hexaflumuron-impregnated timber-based baits. In 2000, all treated devices were removed from the site and replaced with untreated devices. Ongoing monitoring in the 27 other properties within the eradication zone has continued to reveal no evidence of termite activity. Within the treatment zone, activity was last detected in a structural, above-ground timber in August 2000. With no activity detected in any device or substrate for three and a half years, inspection frequencies have now been reduced to two occasions a year (early and late season respectively).

Acknowledgments

We wish to thank the UK Office of the Deputy Prime Minister (ODPM) for financial and logistical support, the program Steering Group for its advice, as well as the members of the project consortium, including: ODPM; Building Research Establishment Ltd.; Imperial College London; Natural History Museum, London; Dow AgroSciences; UK Forestry Commission; and the Natural Resources Institute (University of Greenwich). Further thanks are offered to the residents of Saunton who have cooperated with the program so successfully.

References

Abe, T. (1980) Studies on the distribution and ecological role of termites in a lowland rain forest of West Malaysia. 4. The role of termites in the process of wood decomposition in Pasoh Forest Reserve. *Revue d'Ecologie et de Biologie du Sol*, **17**, 23–40.

Abe, T. & Matsumoto, T. (1979) Studies on the distribution and ecological role of termites in a lowland rain forest of West Malaysia. 3. Distribution and abundance of termites in Pasoh Forest Reserve. *Japanese Journal of Ecology*, **29**, 337–351.

Abensperg-Traun, M. (1993) A comparison of 2 methods for sampling assemblages of subterranean, wood-eating termites (Isoptera). *Australian Journal of Ecology*, **18**, 317–324.

Adis, J. (1987) Extraction of arthropods from neotropical soils with a modified Kempson apparatus. *Journal of Tropical Ecology*, **3**, 131–138.

Atkinson, L. & Adams, E.S. (1997) The origins and relatedness of multiple reproductives in colonies of the termite *Nasutitermes corniger*. *Proceedings of the Royal Society of London Series B: Biological Sciences*, **264**, 1131–1136.

Baroni-Urbani, C., Josens, G., & Peakin, G.J. (1978) Empirical data and demographic parameters. In *Production Ecology of Ants and Termites* (ed. M.V. Brian), pp. 5–44. Cambridge University Press, Cambridge.

Bignell, D.E., Eggleton, P., Nunes, L., & Thomas, K.L. (1997) Termites as mediators of carbon fluxes in tropical forest: budgets for carbon dioxide and methane emissions. In *Forests and Insects* (ed. A.D. Watt, N.E. Stork, & M.D. Hunter), pp. 109–134. Chapman & Hall, London.

Black, H.I.J. & Okwakol, M.J.N. (1997) Agricultural intensification, soil biodiversity and agroecosystem function in the tropics: the role of termites. *Applied Soil Ecology*, **6**, 37–53.

Black, H.I.J. & Wood, T.G. (1989) The effects of cultivation on the vertical distribution of *Microtermes* spp. (Isoptera: Termitinae: Macrotermitinae) in soil at Mokwa, Nigeria. *Sociobiology*, **15**, 133–138.

Buxton, R.D. (1981) Changes in the composition and activities of termite communities in relation to changing rainfall. *Oecologia*, **51**, 371–378.

Clement, J.L., Bagneres, A.G., Uva, P., et al. (2001) Biosystematics of *Reticulitermes termites* in Europe: morphological, chemical and molecular data. *Insectes Sociaux*, **48**, 202–215.

Coaton, W.G.H. & Sheasby, J.L. (1975) National survey of the Isoptera of Southern Africa. 10. The genus *Hodotermes* Hagen (Hodotermitidae). *Cimbebasia*, **3 (10)**, 105–138.

Collins, N.M. (1979a) A comparison of the soil macrofauna of three lowland forest types in Sarawak. *Sarawak Museum Journal*, **27**, 267–282.

Collins, N.M. (1979b) Observations on the foraging activity of *Hospitalitermes umbrinus* (Haviland), (Isoptera: Termitidae) in the Gunung Mulu National Park, Sarawak. *Ecological Entomology*, **4**, 231–238.

Collins, N.M. (1983) Termite populations and their role in litter removal in Malaysian rain forests. In *Tropical Rain Forest: Ecology and Management* (ed. S.L. Sutton, T.C. Whitmore, & A.C. Chadwick), pp. 311–325. Blackwell, Oxford.

Collins, N.M. (1984) The termites (Isoptera) of the Gunung Mulu National Park, with a key to the genera known from Sarawak. *Sarawak Museum Journal*, **30**, 65–87.

Collins, N.M. (1989) Termites. In *Tropical Rain Forest Ecosystems: Biogeographical and Ecological Studies* (ed. H. Lieth & M.J.A. Werger), pp. 455–471. Elsevier, Amsterdam.

Dangerfield, J.M. & Mosugelo, D.K. (1997) Termite foraging on toilet roll baits in semi-arid savanna, south-east Botswana (Isoptera: Termitidae). *Sociobiology*, **30**, 133–143.

Darlington, J.P.E.C. (1982) The underground passages and storage pits used in foraging by a nest of the termite *Macrotermes michaelseni* in Kajiado, Kenya. *Journal of Zoology (London)*, **198**, 237–247.

Darlington, J.P.E.C. (1984) A method for sampling the population of large termite nests. *Annals of Applied Biology*, **104**, 427–436.

Darlington, J.P.E.C. (1990) Populations in nests of the termite *Macrotermes subhyalinus* in Kenya. *Insectes Sociaux*, **37**, 158–168.

Darlington, J.P.E.C. (1991) Relationship of individual weights to nest parameters in termites of the genus *Macrotermes* (Isoptera, Macrotermitinae). *Sociobiology*, **18**, 167–176.

Darlington, J.P.E.C. & Dransfield, R.D. (1987) Size relationships in the nest populations and mound parameters in the termite *Macrotermes michaelseni* in Kenya. *Insectes Sociaux*, **34**, 165–180.

Davies, R.G., Eggleton, P., Jones, D.T., Gathorne-Hardy, F.J., & Hernández, L.M. (2003) Evolution of termite functional diversity: analysis and synthesis of local ecological and regional influences on local species richness. *Journal of Biogeography*, **30**, 847–877.

Dawes-Gromadzki, T. (2003) Sampling subterranean termite species diversity and activity in tropical savannas: an assessment of different bait choices. *Ecological Entomology*, **28**, 397–404.

Dejean, A. & Ruelle, J.E. (1995) Importance of *Cubitermes* territaries as shelter for alien incipient termite societies. *Insectes Sociaux*, **42**, 129–136.

de Souza, O.F.F. (1993) Effects of habitat fragmentation on termites in Cerrado. PhD thesis, Imperial College, University of London.

Dibog, L. (1998) Biodiversity and ecology of termites (Isoptera) in a humid tropical forest, southern Cameroon. PhD thesis, Imperial College, University of London.

Dibog, L., Eggleton, P., & Forzi, F. (1998) Seasonality of the soil termites in a humid tropical forest, Mbalmayo, southern Cameroon. *Journal of Tropical Ecology*, **14**, 841–850.

Donovan, S.E., Eggleton, P., & Bignell, D.E. (2001a) Gut content analysis and a new feeding group classification of termites. *Ecological Entomology*, **26**, 356–366.

Donovan, S.E., Eggleton, P., Dubbin, W., Batchelder, M., & Dibog, L. (2001b) The effect of a soil-feeding termite, *Cubitermes fungifaber* (Isoptera: Termitidae) on soil properties: termites may be an important source of soil microhabitat heterogeneity in tropical forests. *Pedobiologia*, **45**, 1–11.

Easey, J.F. & Holt, J.A. (1989) Population estimation of some mound building termites (Isoptera: Termitidae) using radioisotope methods. *Material und Organismen*, **24**, 81–91.

Eggleton, P. (2000) Global patterns of termite diversity. In *Termites: Evolution, Sociality, Symbiosis, Ecology* (ed. T. Abe, D.E. Bignell, & M. Higashi), pp. 25–51. Kluwer, Dordrecht.

Eggleton, P. & Bignell, D.E. (1995) Monitoring the response of tropical insects to changes in the environment: troubles with termites. In *Insects in a Changing Environment* (ed. R. Harrington & N.E. Stork), pp. 473–497. Academic Press, London.

Eggleton, P. & Bignell, D.E. (1997) Secondary occupation of epigeal termite (Isoptera) mounds by other termites in the Mbalmayo Forest Reserve, southern Cameroon, and its biological significance. *Journal of African Zoology*, **111**, 489–498.

Eggleton, P.E., Williams, P.H., & Gaston, K.J. (1994) Explaining global termite diversity: productivity or history? *Biodiversity and Conservation*, **3**, 318–330.

Eggleton, P., Bignell, D.E., Sands, W.A., Waite, B., Wood, T.G., & Lawton, J.H. (1995) The species richness of termites (Isoptera) under differing levels of forest disturbance in the Mbalmayo Forest Reserve, southern Cameroon. *Journal of Tropical Ecology*, **11**, 85–98.

Eggleton, P., Bignell, D.E., Sands, W.A., et al. (1996) The diversity, abundance, and biomass of termites under differing levels of disturbance in the Mbalmayo Forest Reserve, southern Cameroon. *Philosophical Transactions of the Royal Society of London, Series B*, **351**, 51–68.

Eggleton, P., Homathevi, R., Jones, D.T., et al. (1999) Termite assemblages, forest disturbance and greenhouse gas fluxes in Sabah, East Malaysia. *Philosophical Transactions of the Royal Society of London, Series B*, **354**, 1791–1802.

Ellwood, M.D.F., Jones, D.T., & Foster, W.A. (2002) Canopy ferns in lowland dipterocarp forest support a prolific abundance of ants, termites and other invertebrates. *Biotropica*, **34**, 575–583.

Evans, T.A. (1997) Evaluation of markers for Australian subterranean termites (Isoptera: Rhinotermitidae & Termitidae). *Sociobiology*, **29**, 277–292.

Evans, T.A., Lenz, M., & Gleeson, P.V. (1998) Testing assumptions of mark–recapture protocols for estimating population size using Australian mound-building, subterranean termites. *Ecological Entomology*, **23**, 139–159.

Ferrar, P. (1982) Termites of a South African savanna. III. Comparative attack on toilet roll baits in subhabitats. *Oecologia*, **52**, 139–146.

Forschler, B.T. & Townsend, M.L. (1996) Mark–recapture estimates of *Reticulitermes* spp. (Isoptera: Rhinotermitidae) colony foraging populations from Georgia, USA. *Environmental Entomology*, **25**, 952–962.

Fuchs, A., Schreyer, A., Feuerbach, S., & Korb, J. (2004) A new technique for termite monitoring using computer tomography and endoscopy. *International Journal of Pest Management*, **50**, 63–66.

Gathorne-Hardy, F., Syaukani, & Eggleton, P. (2001) The effects of altitude and rainfall on the composition of the termites (Isoptera) of the Leuser Ecosystem (Sumatra, Indonesia). *Journal of Tropical Ecology*, **17**, 379–393.

Gathorne-Hardy, F.J., Syaukani, Davies, R.G., Eggleton, P., & Jones, D.T. (2002) Quaternary rainforest refugia in Southeast Asia: using termites (Isoptera) as indicators. *Biological Journal of the Linnean Society*, **75**, 453–466.

Greaves, T. (1967). Experiments to determine the populations of tree-dwelling colonies of termites (*Coptotermes acinaciformis* (Froggatt) and *C. frenchi* Hill). In *Termites of Australian Forest Trees*. Division of Entomology Technical Paper No. 7. Commonwealth Scientific and Industrial Research Organization, Australia.

Haagsma, K. & Bean, J. (1998) Evaluation of a hexaflumuron-based bait to control subterranean termites in southern California (Isoptera: Rhinotermitidae). *Sociobiology*, **31**, 363–369.

Harris, W.V. (1950) Dry-wood termites. *The East African Agricultural Journal*, **16**, 50–52.

Haverty, M.I. & Nutting, W.L. (1974) Natural wood-consumption rates and survival of a dry-wood and a subterranean termite at constant temperatures. *Annals of the Entomological Society of America*, **67**, 153–157.

Haverty, M.I., Nutting, W.L., & LaFage, J.P. (1975) Density of colonies and spatial distribution of foraging territories of the desert subterranean termite, *Heterotermes aureus* (Snyder). *Environmental Entomology*, **4**, 105–109.

Haverty, M.I., Nutting, W.L., & LaFage, J.P. (1976) A comparison of two techniques for determining abundance of subterranean termites in an Arizona desert grassland. *Insectes Sociaux*, **23**, 175–178.

Haverty, M.I., Getty, G.M., Copren, K.A., & Lewis, V.R. (1999) Seasonal foraging and feeding behaviour of Reticulitermes spp. (Isoptera: Rhinotermitidae) in a wildland and a residential location in northern California. *Environmental Entomology*, **26**, 1077–1084.

Henderson, G. (1996) Alate production, flight phenology, and sex ratio in *Coptotermes formosanus* Shiraki, and introduced subterranean termite in New Orleans, Louisiana. *Sociobiology*, **28**, 319–326.

Hinze, B. & Leuthold, R.H. (1999) Age related polyethism and activity rhythms in the nest of the termite *Macrotermes bellicosus* (Isoptera, Termitidae). *Insectes Sociaux*, **46**, 392–397.

Hoare, A. & Jones, D.T. (1998) Notes on the foraging behaviour and taxonomy of the southeast Asian termite *Longipeditermes longipes* (Termitidae: Nasutitermitinae). *Journal of Natural History*, **32**, 1357–1366.

Holt, J.A. & Easey, J.F. (1985) Polycalic colonies of some mound building termites (Isoptera: Termitidae) in northeastern Australia. *Insectes Sociaux*, **32**, 61–69.

Holt, J.A. & Easey, J.F. (1993) Numbers and biomass of mound-building termites (Isoptera) in a semi-arid tropical woodland near Charters Towers, north Queensland, Australia. *Sociobiology*, **21**, 281–286.

Holt, J.A. & Lepage, M. (2000) Termites and soil properties. In *Termites: Evolution, Sociality, Symbioses, Ecology* (ed. T. Abe, D.E. Bignell, & M. Higashi), pp. 389–407. Kluwer, Dordrecht.

Husseneder, C., Brandl, R., Epplen, C., Epplen, J.T., & Kaib, M. (1998) Variation between and within colonies in the termite: morphology, genomic DNA, and behaviour. *Molecular Ecology*, **7**, 983–990.

Jenkins, T.M., Dean, R.E., Verkerk, R.H.J., & Forschler B.T. (2001) Phylogenetic analyses of two mitochondrial genes and one nuclear intron region illuminate European subterranean termite (Isoptera : Rhinotermitidae) gene flow, taxonomy, and introduction dynamics. *Molecular Phylogenetics and Evolution*, **20**, 286–293.

Jones, D.T. (1996) A quantitative survey of the termite assemblage and its consumption of food in lowland mixed dipterocarp forest in Brunei Darussalam. In *Tropical Rainforest Research: Current Issues* (ed. D.S. Edwards, W.E. Booth, & S. Choy), pp. 297–305. Kluwer, Dordrecht.

Jones, D.T. & Eggleton, P. (2000) Sampling termite assemblages in tropical forests: testing a rapid biodiversity assessment protocol. *Journal of Applied Ecology*, **37**, 191–203.

Jones, D.T. & Gathorne-Hardy, F. (1995) Foraging activity of the processional termite *Hospitalitermes hospitalis* (Termitidae: Nasutitermitinae) in the rain forest of Brunei, north-west Borneo. *Insectes Sociaux*, **42**, 359–369.

Jones, J.A. (1990) Termites, soil fertility and carbon cycling in dry tropical Africa: a hypothesis. *Journal of Tropical Ecology*, **6**, 291–305.

Kambhampati, S. & Eggleton, P. (2000) Phylogenetics and taxonomy. In *Termites: Evolution, Sociality, Symbioses, Ecology* (ed. T. Abe, D.E. Bignell, & M. Higashi), pp. 1–23. Kluwer, Dordrecht.

Kirton, L.G., Brown, V.K., & Azmi, M. (1998) A new method of trapping subterranean termites of the genus *Coptotermes* (Isoptera: Rhinotermitidae) for field and laboratory experimental studies. *Sociobiology*, **32**, 451–458.

Lawton, J.H., Bignell, D.E., Bloemers, G.F., Eggleton, P., & Hodda, M.E. (1996) Carbon flux and diversity of nematodes and termites in Cameroon forest soils. *Biodiversity and Conservation*, **5**, 261–273.

Lee, K.E. & Wood, T.G. (1971) *Termites and Soils*. Academic Press, London.

Lenz, M., Creffield, J.W., & Barrett, R.A. (1992) An improved field method for assessing the resistance of woody and non-woody materials to attack by subterranean termites. *Material und Organismen*, **27**, 89–115.

Lewis, V.R., Haverty, M.I., Getty, G.M., Copren, K.A., & Fouche, C. (1998) Monitoring station for studying populations of *Reticulitermes* (Isoptera: Rhinotermitidae) in California. *Pan-Pacific Entomologist*, **74**, 121–133.

Lobry de Bruyn, L.A. & Conacher, A.J. (1990) The role of termites and ants in soil modification: a review. *Australian Journal of Soil Research*, **28**, 55–93.

Madge, D.S. (1969) Litter disappearance in forest and savanna. *Pedobiologia*, **9**, 288–299.

Maki, K. & Abe, T. (1986) Proportion of soldiers in the colonies of a dry wood termite, *Neotermes koshunensis* (Kalotermitidae, Isoptera). *Physiology and Ecology Japan*, **23**, 109–117.

Mankin, R.W., Osbrink, W.L., Oi, F.M., & Anderson, J.B. (2002) Acoustic detection of termite infestations in urban trees. *Journal of Economic Entomology*, **95**, 981–988.

Martius, C., Bandeira, A.G., & Medeiros, L.G.S. (1996) Variation in termite alate swarming in rain forests of central Amazonia. *Ecotropica*, **2** (1), 1–11.

Mathot, G. (1967). Premier essai de détermination de facteurs écologiques correlatifs à la distribution et l'abondance de *Cubitermes sankurensis* (Isoptera, Termitidae, Termitinae). *Proceedings, 5th Congress of the International Union for the Study of Social Insects*, Toulouse, 1965, pp. 117–129.

Matsumoto, T. & Abe, T. (1979) The role of termites in an equatorial rain forest ecosystem of west Malaysia. 2. Leaf litter consumption on the forest floor. *Oecologia*, **38**, 261–274.

Medeiros, L.G.S., Bandeira, A.G., & Martius, C. (1999) Termite swarming in the northeastern Atlantic rain forest of Brazil. *Studies non Neotropical Fauna and Environment*, **34**, 76–87.

Meyer, V.W., Braack, L.E.O., Biggs, H.C., & Ebersohn, C. (1999) Distribution and density of termite mounds in the northern Kruger National Park, with specific reference to those constructed by *Macrotermes* Holmgren (Isoptera: Termitidae). *African Entomology*, **7**, 123–130.

Mill, A.E. (1983) Observations on Brazilian termite alate swarms and some structures used in the dispersal of reproductives (Isoptera: Termitidae). *Journal of Natural History*, **17**, 309–320.

Miller, L.R. (1993) Fluorescent dyes as markers in studies of foraging biology of termite colonies (Isoptera). *Sociobiology*, **23**, 127–134.

Miura, T. & Matsumoto, T. (1998) Foraging organization of the open-air processional lichen-feeding termite *Hospitalitermes* (Isoptera, Termitidae) in Borneo. *Insectes Sociaux*, **45** (1), 17–32.

Myles, T.G. (1996) Development and evaluation of a transmissible coating for control of subterranean termites. *Sociobiology*, **28**, 373–457.

Noirot, C. (1970) The nests of termites. In *Biology of Termites*, vol. 2 (ed. K. Krishna & F.M. Weesner), pp. 73–125. Academic Press, New York.

Nutting, W.L. (1969) Flight and colony foundation. In *Biology of Termites*, vol. 1 (ed. K. Krishna & F.M. Weesner), pp. 233–282. Academic Press, New York.

Nutting, W.L. & Jones, S.C. (1990) Methods for studying the ecology of subterranean termites. *Sociobiology*, **17**, 167–189.

Ohiagu, C.E. (1979) Nest and soil populations of *Trinervitermes* spp. with particular reference to *T. geminatus* (Wasmann), (Isoptera), in southern Guinea savanna near Mokwa, Nigeria. *Oecologia*, **40**, 167–178.

Pearce, M.J. (1990) A new trap for collecting termites and assessing their foraging activity. *Tropical Pest Management*, **36**, 310–311.

Pearce, M.J. (1997) *Termites: Biology and Pest Management*, CAB International, Wallingford.

Pearce, M.J., Cowie, R.H., Pack, A.S., & Reavey, D. (1990) Intraspecific aggression, colony identity and foraging distances in Sudanese *Microtermes* spp. (Isoptera, Termitidae, Macrotermitinae). *Ecological Entomology*, **15**, 71–77.

Peters, B.C. & Fitzgerald, C.J. (1999) Field evaluation of the effectiveness of three timber species as bait stakes and the bait toxicant hexaflumuron in eradicating *Coptotermes acinaciformis* (Froggatt) (Isoptera: Rhinotermitidae). *Sociobiology*, **33**, 227–238.

Peters, B.C. & Fitzgerald, C.J. (2003) Field evaluation of the bait toxicant chlorfluazuron in eliminating *Coptotermes acinaciformis* (Froggatt) (Isoptera : Rhinotermitidae). *Journal of Economic Entomology*, **96**, 1828–1831.

Pomeroy, D.E. (1977) The distribution and abundance of large termite mounds in Uganda. *Journal of Applied Ecology*, **14**, 465–475.

Potter, M.F., Eliason, E.A., Davis, K., & Bessin, R.T. (2001) Managing subterranean termites (Isoptera : Rhinotermitidae) in the Midwest with a hexaflumuron bait and placement considerations around structures. *Sociobiology*, **38**, 565–584.

Rebello, A.M.C. & Martius, C. (1994) Dispersal flights of termites in Amazonian forests (Isoptera). *Sociobiology*, **24**, 127–146.

Roisin, Y. & Pasteels, J.M. (1986) Reproductive mechanisms in termites: polycalism and polygyny in *Nasutitermes polygynus* and *N. costalis*). *Insectes Sociaux*, **33**, 149–167.

Salt, G. (1952) The arthropod population of the soil in some East African pastures. *Bulletin of Entomological Research*, **43**, 203–220.

Sands, W.A. (1961) Nest structure and size distribution in the genus *Trinervitermes* (Isoptera, Termitidae, Nasutitermitinae) in West Africa. *Insectes Sociaux*, **8**, 177–188.

Sands, W.A. (1965) Mound population movements and fluctuations in *Trinervitermes ebenerianus* Sjöstedt. *Bulletin de l'Union Internationale pour l'étude des insectes sociaux*, **12**, 49–58.

Sands, W.A. (1972) Problems in attempting to sample tropical subterranean termite populations. *Ekologia Polska*, **20** (3), 23–31.

Schuurman, G. & Dangerfield, J.M. (1997) Dispersion and abundance of *Macrotermes michaelseni* colonies: a limited role for intraspecific competition. *Journal of Tropical Ecology*, **13**, 39–49.

Silva, E.G. & Martius, C. (2000) Termite (Isoptera) sampling from soil: handsorting or Kempson extraction? *Sociobiology*, **36**, 209–216.

Sleaford, F., Bignell, D.E., & Eggleton, P. (1996) A pilot analysis of gut contents in termites from the Mbalmayo Forest Reserve, Cameroon. *Ecological Entomology*, **21**, 279–288.

Spragg, W.T. & Paton, R. (1977) Tracing, trophallaxis and population measurement of colonies of subterranean termites (Isoptera) using a radioactive tracer. *Annals of the Entomological Society of America*, **73**, 708–714.

Strickland, A.H. (1944) The arthropod fauna of some tropical soils, with notes on the techniques applicable to entomological soil surveys. *Tropical Agriculture*, **21**, 107–114.

Su, N.Y. (2002) Dimensionally stable sensors for a continuous monitoring program to detect subterranean termite (Isoptera: Rhinotermitidae) activity. *Journal of Economic Entomology*, **95**, 975–980.

Su, N.Y. & Scheffrahn, R.H. (1996) Fate of subterranean termite colonies (Isoptera) after bait applications: an update and review. *Sociobiology*, **27**, 253–275.

Su, N.Y., Scheffrahn, R.H., & Ban, P.M. (1988) Retention time and toxicity of a dye marker, Sudan red 7B, on formosan and eastern subterranean termites (Isoptera: Rhinotermitidae). *Journal of Entomological Science*, **23**, 235–239.

Su, N.Y., Ban, P.M., & Scheffrahn, R.H. (1991) Evaluation of twelve dye markers for population studies of the eastern and Formosan subterranean termite (Isoptera: Rhinotermitidae). *Sociobiology*, **19**, 349–362.

Su, N.Y., Ban, P.M., & Scheffrahn, R.H. (1993) Foraging populations and territories of the eastern subterranean termite (Isoptera: Rhinotermitidae) in southeastern Florida. *Environmental Entomology*, **22**, 1113–1117.

Sugimoto, A., Bignell, D.E., & MacDonald, J.A. (2000) Global impact of termites on the carbon cycle and atmospheric trace gases. In *Termites: Evolution, Sociality, Symbiosis, Ecology* (ed. T. Abe, D.E. Bignell, & M. Higashi), pp. 409–435. Kluwer, Dordrecht.

Sugio, K. (1995) Trunk trail foraging of the fungus-growing termite *Macrotermes carbonarius* (Hagen) in southeastern Thailand. *Tropics*, **4**, 211–222.

Takematsu, Y., Inoue, T., Hyodo, F., Sugimoto, A., Kirtebutr, N., & Abe, T. (2003) Diversity of nest types in *Microcerotermes crassus* (Termitinae, Termitidae, Isoptera) in a dry evergreen forest of Thailand. *Sociobiology*, **42**, 587–596.

Tayasu, I., Abe, T., Eggleton, P., & Bignell, D.E. (1997) Nitrogen and carbon isotope ratios in termites: an indicator of trophic habit along the gradient from wood-feeding to soil-feeding. *Ecological Entomology*, **22**, 343–351.

Taylor, H.S., MacKay, W.P., Herrick, J.E., Guerra, R.A., & Whitford, W.G. (1998) Comparison of field methods to detect termite activity in the Northern Chihuahuan Desert. *Sociobiology*, **32**, 1–15.

Thoms, E.M. (2000) Use of an acoustic emissions detector and intragallery injection of spinosad by pest control operators for remedial control of drywood termites (Isoptera: Kalotermitidae). *Florida Entomologist*, **83**, 64–74.

Thorne, B.L., Russek-Cohen, E., Forschler, B.T., Breisch, N.L., & Traniello, J.F.A. (1996) Evaluation of mark–release–recapture methods for estimating forager population-size of subterranean termite (Isoptera: Rhinotermitidae) colonies. *Environmental Entomology*, **25**, 938–951.

Usher, M.B. & Ocloo, J.K. (1974) An investigation of stake size and shape in "graveyard" field tests for termite resistance. *Journal of the Institute of Wood Science*, **9**, 32–36.

Verkerk, R.H.J. (1990) *Building Out Termites: an Australian Manual for Environmentally Responsible Control*, Pluto Press, Leichhardt, Australia.

Verkerk, R.H.J & Bravery, A.F. (2001) The UK termite eradication programme: justification and implementation. *Sociobiology*, **37**, 351–360.

Watt, A.D., Stork, N.E., Eggleton, P., et al. (1997) Impact of forest loss and regeneration on insect abundance and diversity. In *Forests and Insects* (ed. A.D. Watt, N.E. Stork, & M.D. Hunter), pp. 273–286. Chapman & Hall, London.

Weissling, T.J. & Thoms, E.M. (1999) Use of an acoustic emission detector for locating Formosan subterranean termite (Isoptera: Rhinotermitidae) feeding activity when installing and inspecting aboveground termite bait stations containing hexaflumuron. *Florida Entomologist*, **82**, 60–71.

Wilson, E.O. (1992) The effects of complex social life on evolution and biodiversity. *Oikos*, **63**, 13–18.

Wood, T.G. (1978) Food and feeding habits of termites. In *Production Ecology of Ants and Termites* (ed. M.V. Brian), pp. 55–80. Cambridge University Press, Cambridge.

Wood, T.G. (1996) The agricultural importance of termites in the tropics. *Agricultural Zoology Reviews*, **7**, 117–155.

Wood, T.G. & Johnson, R.A. (1978) Abundance and vertical distribution in soil of *Microtermes* (Isoptera: Termitidae) in savanna woodland and agricultural ecosystems at Mokwa, Nigeria. *Memorabilia Zoolgie*, **29**, 203–213.

Wood, T.G. & Lee, K.E. (1971) Abundance of mounds and competition among colonies of some Australian termite species. *Pedobiologia*, **11**, 341–366.

Wood, T.G. & Sands, W.A. (1978) The role of termites in ecosystems. In *Production Ecology of Ants and Termites* (ed. M.V. Brian), pp. 245–292. Cambridge University Press, Cambridge.

Wood, T.G., Johnson, R.A., Ohiagu, C.E., Collins, N.M., & Longhurst, C. (1977). *Ecology and Importance of Termites in Crops and Pastures in Northern Nigeria*. Project report 1973–76, Centre for Overseas Pest Research, London.

Wood, T.G., Johnson, R.A., Bacchus, S., Shittu, M.O., & Anderson, J.M. (1982) Abundance and distribution of termites (Isoptera) in a riparian forest in the Southern Guinea savanna zone of Nigeria. *Biotropica*, **14**, 25–39.

Index of methods and approaches

Methodology	Topics addressed	Comments
Estimating colony density		
Plot surveys	Visual survey of plots to count the number of mounds and/or arboreal nests.	Only possible with conspicuous colonies. Aerial photography can be used to survey larger mounds in open habitats. Often difficult to see arboreal nests in forest canopy, even using binoculars. Many arboreal nests may be hidden inside wood or large epiphytes.
	Recording the colony coordinates within plots also allows estimates of dispersion.	
Detection of acoustic emissions	Hand-held device for detecting the acoustic emissions from termite colonies hidden inside wood.	This has been used for the detection of pest species in buildings and urban trees. The sampling methodology of this relatively new technology has not yet been fully developed, nor has it been tested on non-pest species in natural habitats.

Continued

Methodology	Topics addressed	Comments
Estimating population density		
Direct sampling of nests	Estimation of population in a single colony by destructive sampling of entire nest, or sub-sampling of part of the nest.	Termites can be extracted from nest material by hand-sorting or flotation methods. Sampling an entire nest can be very labor-intensive and time-consuming. Fumigation of nest to kill the colony before sampling gives a much more accurate population estimate.
Mark–recapture techniques	Estimation of colony population size by mark–recapture protocols using colored stains, dyes, or radioisotopes to mark individual termites.	Population estimates can be very inaccurate compared with direct population counts because the assumptions of the protocol are often violated.
Soil pits, monoliths, or trenches	Estimation of subterranean termite population by digging pits, monoliths, or trenches, and then removing the termites by hand-sorting the soil. Digging very large pits or long trenches can also give estimates of hypogeal colony density.	Very labor-intensive and/or time-consuming. Density estimates often have high variances due to the very patchy distribution of termites in soil. Less effective in savannas, where termites often migrate further down the soil profile compared with termites in forest soils.
Soil cores	Estimation of subterranean termite population by taking soil cores, and then removing the termites by hand-sorting, flotation or using an automatic extraction device. This method does not give accurate estimates of the density of hypogeal colonies.	Collecting soil samples with a sharp-edged corer is quicker than digging pits, monoliths or trenches. Removing the termites by hand-sorting the soil is considered to be more efficient than using flotation methods or extraction devices such as Berlese–Tullgren funnels.
Sampling dead wood	Estimation of total population by splitting open all items of dead wood collected within quadrats, and hand-sorting the termites. Populations in larger items of wood can be estimated by sampling sub-units of known volume or weight. Alternatively, the wood can be split open and	Very labor-intensive and/or time-consuming because all items of wood must be split lengthways and individual termites extracted by hand. The semi-quantitative method of visual estimation has never been calibrated against direct counts to test its accuracy.

Continued

Methodology	Topics addressed	Comments
	the number of individuals can be estimated visually by counting in units of 10s, 100s, or 1000s.	
Sampling regime for estimating population density of a local termite assemblage	A plot-based sampling regime for estimating the population density of the local assemblage in tropical forest by sampling termites in mounds, nests, dead wood, and soil.	A standardized sampling regime that combines plot surveys of mounds and nests with soil pits and dead wood from quadrats. Very labor-intensive and time-consuming. Excludes arboreal termites occurring at more than 2 m above ground level.
Assessing species composition		
Transect protocol	A rapid sampling method for assessing the species composition of a local assemblage in a tropical forest. This protocol standardizes the amount of sampling effort and area along a belt transect.	This protocol has been shown to produce samples that represent accurately the taxonomic and functional composition of the local assemblage. No sampling is conducted at more than 2 m above ground level but many arboreal-nesting species are collected as they forage on the ground.
Baiting methods	Estimation of local species richness by attracting foraging termites to cellulose baits. A range of materials can be used as baits, including wood, litter bags, toilet rolls, corrugated cardboard, dungpats, and sawdust.	Not all wood-feeding species are attracted to cellulose baits. Species show preferences for different bait material, so a combination of bait types is recommended to maximize the number of species attracted.
Light trapping	Estimation of local species richness by attracting alates to light traps at night.	Light traps fail to capture many species because not all show a clear preference for nocturnal swarming, and the alates of many species have a poor dispersal range. Even running traps throughout the year may only capture about half of the known local species.
Studying termite activity		
Observing termites that forage in the open air	Estimation of the number of individuals involved in foraging by direct counts in real time, or from close-up photographs taken of the	Only a relatively small number of species send out processional columns of foragers that move in the open

Continued

Methodology	Topics addressed	Comments
	foraging column at regular intervals. The amount of food collected by the colony can be estimated if the workers return with visible balls of forage held in their mandibles. The foraging range and frequency of individual colonies can be studied easily.	air. Many of these only forage at night, making it more difficult to quantify their activity.
Baiting experiments	Studying rates of consumption, food preference and size of foraging territory. Can also be used to study inter- and intraspecific foraging activity.	Ensure that timber purchased as baits has not been treated with pesticide, and that toilet rolls are unscented as only foragers are attracted to baits, this gives a measure of relative intensity of foraging activity rather than relative population density.
Mark–recapture	Studying foraging activity and territory size by using colored stains, dyes, or radioisotopes to mark individual termites.	Provided sufficient foragers are marked, it is easier to produce reliable estimates of foraging activity and territory size than it is to use mark–recapture techniques to estimate population density.
Light trapping	Studying temporal patterns of swarming by attracting alates to light traps.	Local assemblages show little interspecific synchronicity of swarming, so traps operating over short periods will fail to capture many species. Even running traps throughout the year may only capture about half of the known local species.
		In addition, not all species show a clear preference for nocturnal swarming, and the alates of many species have a poor dispersal range.
Detection of acoustic emissions	Detecting infestations of termites hidden in wood by using a handheld device.	This has been used for the detection of pest species in buildings and urban trees. The sampling methodology of this relatively new technology has not yet been fully developed, nor has it been tested on non-pest species in natural habitats.

CHAPTER 11

Parasitoids and predators

NICK MILLS

Introduction

Insect parasitoids and predators are major contributors to the third trophic level of terrestrial plant-based food webs, and pose some interesting and unique constraints for sampling. For example, as parasitoids spend a large part of their life cycle in intimate association with a host, sampling is often based on the host population with few opportunities to estimate absolute densities of the free-living adult stage. In contrast, predators are always free-living and predator populations or communities can readily be sampled independently of prey populations. In addition, the third trophic level imposes constraints on the abundance of parasitoid and predator populations relative to their host populations, and thus the accuracy of estimates of abundance or diversity are more dependent on large sample size. Finally, although the size of insect parasitoids and predators varies from the larger rhyssine ichneumonids and carabid beetles to the smallest mymarid egg parasitoids and coccinellids, the small size of many entomophagous species can make them rather more difficult to sample than other insects using standard sampling techniques.

There are many reasons why researchers may need to sample insect parasitoid and predator populations, but the three most frequent can be categorized as evaluating (i) the composition of the entomophagous assemblage associated with a particular host species, (ii) the impact of entomophagous species on the dynamics of a particular host population, and (iii) the biodiversity of entomophagous species within a local or regional community. In this chapter I will discuss the issues that pertain to the sampling of insect parasitoids and predators for each of these purposes, and indicate how they build upon the standard sampling techniques discussed in other chapters.

Composition of entomophagous insect assemblages

Tremendous care needs to be exercised in the sampling of entomophagous insect assemblages, as the literature contains numerous errors in predator–prey and host–parasitoid associations (Askew & Shaw 1986, Hodek 1993, Shaw 1994, 1997). Mistakes can arise as a result of either misidentification or incorrect association. As the majority of parasitoids and predators attack juvenile stages of their host, it can be difficult to correctly identify the immature host, particularly when several host species occur together in the same location. Similarly, the majority of insect predators are predaceous during their immature stages, and correct identification of juvenile predators can be challenging.

In many cases it is necessary to rear the immature parasitoids and predators through to the adult stage for correct identification, and it is surprising how frequently other insects can be introduced into a rearing and result in incorrect associations. This is less of a problem for insect predators, but of particular concern for insect parasitoids as the food plant used in rearing may support additional small or cryptic species that produce parasitoids of their own. This is particularly noticeable when parasitoids recovered from a rearing include species that belong to a taxonomic group that has absolutely no association with the host insect of interest. For example, it is not uncommon for aphid parasitoids (Braconidae: Aphidiinae) to be recorded from rearings of lepidopteran larvae (Herting 1975), due to the presence of parasitized aphids or aphid mummies on the foliage or bark of the food plant. As aphidiine parasitoids have never been reared from Lepidoptera, but instead have a strong and close association with the Aphididae (Stary 1970), the error can easily be detected. However, when contaminating hosts are more closely aligned to the host in question, such as a buprestid feeding under the bark of a log primarily infested with scolytid beetles, or a twig-boring lepidopteran in the branches of foliage fed to leaf-chewing Lepidoptera, the incorrect associations are far more difficult to detect.

Parasitoid assemblages

The intimacy of contact between parasitoid and host requires correct associations to be determined by rearing. Direct observation of parasitoid species probing a particular host provides an indication of potential association, but such observations cannot be relied upon as the parasitoid may either not accept the host for oviposition (Kitt & Keller 1998), or eggs may be consistently encapsulated by the host to prevent parasitism (Blumberg & Van Driesche 2001).

Contamination of host rearings by other host insects or parasitoids can be minimized by use of the following procedures (after Shaw 1997):

• Search the food plant thoroughly to remove contaminants, not only from foliage but also from the bark, stem, flowers, or fruit.

• Never rear more than one host species together in the same rearing container, even if they are easily distinguishable.

• Count the host individuals and parasitoid clutches to more accurately associate host mortality with the presence of parasitoid cocoons, pupae, or puparia.

• Keep exited parasitoids as individual clutches whenever possible to accurately associate adults with immature stages.

- Preserve the host remains and parasitoid cocoons/puparia with the adult parasitoid to provide a means of checking host identity and associating immature stages of the parasitoids with the adult.

To determine the complete parasitoid assemblage of a host species it is necessary to sample all stages of the life cycle of the host insect, as parasitoids have varied, and sometimes very narrow, windows of parasitism of a host (Mills 1994). Larval and nymphal stages of host insects have frequently been sampled for parasitism, but we know far less about parasitoids confined to the egg, pupal, or adult stages of their hosts. The most important factor influencing the species richness of parasitoid assemblages is sample size (Hawkins 1994). The number of hosts that must be collected from a local host population is influenced by two factors, the probability of a host individual being parasitized, and the degree of heterogeneity in the distribution of the parasitoid species within the host population. On average, the number of parasitoid species associated with holometabolous hosts in the UK is approximately seven, but in individual instances can be as high as 25 (Hawkins 1988). Nonetheless, to be able to collect the majority of species in a local parasitoid assemblage requires a sample size of 1000 individuals or more (Hawkins 1994). In some cases, surveys of parasitoid assemblages for readily reared hosts can be facilitated by use of trap hosts to increase sample size (e.g. Floate et al. 1999). In contrast, the parasitoid assemblages of hemimetabolous hosts contain far fewer species. For example, aphids support an average of about two parasitoid species (Porter & Hawkins 1998, Stadler 2002), although individual aphid species may support as many as eight (Muller et al. 1999). The lower richness of hemimetabolous parasitoid assemblages should reduce the influence of heterogeneity (Keating & Quinn 1998) and facilitate the estimation of local assemblage richness from smaller sample sizes. In this regard, the biodiversity software package EstimateS (Colwell 1997) can be used to indicate the sufficiency of sample size, and to estimate the total species richness of the complete parasitoid assemblage, as used by Stireman and Singer (2002) in studying the parasitoid assemblage of a polyphagous caterpillar.

Predator assemblages

In contrast to parasitoid assemblages, the lack of intimacy in the interaction between predators and their prey poses interesting challenges for correct identification of the range of predators associated with a particular prey species. One of three approaches can be used, direct observation, indirect sampling, or gut detection.

Direct observation

This is the most effective sampling strategy as it is the only technique that unequivocally ensures the validity of the predator–prey association. Qualitative sampling is accomplished for any readily observable prey species, by visual search, collection, and correct identification of all predators seen to be actively feeding on the prey species. Direct observation is time consuming, but has the advantages that it can also be used to quantify predator abundance, and that it provides the observer with an intimate knowledge of the activity and behavior of the different predator species. As predators are most frequently observed in their immature stages, correct identification is essential. This can be achieved by building a reference set of close-up photographs showing the successive stages of development as individuals are reared through to maturity. In addition, valuable keys to the immature stages of some of the more frequent taxa of insect predators include those for the Coccinellidae (Hodek 1973, Gordon & Vandenberg 1995, Rhoades 1996), Chrysopidae (Diaz-Aranda & Monserrat 1995, Tauber et al. 2000, Monserrat et al. 2001), and Syrphidae (Rotheray 1988, 1993, Rotheray & Gilbert 1989, 1999).

Indirect sampling

The most frequent approach for evaluating predator assemblages is indirect sampling through use of either suction samplers (e.g. Parajulee & Slosser 1999), sweep nets (e.g. Elliott & Kieckhefer 1990), beating trays (e.g. Wyss 1996), or yellow pan traps (e.g. Bowie et al. 1999). Although these sampling techniques provide greater numbers of entomophagous species than direct observation, the problem of accurate association is paramount, and so caution should be exercised when using indirect sampling to identify the predator assemblage of a particular prey species. The associations should at least be tested in feeding trials, using simple laboratory arenas with predators at different stages in their development. Care must also be taken to distinguish essential prey, which fully support growth and reproduction of the predator, from alternative prey, which can sustain predators for short periods of time (Hodek 1993). Indirect sampling is best used as a technique to quantify the absolute or relative abundance of members of a predator assemblage that are known to be true associates from either direct observation or gut detection. Both the choice of sampling method and the time of day that samples are collected can have an important influence on the types of predators collected. For example, Brown and Schmitt (2001) found that chrysopids, clerids, and Thysanoptera were more effectively sampled from apple trees by beating at night. Similarly, Costello and Daane (1997) found that drop cloths were more efficient than funnels or suction samplers for monitoring spider assemblages in vines.

Gut detection

This last approach makes use of either artificial (elemental and immunological) or natural (immunological and PCR) markers for the prey and methods to screen predators, collected through indirect sampling, for the presence of markers in their gut. Artificial markers include rubidium, which must be fed to prey via the food plant or diet (Akey & Burns 1991, Johnson & Reeves 1995), and readily available vertebrate immunoglobulins, which can either be fed to prey or applied topically (Hagler & Durand 1994). In contrast, natural markers are present naturally in all prey populations and take the form of monoclonal antibodies (currently available for only a small number of prey species; Green-stone 1996) and PCR products (Chen et al. 2000). Standard techniques are available to detect the presence or absence of all four forms of marker in the guts of individual predators (see individual references cited above). Two key difficulties with gut detection are that they do not distinguish necrophagy, saprophagy, and higher-order predation from primary predation, and that detectability and reliability vary with predator species and size, with meal size and time since feeding, and in some cases with the size of the marker. However, the relative ease with which PCR primers can be used to develop specific markers for individual prey species is likely to lead to significant advancements in gut detection in the near future and its more widespread use in the characterization of predator assemblages.

Regional variation

For both parasitoids and predators, the regional species richness of an assemblage is likely to be influenced by the rate of turnover or beta-diversity of parasitoid or predator species across the geographic distribution of the host. Thus it is important to consider how much sampling must be conducted a single site, and how many sites should be sampled in order to provide an accurate estimate of a regional assemblage. For example, Hawkins (1988) clearly shows how the size of the species-rich parasitoid assemblages of Lepidoptera in the UK increases with spatial scale. In contrast, the size of the parasitoid assemblages of grass-feeding chalcid wasps remains more or less constant throughout the UK, with little evidence of regional variation (Dawah et al. 1995). The number of sites sampled will be of greatest importance in assessing regional diversity, as shown by Gaston and Gauld (1993) for the pimpline ichneumonid fauna of Costa Rica, but it remains unclear how large a sample is needed at each locality.

Sampling to estimate the impact of entomophagous species

The most frequent need for sampling entomophagous insects is to estimate their impact on the dynamics of a particular host population (Luck et al. 1988, Sunderland 1988, Mills 1997). As both parasitoids and predators are important sources of mortality at all stages in the life cycle of phytophagous insects we often want to know which entomophagous species play the greatest role in reducing or regulating host abundance. The impact of entomophagous species can be analyzed through life table analysis (Bellows et al. 1992, Yamamura

1999) or through simulation modeling (Gutierrez 1996, van Lenteren & van Roermund 1999, Kean & Barlow 2001). Both approaches require data from the regular sampling of both host and entomophagous insect populations and present many challenges to the field ecologist. As parasitoids have an intimate relationship with their host that lasts many days, the impact of parasitoids has focused on sampling hosts to directly estimate percent parasitism. In contrast, as predation events are so brief in time it is extremely difficult to obtain a direct estimate of predation mortality, and thus sampling has focused on monitoring the abundance and consumption rate of predators.

Percent parasitism

Percent parasitism is commonly estimated from a sample of hosts as the ratio of the number parasitized to the total number of hosts in the sample. In general, reports of percent parasitism in the literature are based either on the peak or the mean level of parasitism from a series of samples collected at intervals from the same location. Van Driesche (1983) points out the gross inaccuracies in this common practice, and Van Driesche et al. (1991) provide a number of solutions for more effective estimation of percent parasitism. For hosts with discrete generations, the goal is to estimate stage-specific parasitism, the percentage of host individuals recruited to the susceptible stage that are attacked by a particular parasitoid species, representing the generational mortality attributable to parasitism (Table 11.1). In contrast, for hosts with overlapping generations, it is

Table 11.1 A summary of the methodological approaches for estimating the impact of parasitism.

Methodology	Application	Examples
Stage-specific generational parasitism for hosts with discrete generations		
Single sample	Effective only if susceptible host stage is clearly separated from host stage killed	Hill (1988), Gould et al. (1992a)
Death rate analysis	Useful when there is insufficient knowledge of host stages susceptible to parasitism	Gould et al. (1992a)
Recruitment analysis	Unbiased, but requires detailed knowledge and regular sampling	Van Driesche & Bellows (1988), Gould et al. (1992a)
Time-specific rate of parasitism for hosts with discrete or overlapping generations		
Inclusive host stages	Should be applied in all cases to avoid bias caused by non-inclusive stages	
Recruitment analysis	Avoids bias, but requires detailed knowledge of susceptible stages	Lopez and Van Driesche (1989)
Correction for differential duration of development	Should be applied whenever parasitism influences host development time	Russell (1987)

important to estimate a time-specific rate of parasitism, the percentage of hosts that die from parasitism during a specific interval of time, and how that rate changes over a season (Table 11.1). The susceptible stage of the host is the earliest life stage or instar that can be attacked by the parasitoid of interest, a characteristic of either the parasitoid guild (a classification that defines the pattern of host utilization by parasitoids; Mills 1994) or the individual parasitoid species. In some cases, the estimation of percent parasitism also requires knowledge of the earliest host stage or instar killed by the parasitoid, such that the inclusive host stages, those that support the parasitoid during its successive stages of development from oviposition to emergence from the host, can be clearly defined.

Stage-specific generational parasitism

The key to estimation of generational parasitism is to devise a sampling plan that provides accurate input for the ratio estimate of percent parasitism. The numerator of the ratio must be representative of the total number of host individuals in the population parasitized by a specific parasitoid species, and the denominator must be representative of the total number of host individuals in the population recruited to the susceptible stage. There are three ways in which generational parasitism can be estimated, depending on the life history of the parasitoid: single samples, death rate analysis, and recruitment analysis.

Single samples

Under some circumstances the life histories of host and parasitoid can facilitate the sampling of host populations for the estimation of generational parasitism. This occurs when the susceptible host stage is clearly separated from the host stage killed by the parasitoid (Van Driesche 1983, Royama 2001). Several guilds of endoparasitoids of holometabolous hosts are known to attack early in host life cycle and delay their development to kill much later host stages (Mills 1994), such as egg–prepupal parasitoids, early larval parasitoids, and larval–pupal parasitoids. For these parasitoids there is a point in the host life cycle when parasitoid attack is complete, but mortality from parasitism has yet to occur. In other situations, host diapause can also arrest host and parasitoid development at a point intermediate between the susceptible stage and the stage experiencing mortality from parasitism. In both of these cases, a single sample of hosts collected when the population is at an intermediate life stage can provide an effective estimate of generational parasitism (Van Driesche 1983). This approach proved valuable for estimating percent parasitism by larval parasitoids of both armyworm (Hill 1988) and gypsy moth larvae (Gould et al. 1992a). Of course, such estimates can still be biased by any differential mortality of parasitized and healthy hosts that might occur before collection of the host sample or during subsequent rearing.

Additionally, parasitoids often leave distinct emergence holes in the resilient coverings of eggs, pupae and many sessile Homoptera. In these instances, it is possible to monitor generational mortality directly from the percentage of parasitized individuals in a single sample collected after hosts have developed beyond the inclusive stage (Van Driesche 1983). This approach has often been used for egg (e.g. Nakamura & Abbas 1987) and pupal (e.g. Doane 1971) parasitism, but can lead to underestimation if there is a significant level of predation as a contemporaneous mortality factor. In the latter case, both predation and parasitism must be estimated and marginal attack rates calculated (see below).

Death rate analysis

Death rate analysis was developed by Gould et al. (1989, 1992a) to estimate generational parasitism in populations of gypsy moth, and is based on regular monitoring of hosts killed by a specific parasitoid species. A series of host samples must be collected at regular intervals and for each sample the number of individuals killed by the parasitoid is noted while the hosts are held under natural conditions for the interval between two successive samples. The approach differs from the more traditional measurement of parasitism in that the death rate from parasitism is monitored for the interval between samples only rather than keeping each sample until all hosts have either died from parasitism or developed beyond the inclusive stage. The death rate is first converted to a proportion surviving from parasitism, and generational parasitism is subsequently estimated from $(1 - s) * 100$, where s is the product of the proportion surviving from each of the successive samples (Gould et al. 1992a). Death rate analysis has the advantage that it does not require prior knowledge of the host stages susceptible to parasitism or those that are killed by parasitism. However, the series of successive samples must span the complete period during which hosts are killed by the parasitoid, and each sample must be representative of all host stages present at that time. It also assumes that no other fast-acting mortality factor could have intervened in the field during the interval between samples to kill some of the host individuals in each sample before they are killed by parasitism. As a result it is important to use a sample interval that is short enough to reduce the chances of bias due to the exclusion of contemporaneous mortality factors. Death rate analysis has not been used extensively, but has proved valuable for monitoring parasitism of gypsy moth larvae (Gould et al. 1990, 1992a), and a variant of the approach is advocated by Royama (2001).

Recruitment analysis

In some cases, the recruitment of hosts to the susceptible stage and of parasitoids to the host population can be estimated directly. Van Driesche and Bellows (1988) successfully developed this approach to estimate the generational parasitism of imported cabbageworm by the larval parasitoid *Cotesia glomerata*. Recruitment of individuals to the susceptible host stage (young larvae) was measured by removing all larvae from randomly selected collard plants and counting the number of new first and second instar larvae 3–4 days later. This procedure was repeated every 3–4 days over the complete egg-laying period of the host, using new plants each time, and the counts of young larvae summed to estimate total recruitment to the susceptible stage for that generation. The recruitment of parasitized hosts was then measured using a short marker stage method. This involved dissection of the host larvae removed from the collard plants each time to count the number of hosts containing parasitoid eggs (excluding those containing parasitoid larvae), representing only those parasitoids recruited during the previous sampling interval. The total count of hosts with newly recruited parasitoids from the complete set of samples provided an estimate of the sum of hosts parasitized for the generation, and generational parasitism could be estimated from the ratio of the sum of parasitized hosts to the sum of hosts recruited to the susceptible stage.

As an alternative, the recruitment of parasitized hosts can be measured using the trap host method, in which susceptible-stage hosts are exposed to parasitism under natural conditions for a period equivalent to the sample interval. This works best for sessile hosts, as mobile hosts often have a tendency to move away from the location where released. In addition, it is critical to use trap hosts at natural densities, place them in natural settings, and verify that their development rate during exposure is typical, to ensure that they are no more or less susceptible to parasitism than wild hosts. Trap hosts have been used to monitor recruitment of parasitized hosts for imported cabbageworm (Van Driesche 1988), and for larvae (Gould et al. 1992a) and pupae (Gould et al. 1992b) of gypsy moth.

The success of recruitment analysis is dependent upon the biology of the species involved and the ease with which methods can be devised to effectively measure host and parasitoid recruitment. If more than one parasitoid species is to be monitored, different techniques may need to be employed for each, and this can result in a very demanding and time-consuming sampling plan. Nonetheless, when possible this approach provides an unbiased estimate of stage-specific parasitism.

Time-specific rate of parasitism

Although time-specific rates of parasitism are an effective way to monitor changes in parasitism over a season for hosts with overlapping generations, they are also frequently used to estimate the impact of parasitism at specific points in time for hosts with discrete generations. The most important biases in sampling for time-specific rates of parasitism in host populations with overlapping generations are the presence of non-inclusive host stages in the sample, and the duration of time that parasitized individuals spend in the inclusive stages relative to healthy individuals. The presence of non-inclusive host stages in a sample leads to underestimation, and prolonged availability of parasitized hosts leads to overestimation of the rate of parasitism.

Knowledge of the inclusive stages of the host in relation to specific parasitoids can be used to ensure that only those stages that support the parasitoid during its development are sampled for dissection or rearing in the laboratory. By limiting sampling to the inclusive host stages, estimates of rate of parasitism can be greatly improved, but they are still subject to bias caused by differential mortality during development and the intervention of other contemporaneous mortality factors. One approach to avoid these additional sources of bias is recruitment analysis (see above). Although recruitment analysis has been used successfully to monitor parasitism of cabbage aphid populations (Lopez & Van Driesche 1989), it has not been used extensively. Other approaches used for stage-specific parasitism are not appropriate for hosts with overlapping generations due to the disruptive influence of variable rates of recruitment to the susceptible stage and of loss from the inclusive stage for both host and parasitoid populations.

A simple correction factor can be used to improve estimates of rate of parasitism when parasitized and healthy hosts spend different periods of time in the inclusive host stages. Parasitized hosts often develop more slowly than healthy hosts, and as a result become over-represented in samples. Russell (1987) indicates that this can be corrected by elevating the number of healthy hosts in the calculation of percent parasitism by a factor determined from the ratio of the period of time spent in the inclusive stages by parasitized (P) and healthy (H) hosts:

$$\% \text{ parasitism} = \frac{\text{No. of parasitized hosts} * 100}{[\text{No. of healthy hosts} * \text{ P/H}] + \text{No. of parasitized hosts}} \quad (11.1)$$

This correction is most effective when sampling intervals are short relative to the development periods of the host and parasitoid, but can still provide a useful improvement in the estimation of rate of parasitism when sample intervals are longer.

Rates of parasitism over specific intervals of time can also be used to estimate the influence of various environmental factors on parasitism of hosts with either discrete or overlapping generations. In this case, parasitism is best estimated through use of trap hosts exposed to parasitism for short intervals of time under different environmental conditions. This approach has frequently been used to monitor the effectiveness of mass-released parasitoids in inundative biological control programs (e.g. Petersen et al. 1995, Wang et al. 1999), and to assess the influence of landscape diversity on parasitism (e.g. Cappuccino et al. 1998, Menalled et al. 1999).

General sampling concerns

For all these approaches, samples of hosts must be large enough to provide reasonable accuracy in the estimation of a percentage, must be unbiased in their representation of parasitized and healthy hosts, and must be free from cross-contamination after collection. One important source of bias in sampling host populations in the field is the differential spatial distribution of parasitized and healthy individuals. Heterogeneous parasitism can result from random or aggregated attack (Olson et al. 2000), greater parasitoid activity at the edge of a host population (Lopez et al. 1990, Brodmann et al. 1997), or parasitized hosts moving away from the main host population (Ryan 1985, Lopez et al. 1990). The collection of hosts from a greater number of quadrats within the sample site, the use of transects to collect samples, and knowledge of the influence of parasitism on host behavior can all help to minimize bias from heterogeneity. In addition, when hosts are reared to await emergence of parasitoids, considerable care must be taken to prevent cross-contamination among hosts through parasitism by emerging parasitoids. Fortunately, this can readily be avoided by separating hosts into individual containers as they are collected from the field.

Once a sample of hosts has been collected from the field they are either reared to estimate apparent parasitism, the observed percentage of hosts *killed* by a parasitoid, or they are dissected to estimate the marginal attack rate, the percentage of hosts *attacked* by a parasitoid. This is an important distinction, as the two measurements can provide very different estimates of parasitism from the same sample if there is differential mortality of parasitized and healthy hosts or interference between contemporaneous mortality factors (Waage & Mills 1992, Royama 2001). Royama (1981, 2001) and Elkinton et al. (1992) provide detailed discussions of how to separate the marginal death rates, based on rearing data, of contemporaneous mortality factors. Although dissection of hosts, if based on use of short marker stages for parasitism, provides a more direct measure of the marginal death rate, this approach is dependent upon accurate detection of parasitoid eggs or young larvae, which can often be concealed within host tissues. Recent studies using DNA markers, however, indicate that species-specific parasitism can be detected 24 hours after parasitoid oviposition, offering new opportunities in the accuracy of estimating marginal death rates (Tilmon et al. 2000, Zhu et al. 2000, Ratcliffe et al. 2002, Zhu & Williams 2002).

Predation

Predation is rather more difficult to measure than parasitism for two reasons: the interaction between a predator and its prey is very brief, a matter of minutes in most cases, which severely reduces the chances of detection by an observer; and very often there are few, if any, remains of the prey that can be detected after predation has taken place. As a result there are two basic approaches that can be used to estimate predation under natural conditions. In the few cases where remains of prey are detectable after a predation event, percent predation provides a direct estimate of the impact of a predator population. In most cases, however, the impact of predation must be estimated indirectly as a predation rate based on a combination of predator abundance and per capita rate of consumption.

Percent predation

Eggs, pupae, and the coverings of Homoptera often remain intact for a sufficient length of time after predation has occurred to be used to directly quantify percent predation from a single sample, as was the case for percent parasitism. Persistent structural components of the prey can also provide distinctive evidence of the type of predator: for example, a pair of small holes result from predation by Neuroptera, a single hole from Heteroptera, and a peppering of small holes from ants. Andow (1990) differentiated the attack of several different insect predators on eggs of the European corn borer, and examples of successful measurement of predation from prey remains include whitefly nymphs (Heinz & Parrella 1994), European corn borer eggs (Andow 1992), and gypsy moth pupae (Cook et al. 1994). In addition, sessile prey, such as eggs and pupae, can readily be placed out in the field at natural densities to monitor losses from predation, but it is important that the sentinel prey are no more or less susceptible to predation than the wild population. For example, Andow (1992) exposed egg masses of the European corn borer to monitor predation in corn fields under different tillage systems, and Cook et al. (1994, 1995) exposed freeze-dried pupae of gypsy moth in the field to monitor predation by small mammals. Mobile stages of prey, such as larvae or adults, can be exposed to predation for limited time periods by tethering with strong sewing thread. Tethering obviously affects the behavior of the prey, however, and thus predation estimated in this way must carefully correct for the greater susceptibility of the prey (see Weseloh [1990] for an example).

Predation rate

In contrast to percent predation, predation rate (N_a) is a measure of the biomass (or number) of prey eaten per unit area per unit time. Estimating predation rate requires sampling for predator abundance and for per capita rate of consumption by the predator population. The predation rate is then given by:

$$N_a = \sum_{i=1}^{n} P_i R_i (p_i / D_i) \qquad (11.2)$$

where P_i is the density of predators of stage i per unit area, R_i is the per capita rate of consumption of prey by an individual predator of stage i per unit time, p_i is the proportion of predators of stage i that test positive for gut detection of prey, D_i is the detection interval for prey in the gut of a predator of stage i, and n is the number of predator stages. The bracketed term is included only when gut detection is the approach used to estimate predation events (see below).

Predator density (P)

The abundance of predators is most commonly monitored through use of sweep-netting, vacuum sampling, or visual counts of individuals either directly in the field or from destructive plant samples brought into the laboratory for closer examination. Predator individuals must be classified by stage of development (instar) due to change in predation potential with age, and samples must be taken at regular intervals to show changes in age structure of the predator population over the period of interest. In addition, a sampling plan must be devised to optimize the accuracy of predator density estimation, based on the number, size, and cost of samples to be taken (Chapter 1). In general, the accuracy of estimates is enhanced by use of a greater number of smaller-sized samples than fewer large-sized samples, as found by Ellington and Southward (1999) for monitoring the abundance of predators in cotton.

Rate of consumption (R)

There are three approaches to the estimation of the per capita rate of consumption of predators in each of their successive stages of development: laboratory estimation, direct observation, and gut detection (Table 11.2).

Laboratory estimation is based on monitoring the number of prey consumed per unit time by individual predators in artificial arenas held at constant temperature (representing the average temperatures experienced under natural conditions) with excess prey. These estimates represent the stage-specific maximum per capita feeding rates of the predator under ideal conditions, without the need to search for and subdue prey and where metabolic costs are mini-

Table 11.2 A summary of the methodological approaches for estimating the rate of consumption by insect predators.

Methodology	Application	Examples
Laboratory estimation	Used to estimate maximum stage-specific per capita feeding rates under ideal conditions	Tamaki et al. (1974), Ro & Long (1998)
Direct observation	Uses multiple point observations of stage-specific per capita feeding activity under field conditions	Edgar (1970), van den Berg et al. (1997)
Gut detection: dissection	Effective if heavily chitinized parts of the prey are consumed	Sunderland (1975), Breene et al. (1990)
Gut detection: electrophoresis	Esterase bands; easy to develop, low specificity, moderate retention time	Jones & Morse (1995), Solomon et al. (1996)
Gut detection: immunology	Monoclonal antibodies; difficult to develop, high specificity, moderate retention time	Greenstone (1996), Hagler (1998)
Gut detection: PCR	PCR products; easy to develop, high specificity, short retention time	Chen et al. (2000), Greenstone & Shufran (2003)

mized. This approach was developed by Bombosch (1963) and Tamaki et al. (1974) to estimate the impact of predation on aphid populations, and has subsequently been used by Chambers and Adams (1986) and Ro and Long (1998).

Direct observation is not only the best way to assess the range of predators attacking a particular prey population and the diet breadth of individual predator species, but can also provide stage-specific estimates of rate of consumption under natural conditions. Edgar (1970) was among the first to use this approach to estimate predation rates of lycosid spiders from observations of the number of hours a day that the predators are active (t_a), the mean proportion of time that an average predator spends actively feeding (f) and the mean time in hours that a predator takes to fully consume a single prey item (t_c). The rate of consumption (R), or the number of prey consumed by an individual predator per day is given by:

$$R = t_a f / t_c \qquad (11.3)$$

The mean proportion of time spent feeding by an average predator (f) is estimated from a series of point samples of predator activity (proportion observed feeding) taken at intervals during the active hours of a day and throughout the period of interest. As the proportion of time spent feeding will vary with prey density, point samples of predator activity should be repeated across a series of plots representing a gradient of prey abundance. The predator activity period (t_a) is estimated from direct field observation and predator consumption time (t_c) can be estimated either directly in the field or from laboratory observation. The direct observation approach has subsequently been used to estimate spider predation in a variety of crops (Kiritani et al. 1972, Sunderland et al. 1986, Nyffeler et al. 1987) and a variant of this technique has been used to quantify aphid predation (van den Berg et al. 1997). It can readily be applied to many types of predators that can be observed without disturbance under natural conditions, and has even been used to estimate rates of parasitism of bark beetles (Mills 1991).

Gut detection requires positive identification of prey remains in the gut of field-collected predators. In early studies, detection of prey remains relied on dissection and visual inspection (e.g. Sunderland 1975, Hildrew & Townsend 1982, Breene et al. 1990). Subsequently, electrophoretic analysis (Powell et al. 1996, Solomon et al. 1996) was used, allowing predators with extra-oral digestion as well as chewing predators to be studied, and most recently immunological (Greenstone 1996, Hagler 1998) and PCR analysis (Chen et al. 2000, Greenstone & Shufran 2003) have been developed for prey detection. An important caveat associated with gut detection of prey is that a positive for detection can result from saprophagy, necrophagy, and higher-order predation as well as from primary predation, and thus prior knowledge of the feeding habits of the predator is essential. Of the newer techniques available, electrophoretic analysis is the least sensitive, as the same esterase bands may be shared by a specific prey species, the alternative foods of the predator, the prey's food, and even the gut wall of the predator. Recent examples, highlighting some of these difficulties, can be found in Giller (1986), Lister et al. (1987) and Jones and Morse (1995). Immunological analysis is based on the development of monoclonal antibodies for the prey of interest and their detection in the gut of a predator through one of several gut content immunoassays (see Hagler 1998 for specific details of techniques). Monoclonal antibodies are very specific, either for prey species or stages, and are considered better for gut detection than poly-clonal antibodies. They have distinct disadvantages, however, in that they are difficult and costly to produce (hence available for just a handful of prey species; Greenstone 1996). Most recently PCR analysis has shown promise for the detection of prey-specific DNA fragments in the guts of insect predators (Zaidi et al. 1999, Chen et al. 2000, Hoogendorn & Heimpel 2001) and spiders (Agusti et al. 2003, Greenstone & Shufran 2003). The shorter development time, lower cost, and greater certainty of analysis is likely to make PCR analysis of gut detection more widely applicable than immunological analysis in the future.

Whichever method is used for prey detection in the guts of the predators, the results provide an estimate of the proportion of predators (p) that show positive for feeding on the prey species of interest. It must be remembered, however, that these detection events are qualitative rather than quantitative as a predator will show positive whether it has eaten one or several prey items. Thus to esti-mate predation rate, firstly the detection interval (D) must be experimentally determined to allow the observed proportion of predators with positive re-sponses to be corrected for loss of detectable prey remains, and secondly the rate of consumption (R) must be experimentally determined. The latter has either been assumed to be equivalent to just a single prey item (Dempster 1960), or has been estimated using the laboratory estimation approach discussed above. Re-cently, however, Naranjo and Hagler (2001) have recommended incorporating a functional response experiment with gut detection to estimate the relation-ship between R and prey density and thus to improve the estimation of preda-tion rate.

Diversity of entomophagous species

With the increasing need for monitoring ecosystem health and biodiversity in general, predators and occasionally parasitoids have been used as indicator groups. The sensitivity of spider assemblages to environmental conditions has led to their use as an indicator group in the monitoring of restoration land-scapes. For example, spiders provide indicator species for restoration activities in redwood forests (Willett 2001), and have proved valuable in monitoring the reclamation of limestone quarries (Wheater et al. 2000). Similarly, carabid beetle assemblages have been used as indicator species to assess management practices in cereal crops and grasslands (Luff 1996), landscape changes due to human activity (Niemelä et al. 2000), and more generally for habitat disturbance (Rainio & Niemelä 2003). Ant assemblages have also been shown to be valuable bioindicators of biodiversity responses to pollution (Madden & Fox 1999, Andersen et al. 2002).

Parasitoid assemblages have not been used as general indicators of ecosystem health, but braconid parasitoids have been suggested to be a valuable indicator taxon for disturbance in forest stands (Lewis & Whitfield 1999). In addition, parasitoids have proved to be a useful group for investigating latitudinal gradients in species diversity (Janzen 1981, Hawkins 1994, Sime & Brower 1998) and for estimation of global biodiversity (Bartlett et al. 1999, Dolphin & Quicke 2001).

Monitoring the species richness of entomophagous guilds in terrestrial communities has made use of a variety of sampling techniques. The type of habitat is probably the most important determinant of the sampling technique to employ. For forest trees, insecticide fogging is clearly the only way to adequately sample tree canopies (Ozanne et al. 2000, Majer et al. 2001; Chapter 7). In contrast, for epigeal predators, pitfall trapping has been shown to be effective for carabids (Larsen & Williams 1999; Chapter 3), spiders (Brennan et al. 1999), and ants (Miller & New 1997). The type of habitat can also influence the efficacy of different sampling methods for the same taxon of predators. For example, visual searching proved to be the most efficient way to sample spiders in citrus orchards (Amalin et al. 2001), whereas pitfall traps were more effective than visual search for spiders in a heathland landscape (Churchill & Arthur 1999).

In comparing five different sampling methods to study the diversity of parasitoids in the forests of Sulawesi, Noyes (1989) found insecticide fogging to be the most effective, followed by sweep-netting, Malaise traps, yellow water traps and lastly intercept traps. Malaise traps (Chapter 4) have been used extensively in the sampling of parasitoid communities for biodiversity studies. Townes (1972) describes the design of a Malaise trap suitable for sampling parasitoids, which has been used effectively to monitor the larger Ichneumonidae (Gaston & Gauld 1993, Gaasch et al. 1998) and Braconidae (Lewis & Whitfield 1999), as well as smaller taxa such as Mymaridae (Noyes 1989). Malaise traps have also been widely used to monitor the diversity of syrphid predators in a variety of agricultural landscapes (Hondelmann 1998, Salveter 1998). The advantage of Malaise traps is that they can be left *in situ* for longer periods of time, as they can collect directly into preserving materials. However, the placement of the traps can have an important influence on the number of insects trapped (Chapter 7).

Yellow pan traps (Chapter 6) have also proved valuable for sampling the diversity of parasitoids in orchard (Purcell & Messing 1996), forest (Villemant & Andrei-Ruiz 1999), and fen (Finnamore 1994) landscapes. In addition, yellow sticky traps (Chapters 5 & 6) have been used to monitor the diversity of parasitoids on Bahamian islands (Schoener et al. 1995) and of coccinellids in a series of agricultural landscapes (Colunga-Garcia et al. 1997). Other sampling methods used to investigate the diversity of parasitoid and predator communities include sweep-net sampling (Chapter 4) in old field successions (Siemann et al. 1999) and suction sampling (Chapters 4 & 6) in grasslands (Harper et al. 2000). Noyes (1982) describes the construction of a sweep net that has been found to be particularly suitable for sampling smaller insect parasitoids, and Noyes (2003) provides a detailed account of the full range of sampling methods used for collecting smaller parasitoids, particularly the Chalcidoidea.

Duelli et al. (1999) and Duelli and Obrist (2003) provide useful insights for ways in which sampling for biodiversity of entomophagous species can be optimized in cultivated landscapes, and Kitching et al. (2001) discuss the need to use packages of different sampling methods to be effective in the assessment of arthropod biodiversity. One final consideration when sampling entomophagous communities is that many species of parasitoids and predators are rather less common than their hosts or prey. This suggests a need for greater sampling effort when planning the number of sites, number of samples, and size of each sample for an inventory survey.

References

Agusti, N., Shayler, S.P., Harwood, J.D., Vaughan, I.P., Sunderland, K.D., & Symondson, W.O.C. (2003) Collembola as alternative prey sustaining spiders in arable ecosystems: prey detection within predators using molecular markers. *Molecular Ecology*, **12**, 3467–3475.

Akey, D.H. & Burns, D.W. (1991) Analytical consideration and methodologies for elemental determinations in biological samples. *Southwestern Entomologist*, **14** (Suppl.), 25–36.

Amalin, D.M., Pena, J.E., McSorley, R., Browning, H.W., & Crane, J.H. (2001) Comparison of different sampling methods and effect of pesticide application on spider populations in lime orchards in south Florida. *Environmental Entomology*, **30**, 1021–1027.

Andersen, A.N., Hoffmann, B.D. Muller, W.J., & Griffiths A.D. (2002) Using ants as bioindicators in land management: simplifying assessment of ant community responses. *Journal of Applied Ecology* **39**, 8–17.

Andow, D.A. (1990) Characterization of predation on egg masses of *Ostrinia nubilalis* (Lepidoptera: Pyralidae). *Annals of the Entomological Society of America*, **83**, 482–486.

Andow, D.A. (1992) Fate of eggs of first generation *Ostrinia nubilalis* (Lepidoptera: Pyralidae) in three conservation tillage systems. *Environmental Entomology*, **21**, 388–393.

Askew, R.R. & Shaw, M.R. (1986) Parasitoid communities: their size, structure and development. In *Insect Parasitoids* (ed. J. Waage & D. Greathead), pp. 225–264. Academic Press, London.

Bartlett, R., Pickering, J., Gauld, I., & Windsor, D. (1999) Estimating global biodiversity: tropical beetles and wasps send different signals. *Ecological Entomology*, **24**, 118–121.

Bellows, T.S., Van Driesche, R.G., & Elkinton, J.S. (1992) Life-table construction and analysis in the evaluation of natural enemies. *Annual Review of Entomology*, **37**, 587–614.

Blumberg, D. & Van Driesche, R.G. (2001) Encapsulation rates of three encyrtid parasitoids by three mealybug species (Homoptera: Pseudococcidae) found commonly as pests in commercial greenhouses. *Biological Control*, **22**, 191–199.

Bombosch, S. (1963) Untersuchungen zur vermehrung von *Aphis fabae* Scop. in Samenrübenbestanden unter besonderer Berücksichtigung der Schwebfliegen (Diptera: Syrphidae). *Zeitschrift für Angewandte Entomologie*, **52**, 105–141.

Bowie, M.H., Gurr, G.M., Hossain, Z., Baggen, L.R., & Frampton, C.M. (1999) Effects of distance from field edge on aphidophagous insects in a wheat crop and observations on trap design and placement. *International Journal of Pest Management*, **45**, 69–73.

Breene, R.G., Sweet, M.H., & Olson, J.K. (1990) Analysis of the gut contents of naiads of *Enallagma civile* (Odonata: Coenagrionidae) from a Texas pond. *Journal of the American Mosquito Control Association*, **6**, 547–548.

Brennan, K.E.C., Majer, J.D., & Reygaert, N. (1999) Determination of an optimal pitfall trap size for sampling spiders in a Western Australian Jarrah forest. *Journal of Insect Conservation*, **3**, 297–307.

Brodmann, P.A., Wilcox, C.V., & Harrison, S. (1997) Mobile parasitoids may restrict the spatial spread of an insect outbreak. *Journal of Animal Ecology*, **66**, 65–72.

Brown, M.W. & Schmitt, J.J. (2001) Seasonal and diurnal dynamics of beneficial insect populations in apple orchards under different management intensity. *Environmental Entomology*, **30**, 415–424.

Cappuccino, N., Lavertu, D., Bergeron, Y., & Regniere, J. (1998) Spruce budworm impact, abundance and parasitism rate in a patchy landscape. *Oecologia*, **114**, 236–242.

Chambers, R.J. & Adams, T.H.L. (1986) Quantification of the impact of hoverflies (Diptera: Syrphidae) on cereal aphids in winter wheat: an analysis of field populations. *Journal of Applied Ecology*, **23**, 895–904.

Chen, Y., Giles, K.L., Payton, M.E., & Greenstone, M.H. (2000) Identifying key cereal aphid predators by molecular gut analysis. *Molecular Ecology*, **9**, 1887–1898.

Churchill, T.B. & Arthur, J.M. (1999) Measuring spider richness: effects of different sampling methods and spatial and temporal scales. *Journal of Insect Conservation*, **3**, 287–295.

Colunga-Garcia, M., Gage, S.H., & Landis, D. (1997) Response of an assemblage of Coccinellidae (Coleoptera) to a diverse agricultural landscape. *Environmental Entomology*, **26**, 797–804.

Colwell, R.K. (1997) EstimateS: Statistical estimation of species richness and shared species from samples. Version 5. User's guide and application published at http://viceroy.eeb. uconn.edu/estimates.

Cook, S.P., Hain, F.P., & Smith, H.R. (1994) Oviposition and pupal survival of gypsy moth (Lepidoptera: Lymantriidae) in Virginia and North Carolina pine–hardwood forests. *Environmental Entomology*, **23**, 360–366.

Cook, S.P., Smith, H.R., Hain, F.P., & Hastings, F.L. (1995) Predation of gypsy moth (Lepidoptera: Lymantriidae) pupae by invertebrates at low small mammal population densities. *Environmental Entomology*, **24**, 1234–1238.

Costello, M.J. & Daane, K.M. (1997) Comparison of sampling methods used to estimate spider (Araneae) species abundance and composition in grape vineyards. *Environmental Entomology*, **26**, 142–149.

Dawah, H.A., Hawkins, B.A., & Claridge, M.F. (1995) Structure of the parasitoid communities of grass-feeding chalcid wasps. *Journal of Animal Ecology*, **64**, 708–720.

Dempster, J.P. (1960) A quantitative study of the predators of the eggs and larvae of the broom beetle, *Phytodecta olivacea* Forster, using the precipitin test. *Journal of Animal Ecology*, **29**, 149–167.

Diaz-Aranda, L.M. & Monserrat, V.J. (1995) Aphidophagous predator diagnosis: Key to genera of European chrysopid larvae (Neur.: Chrysopidae). *Entomophaga*, **40**, 169–181.

Doane, C.C. (1971) A high rate of parasitization by *Brachymeria intermedia* (Hymenoptera: Chalcididae) on the gypsy moth. *Annals of the Entomological Society of America*, **64**, 753–754.

Dolphin, K. & Quicke, D.L.J. (2001) Estimating the global species richness of an incompletely described taxon: an example using parasitoid wasps (Hymenoptera: Braconidae). *Biological Journal of the Linnean Society*, **73**, 279–286.

Duelli, P. & Obrist, M.K. (2003) Biodiversity indicators: the choice of values and measures. *Agriculture Ecosystems & Environment*, **98**, 87–98.

Duelli, P., Obrist, M.K., & Schmatz, D.R. (1999) Biodiversity evaluation in agricultural landscapes: above-ground insects. *Agriculture Ecosystems & Environment*, **74**, 33–64.

Edgar, W.D. (1970) Prey and feeding behaviour of adult females of the wolf spider *Pardosa amentata* (Clerk). *Netherlands Journal of Zoology*, **20**, 487–491.

Elkinton, J.S., Buonaccorsi, J.P., Bellows, T.S., & Van Driesche, R.G. (1992) Marginal attack rate, k-values and density dependence in the analysis of contemporaneous mortality factors. *Researches on Population Ecology*, **34**, 29–44.

Ellington, J. & Southward, M. (1999) Quadrat sample precision and cost with a high-vacuum insect sampling machine in cotton ecosystems. *Environmental Entomology*, **28**, 722–728.

Elliott, N.C. & Kieckhefer, R.W. (1990) Dynamics of aphidophagous coccinellid assemblages in small grain fields in eastern South Dakota. *Environmental Entomology*, **19**, 1320–1329.

Finnamore, A.T. (1994) Hymenoptera of the Wagner Natural Area, a boreal spring fen in central Alberta. *Memoirs of the Entomological Society of Canada*, **169**, 181–220.

Floate, K., Khan, B., & Gibson, G. (1999) Hymenopterous parasitoids of filth fly (Diptera: Muscidae) pupae in cattle feedlots. *Canadian Entomologist*, **131**, 347–362.

Gaasch, C.M., Pickering, J., & Moore, C.T. (1998) Flight phenology of parasitic wasps (Hymenoptera: Ichneumonidae) in Georgia's Piedmont. *Environmental Entomology*, **27**, 606–614.

Gaston, K.J. & Gauld, I.D. (1993) How many species of pimplines (Hymenoptera: Ichneumonidae) are there in Costa Rica? *Journal of Tropical Ecology*, **9**, 491–499.

Giller, P.S. (1986) The natural diet of the Notonectidae: field trials with electrophoresis. *Ecological Entomology*, **11**, 163–172.

Gordon, R.D. & Vandenberg, N. (1995) Larval systematics of North American *Coccinella* L. (Coleoptera: Coccinellidae). *Entomologica Scandinavica*, **26**, 67–86.

Gould, J.R., Van Driesche, R.G., Elkinton, J.S., & Odell, T.M. (1989) A review of techniques for measuring the impact of parasitoids of lymantriids. In *The Lymantriidae: a Comparison of Features of New and Old World Tussock Moths* (ed. W.E. Wallner), pp. 517–531. USDA Forest Service General Technical Report NE-123.

Gould, J.R., Elkinton, J.S., & Wallner, W.E. (1990) Density-dependent suppression of experimentally created gypsy moth, *Lymantria dispar* (Lepidoptera: Lymantriidae), populations by natural enemies. *Journal of Animal Ecology*, **59**, 213–234.

Gould, J.R., Elkinton, J.S., & Van Driesche, R.G. (1992a) Suitability of approaches for measuring parasitoid impact on *Lymantria dispar* (Lepidoptera: Lymantriidae) populations. *Environmental Entomology*, **21**, 1035–1045.

Gould, J.R., Elkinton, J.S., & Van Driesche, R.G. (1992b) Assessment of potential methods of measuring parasitism by *Brachymeria intermedia* (Nees) (Hymenoptera: Chalcididae) of pupae of the gypsy moth. *Environmental Entomology*, **21**, 394–400.

Greenstone, M.H. (1996) Serological analysis of arthropod predation: past, present and future. In *The Ecology of Agricultural Pests: Biochemical Approaches* (ed. W.O.C. Symondson & J.E. Liddell), pp. 265–300. Chapman & Hall, London.

Greenstone, M.H. & Shufran, K.A. (2003) Spider predation: species-specific identification of gut contents by polymerase chain reaction. *Journal of Arachnology*, **31**, 131–134.

Gutierrez, A.P. (1996) *Applied Population Ecology*. Wiley, New York.

Hagler, J.R. (1998) Variation in the efficacy of several predator gut content immunoassays. *Biological Control*, **12**, 25–32.

Hagler, J.R. & Durand, C.M. (1994) A new method for immunologically marking prey and its use in predation studies. *Entomophaga*, **39**, 257–265.

Harper, M.G., Dietrich, C.H., Larimore, R.L., & Tessene, P.A. (2000) Effects of prescribed fire on prairie arthropods: an enclosure study. *Natural Areas Journal*, **20**, 325–335.

Hawkins, B A. (1988) Species diversity in the third and fourth trophic levels patterns and mechanisms. *Journal of Animal Ecology*, **57**, 137–162.

Hawkins, B.A. (1994) *Pattern and Process in Host–Parasitoid Interactions*. Cambridge University Press, Cambridge.

Heinz, K.M. & Parrella, M.P. (1994) Biological control of *Bemisia argentifolii* (Homoptera: Aleyrodidae) infesting *Euphorbia pulcherrima*: evaluations of releases of *Encarsia luteola* (Hymenoptera: Aphelinidae) and *Delphastus pusillus* (Coleoptera: Coccinellidae). *Environmental Entomology*, **23**, 1346–1353.

Herting, B. (1975) *A Catalogue of Parasites and Predators of Terrestrial Arthropods. Volume VI, Part 1.* Commonwealth Agricultural Bureaux, Farnham Royal.

Hildrew, A.G. & Townsend, C.R. (1982) Predators and prey in a patchy environment: a fresh-water study. *Journal of Animal Ecology*, **51**, 797–815.

Hill, M.G. (1988) Analysis of the biological control of *Mythimna separata* (Lepidoptera: Noctuidae) by *Apanteles ruficrus* (Braconidae: Hymenoptera) in New Zealand. *Journal of Applied Ecology*, **25**, 197–208.

Hodek, I. (1973) *Biology of Coccinellidae*. Junk, The Hague.

Hodek, I. (1993) Prey and habitat specificity in aphidophagous predators (a review). *Biocontrol Science and Technology*, **3**, 91–100.

Hondelmann, P. (1998) Hoverflies (Diptera: Syrphidae) of agrarian ecosystems: study of the Loess Plains in southern Lower Saxony. *Drosera*, **98**, 113–122.

Hoogendoorn, M. & Heimpel, G.E. (2001) PCR-based gut content analysis of insect predators: using ribosomal ITS-1 fragments from prey to estimate predation frequency. *Molecular Ecology*, **10**, 2059–2067.

Janzen, D.H. (1981) The peak in North American ichneumonid species richness lies between 38° and 42°N. *Ecology*, **62**, 532–537.

Johnson, P.C. & Reeves, R.M. (1995) Incorporation of the biological marker rubidium in gypsy moth (Lepidoptera: Lymantriidae) and its transfer to the predator *Carabus nemoralis* (Coleoptera: Carabidae). *Environmental Entomology*, **24**, 46–51.

Jones, S.A. & Morse, J.G. (1995) Use of isoelectric focusing electrophoresis to evaluate citrus thrips (Thysanoptera: Thripidae) predation by *Euseius tularensis* (Acari: Phytoseiidae). *Environmental Entomology*, **24**, 1040–1051.

Kean, J.M. & Barlow, N.D. (2001) A spatial model for the successful biological control of *Sitona discoideus* by *Microctonus aethiopoides*. *Journal of Applied Ecology*, **38**, 162–169.

Keating, K.A. & Quinn, J.F. (1998) Estimating species richness: the Michaelis–Menten model revisted. *Oikos*, **81**, 411–416.

Kiritani, K., Kawahara, S., Sasaba, T., & Nakasuji, F. (1972) Quantitative evaluation of predation by spiders on the green rice leafhopper, *Nephotettix cincticeps*, by a sight-count method. *Researches on Population Ecology*, **13**, 187–200.

Kitching, R.L., Li, D., & Stork, N.E. (2001) Assessing biodiversity "sampling packages": how similar are arthropod assemblages in different tropical rainforests? *Biodiversity and Conservation*, **10**, 793–813.

Kitt, J.T. & Keller, M.A. (1998) Host selection by *Aphidius rosae* Haliday (Hym., Braconidae) with respect to assessment of host specificity in biological control. *Journal of Applied Entomology*, **122**, 57–63.

Larsen, K.J. & Williams, J.B. (1999) Influence of fire and trapping effort on ground beetles in a reconstructed tallgrass prairie. *Prairie Naturalist*, **31**, 75–86.

Lewis, C.N. & Whitfield, J.B. (1999) Braconid wasp (Hymenoptera: Braconidae) diversity in forest plots under different silvicultural methods. *Environmental Entomology*, **28**, 986–997.

Lister, A., Usher, M.B., & Block, W. (1987) Description and quantification of field attack rates by predatory mites: an example using an electrophoresis method with a species of Antarctic mite. *Oecologia*, **72**, 185–191.

Lopez, E.R. & Van Driesche, R.G. (1989) Direct measurement of host and parasitoid recruitment for assessment of total losses due to parasitism in a continuously breeding species, the cabbage aphid, *Brevicoryne brassicae* (L.) (Hemiptera: Aphididae). *Bulletin of Entomological Research*, **79**, 47–60.

Lopez, E.R., Van Driesche, R.G., & Elkinton, J.S. (1990) Rates of parasitism by *Diaeretiella rapae* (Hymenoptera: Braconidae) for cabbage aphids (Homoptera: Aphididae) in and outside of colonies: why do they differ? *Journal of the Kansas Entomological Society*, **63**, 158–165.

Luck, R.F., Shepard, B.M., & Kenmore, P.E. (1988) Experimental methods for evaluating arthropod natural enemies. *Annual Review of Entomology*, **33**, 367–391.

Luff, M.L. (1996) Use of carabids as environmental indicators in grasslands and cereals. *Annales Zoologici Fennici*, **33**, 185–195.

Madden, K.E. & Fox, B.J. (1999) Arthropods as indicators of the effects of fluoride pollution on the succession following sand mining. *Journal of Applied Ecology*, **34**, 1239–1256.

Majer, J.D., Kitching, R.L., Heterick, B.E., Hurley, K., & Brennan, K.E.C. (2001) North–south patterns within arboreal ant assemblages from rain forests in Eastern Australia. *Biotropica*, **33**, 643–661.

Menalled, F.D., Marino, P.C., Gage, S.H., & Landis, D.A. (1999) Does agricultural landscape structure affect parasitism and parasitoid diversity? *Ecological Applications*, **9**, 634–641.

Miller, L.J. & New, T.R. (1997) Mount Piper grasslands: pitfall trapping of ants and interpretation of habitat variability. *Memoirs of the Museum of Victoria*, **56**, 377–381.

Mills, N.J. (1991) Searching strategies and attack rates of parasitoids of the ash bark beetle (*Leperisinus varius*) and its relevance to biological control. *Ecological Entomology*, **16**, 461–470.

Mills, N.J. (1994) Parasitoid guilds: defining the structure of the parasitoid communities of endopterygote insect hosts. *Environmental Entomology*, **23**, 1066–1083.

Mills, N.J. (1997) Techniques to evaluate the efficacy of natural enemies. In *Methods in Ecological and Agricultural Entomology* (ed. D.R. Dent & M.P. Walton), pp. 271–291. CAB International, Wallingford.

Monserrat, V.J., Oswald, J.D., Tauber, C.A., & Diaz-Aranda, L.M. (2001) Recognition of larval Neuroptera. In *Lacewings in the Crop Environment* (ed. P.K. McEwen, T.R. New, & A.E. Whittington), pp. 43–81. Cambridge University Press, Cambridge.

Muller, C.B., Adriaanse, I.C.T., Belshaw, R., & Godfray, H.C.J. (1999) The structure of an aphid–parasitoid community. *Journal of Animal Ecology*, **68**, 346–370.

Nakamura, K. & Abbas, I. (1987) Preliminary life table of the spotted tortoise beetle, *Aspidomorpha millaria* (Coleoptera: Chrysomelidae) in Sumatra. *Researches on Population Ecology*, **29**, 229–236.

Naranjo, S.E. & Hagler, J.R. (2001) Toward the quantification of predation with predator gut immunoassays: a new approach integrating functional response behavior. *Biological Control*, **20**, 175–189.

Niemelä, J., Kotze, J., Ashworth, A., et al. (2000) The search for common anthropogenic impacts on biodiversity: a global network. *Journal of Insect Conservation* **4**, 3–9.

Noyes, J.S. (1982) Collecting and preserving chalcid wasps (Hymenoptera: Chalcidoidea). *Journal of Natural History*, **16**, 315–334.

Noyes, J.S. (1989) A study of five methods of sampling Hymenoptera (Insecta) in a tropical rainforest, with special reference to the Parasitica. *Journal of Natural History*, **23**, 285–298.

Noyes, J.S. (2003) Universal Chalcidoidea database: collecting and preserving chalcidoids. http://www.nhm.ac.uk/entomology/chalcidoids/collecting1.html [accessed May 6, 2004].

Nyffeler, M., Dean, D.A., & Sterling, W.L. (1987) Evaluation of the importance of the striped lynx spider, *Oxyopes salticus* (Araneae: Oxyopidae), as a predator in Texas cotton. *Environmental Entomology*, **16**, 1114–1123.

Olson, A.C., Ives, A.R., & Gross, K. (2000) Spatially aggregated parasitism on pea aphids, *Acyrthosiphon pisum*, caused by random foraging behavior of the parasitoid *Aphidius ervi*. *Oikos*, **91**, 66–76.

Ozanne, C.M.P., Speight, M.R., Hambler, C., & Evans, H.F. (2000) Isolated trees and forest patches: patterns in canopy arthropod abundance and diversity in *Pinus sylvestris* (Scots Pine). *Forest Ecology and Management*, **137**, 53–63.

Parajulee, M.N. & Slosser, J.E. (1999) Evaluation of potential relay strip crops for predator enhancement in Texas cotton. *International Journal of Pest Management*, **45**, 275–286.

Petersen, J.J., Watson, D.W., & Cawthra, J.K. (1995) Comparative effectiveness of three release rates for a pteromalid parasitoid (Hymenoptera) of house flies (Diptera) in beef cattle feedlots. *Biological Control*, **5**, 561–565.

Porter, E.E. & Hawkins, B.A. (1998) Patterns of diversity for aphidiine (Hymenoptera: Braconidae) parasitoid assemblages on aphids (Homoptera). *Oecologia*, **116**, 234–242.

Powell, W., Walton, M.P., & Jervis, M.A. (1996) Populations and communities. In *Insect Natural Enemies. Practical Approaches to Their Study and Evaluation* (ed. M. Jervis & N. Kidd), pp. 223–292. Chapman & Hall, London.

Purcell, M.F. & Messing, R.H. (1996) Ripeness effects of three vegetable crops on abundance of augmentatively released *Psyttalia fletcheri* (Hym.: Braconidae): improved sampling and release methods. *Entomophaga*, **41**, 105–116.

Rainio, J. & Niemelä, J. (2003) Ground beetles (Coleoptera: Carabidae) as bioindicators. *Biodiversity and Conservation* **12**, 487–506.

Ratcliffe, S.T., Robertson, H.M., Jones, C.J., Bollero, G.A, & Weinzierl, R.A. (2002) Assessment of parasitism of house fly and stable fly (Diptera: Muscidae) pupae by pteromalid (Hymenoptera: Pteromalidae) parasitoids using a polymerase chain reaction assay. *Journal of Medical Entomology*, **39**, 52–60.

Rhoades, M.H. (1996) Key to first and second instars of six species of Coccinellidae (Coleoptera) from alfalfa in southwest Virginia. *Journal of the New York Entomological Society*, **104**, 83–88.

Ro, T.H. & Long, G.E. (1998) Population dynamics pattern of green peach aphid (Homoptera: Aphididae) and its predator complex in a potato system. *Korean Journal of Biological Sciences*, **2**, 217–222.

Rotheray, G.E. (1988) Third stage larvae of six species of aphidophagous Syrphidae Diptera. *Entomologist's Gazette*, **39**, 153–159.

Rotheray, G.E. (1993) *Colour Guide to Hoverfly Larvae (Diptera, Syrphidae) in Britain and Europe*. D. Whiteley, Sheffield.

Rotheray, G.E. & Gilbert, F.S. (1989) The phylogeny and systematics of European predacious Syrphidae (Diptera) based on larval and puparial stages. *Zoological Journal of the Linnean Society*, **95**, 29–70.

Rotheray, G.E. & Gilbert, F.S. (1999) Phylogeny of Palaearctic Syrphidae (Diptera): evidence from larval stages. *Zoological Journal of the Linnean Society*, **127**, 1–112.

Royama, T. (1981) Evaluation of mortality factors in insect life table analysis. *Ecological Monographs*, **5**, 495–505.

Royama, T. (2001) Measurement, analysis, and interpretation of mortality factors in insect survivorship studies, with reference to the spruce budworm, *Choristoneura fumiferana* (Clem.) (Lepidoptera: Tortricidae). *Population Ecology*, **43**, 157–178.

Russell, D.A. (1987) A simple method for improving estimates of percentage parasitism by insect parasitoids from field sampling of hosts. *New Zealand Entomologist*, **10**, 38–40.

Ryan, R.B. (1985) A hypothesis for decreasing parasitization of larch casebearer (Lepidoptera: Coleophoridae) on larch foliage by *Agathis pumila*. *Canadian Entomologist*, **117**, 1573–1574.

Salveter, R. (1998) Habitat use of adult syrphid flies (Diptera: Syrphidae) in a highly diversified agricultural landscape. *Mitteilungen der Schweizerischen Entomologischen Gesellschaft*, **71**, 49–71.

Schoener, T.W., Spiller, D.A., & Morrison, L.W. (1995) Variation in the hymenopteran parasitoid fraction on Bahamian islands. *Acta Oecologica*, **16**, 103–121.

Shaw, M.R. (1994) Parasitoid host ranges. In *Parasitoid Community Ecology* (ed. B.A. Hawkins & W. Sheehan), pp. 111–144. Oxford University Press, Oxford.

Shaw, M.R. (1997) Rearing parasitic Hymenoptera. *The Amateur Entomologist*, **25**, 1–46.

Siemann, E., Haarstad, J., & Tilman, D. (1999) Dynamics of plant and arthropod diversity during old field succession. *Ecography*, **22**, 406–414.

Sime, K.R. & Brower, A.V.Z. (1998) Explaining the latitudinal gradient anomaly in ichneumonoid species richness: evidence from butterflies. *Journal of Animal Ecology*, **67**, 387–399.

Solomon, M.G., Fitzgerald, J.D., & Murray, R.A. (1996) Electrophoretic approaches to predator–prey interactions. In *The Ecology of Agricultural Pests: Biochemical Approaches* (ed. W.O.C. Symondson & J.E. Liddell), pp. 457–468. Chapman & Hall, London.

Stadler, B. (2002) Determinants of the size of aphid–parasitoid assemblages. *Journal of Applied Entomology* 126, 258–264.

Stary, P. (1970) *Biology of Aphid Parasites (Hymenoptera: Aphidiidae) with Respect to Integrated Control*. Junk, The Hague.

Stireman J.O. & Singer M.C. (2002) Spatial and temporal variation in the parasitoid assemblage of an exophytic polyphagous caterpillar. *Ecological Entomology*, **27**, 588–600.

Sunderland, K.D. (1975) The diet of some predatory arthropods in cereal crops. *Journal of Applied Ecology*, **12**, 507–515.

Sunderland, K.D. (1988) Quantitative methods for detecting invertebrate predation occurring in the field. *Annals of Applied Biology*, **112**, 201–224.

Sunderland, K.D., Fraser, A.M., & Dixon, A.F.G. (1986) Field and laboratory studies on money spiders (Linyphiidae) as predators of cereal aphids. *Journal of Applied Ecology*, **23**, 433–447.

Tamaki, G., McGuire, J.U., & Turner, J.E. (1974) Predator power and efficiency: a model to evaluate their impact. *Environmental Entomology*, **3**, 625–630.

Tauber, C.A., De Leon, T., Penny, N.D., & Tauber, M.J. (2000) The genus *Ceraeochrysa* (Neuroptera: Chrysopidae) of America north of Mexico: larvae, adults, and comparative biology. *Annals of the Entomological Society of America*, **93**, 1195–1221.

Tilmon, K.J., Danforth, B.N., Day, W.H., & Hoffmann, M.P. (2000) Determining parasitoid species composition in a host population: a molecular approach. *Annals of the Entomological Society of America*, **93**, 640–647.

Townes, H. (1972) A light-weight Malaise trap. *Entomological News*, **83**, 239–247.

van den Berg, H., Ankash, D., Muhammad, A., et al. (1997) Evaluating the role of predation in population fluctuations of the soybean aphid *Aphis glycines* in farmers' fields in Indonesia. *Journal of Applied Ecology*, **34**, 971–984.

Van Driesche, R. (1983) Meaning of "percent parasitism" in studies of insect parasitoids. *Environmental Entomology*, **12**, 1611–1622.

Van Driesche, R. (1988) Field measurement of population recruitment of *Apanteles glomeratus* (L.) (Hymenoptera: Braconidae), a parasitoid of *Pieris rapae* (L.) (Lepidoptera: Pieridae), and factors influencing adult parasitoid foraging success in kale. *Bulletin of Entomological Research*, **78**, 199–208.

Van Driesche, R. & Bellows, T. (1988) Use of host and parasitoid recruitment in quantifying losses from parasitism in insect populations. *Ecological Entomology*, **13**, 215–222.

Van Driesche, R., Bellows, T.S., Elkinton, J.S., Gould, J.R., & Ferro, D.N. (1991) The meaning of percent parasitism revisited: solutions to the problem of accurately estimating total losses from parasitism. *Environmental Entomology*, **20**, 1–7.

van Lenteren, J.C. & van Roermund, J.W. (1999) Why is the parasitoid *Encarsia formosa* so successful in controlling whiteflies. In *Theoretical Approaches to Biological Control* (ed. B.A. Hawkins & H.V. Cornell), pp. 116–130. Cambridge University Press, Cambridge.

Villemant, C. & Andrei-Ruiz, M.-C. (1999) Diversity and spatial distribution of parasitoid hymenoptera in the green oak forest of the Fango valley (Corsica). *Annales de la Société Entomologique de France*, **35** (Suppl.), 259–262.

Waage, J.K. and Mills, N.J. (1992) Understanding and measuring the impact of natural enemies on pest populations. In *Biological Control Manual. Volume I. Principles and Practice of Biological Control* (ed. R.H. Markham, A. Wodageneh, & S. Agboola), pp. 84–114. International Institute of Tropical Agriculture, Cotonou, Benin.

Wang, B., Ferro, D.N., & Hosmer, D.W. (1999) Effectiveness of *Trichogramma ostriniae* and *T. nubilale* for controlling the European corn borer *Ostrinia nubilalis* in sweet corn. *Entomologia Experimentalis et Applicata*, **91**, 297–303.

Weseloh, R.M. (1990) Estimation of predation rates of gypsy moth larvae by exposure of tethered caterpillars. *Environmental Entomology*, **19**, 448–455.

Wheater, C.P., Cullen, W.R., & Bell, J.R. (2000) Spider communities as tools in monitoring reclaimed limestone quarry landforms. *Landscape Ecology*, **15**, 401–406.

Willett, T.R. (2001) Spiders and other arthropods as indicators in old-growth versus logged redwood stands. *Restoration Ecology*, **9**, 410–420.

Wyss, E. (1996) The effects of artificial weed strips on diversity and abundance of the arthropod fauna in a Swiss experimental apple orchard. *Agriculture Ecosystems and Environment*, **60**, 47–59.

Yamamura, K. (1999) Key-factor/key-stage analysis for life table data. *Ecology*, **80**, 533–537.

Zaidi, R.H., Jaal, Z., Hawkes, N.J., Hemingway, J., & Symondson, W.O.C. (1999) Can multiple-copy sequences of prey DNA be detected amongst the gut contents of invertebrate predators? *Molecular Ecology*, **8**, 2081–2087.

Zhu, Y.C., & Williams, L. (2002) Detecting the egg parasitoid *Anaphes iole* (Hymenoptera: Mymaridae) in tarnished plant bug (Heteroptera: Miridae) eggs by using a molecular approach. *Annals of the Entomological Society of America*, **95**, 359–365.

Zhu, Y.C., Burd, J.D., Elliott, N.C., & Greenstone, M.H. (2000) Specific ribosomal DNA marker for early polymerase chain reaction detection of *Aphelinus hordei* (Hymenoptera: Aphelinidae) and *Aphidius colemani* (Hymenoptera: Aphidiidae) from *Diuraphis noxia* (Homoptera: Aphididae). *Annals of the Entomological Society of America*, **93**, 486–491.

Index of methods and approaches

Methodology	Topics addressed	Comments
The composition of an entomophagous assemblage		
Sampling for parasitoids	Composition of the parasitoid complex of a specific host species.	Minimize contamination; sample all host stages and multiple sites.
Sampling for predators	Composition of the predator assemblage associated with a specific prey species.	Direct observation has many advantages over general sampling or gut detection.

Methodology	Topics addressed	Comments
The impact of an entomophagous species		
Percent parasitism	Stage-specific generational parasitism of a host with discrete generations.	Devise a sampling plan that provides accurate input for the ratio of hosts parasitized to susceptible hosts.
	Time-specific rate of parasitism for hosts with either discrete or continuous generations.	Easily biased by inclusion of instars in host samples that do not support parasitism or by an influence of parasitism on host development time.
Percent predation	Stage-specific generational predation of a prey with discrete generations.	Best with sentinel prey that have persistent structural components (eggs, pupae, scales).
Predation rate	Time-specific rate of prey consumption for prey with either discrete or continuous generations.	Requires regular sampling for predator abundance and accurate estimation of per capita rate of prey consumption by predator instars.
The biodiversity of entomophagous species		
Sampling for biodiversity	Indicators of ecosystem health.	Ants, carabid beetles, and spiders can be used as sensitive bioindicators.
	Latitudinal gradients of species richness.	Parasitoid taxa show contrasting latitudinal gradients of species richness.
	Global biodiversity.	Parasitoids can be used to estimate the richness of only partially known taxa.

Index

Note: Page numbers in *italics* refer to figures and those in **bold** to tables

Abbas, I. 261
Abbott, I. **150**, 156
Abe, T. 221, 225, 228–31
Abensperg-Traun, M. 41, 231
Abies balsamea 83
absolute density 186
abundance
  estimation of 3
  indices of 192
  see also activity–abundance concept
abundance index *see* relative abundance
abundance–frequency distribution 211, *211*, 212
Acanthotermes 223
Acarina 158, **167**
acetic acid 18
  in traps **43**–4
Acidnotermes 223
acoustic emission 233
acoustic monitors 238
Acrididae **78**
active traps 5, 66, 150
activity patterns, diel 45
activity–abundance concept 49–50, **57**
Adaiphrotermes 223
Adalia
  bipunctata 2
  decempunctata 2
Adam, E.E. 38, 42, 44
Adams, E.S. 224
Adams, T.H.L. 267
Addicott, J.F. 182
Adelges
  abietis 80–1, *81*
  tsugae 84

Adelgidae **78**, 80, 83–4
Aderitotermes 223
Adis, J. 44, 147, 150–1, 153, 229
Aedes 132
Aeolus mellillus 20
Africa 89, 138, 147, 226, 228
Agalychnis callidryas 174
Agassiz, D. 96
Agromyzidae **67, 76**
Agusti, N. 268
Aizen, M.A. 146
Akey, D.H. 258
alcohol 215
  in traps **43**–4
aldehydes 133
Aleutherocanthus woglumi 131
All, J.N. 23
Allsopp, P.G. 22
Alyscotermes 223
Alzugaray, M.D.R. 25
Amalin, D.M. 269
Amalotermes 223
America 96, 133
  see also Central America; North America; Panama; South America; United States
Amicotermes 223
Ammer, U. 147
ammonia 18
anaesthetic, aerosol 191
analysis of variance (ANOVA) 180
Andersen, A.N. 269
Anderson, J.R. 133, 179
Andis, M.D. 188
Andow, D.A. 265
Andrei-Ruiz, M.-C. 269

Andrena 128
Andricus **79**, 81
Anenteotermes 223
Anobiidae **78**, 110
Anobium **80**
ants 37, 41, 45, 49, 132
  associated with aphids 17–18
  as bioindicators 269
  ponerine 172
  predation by 265
Aphid Bulletin 121
Aphididae 16, **78**, 83, 255
aphids 2, 6, 11, 16, 30, **78**
  alate 61
  apterous 61
  balsam twig 83
  birch 13
  bird cherry 4
  cabbage 263
  cereal 8, **67**, **76**
  comparative catch study on 131
  European movement of 121–2
  extraction of *28*
  non-cereal **67**, **76**
  parasitoids of 255–6
  pest species 122
  predation of 267
  reproduction during sampling 27
  Rothamsted dataset for 122
  subterranean 17–18
  suction traps for 60–1, 120–1
  true 83
  woolly **78**, 83
    gall woolly **79**
    hemlock woolly 84, *85*
    pine woolly 77
    spruce woolly 80
Aphis fabae 122, 131
Aphodius 132
Apilitermes 223
Aporusa lagenocarpa **150**
Araneae 37, 58, **167**
Arctica 117
Armstrong, G. 45, 128, 137
armyworm 260
  beet 11
Arnold, A.J. 17
arthropods 24–5, 97–8, 150, 157
  arboreal 151
  canopy 146–7

  diel movement of 155
  flightless 157–8, **166**
  suspended-soil 146
Arthur, J.M. 269
Asia, South/Southeast 89, 147, 221–2
Askew, R.R. 254
aspirator (pooter) 69
Asplenium nidus 230
Astalotermes 223
Atchley, W.R. 206
Ateuchotermes 223
Atkinson, L. 224
Attfield, B. 83
Ausden, M. 26
Austin, A.D. 82
Australia 21, 82, 89, **103**, 147–8, 232, 239
autoradiography (δ^{13}C and ^{32}P) 29

Baars, M.A. 41–2, 44–5, 47–8
Badenhausser, I. 22
Baert, L. 49
bagging and clipping, index of methods and
        approaches **166**
Bahamas 269
baited traps 67, 121, 132–6
  animals as bait 133–4
  designs and modifications 132
  index of methods and approaches **145**
  wind effect on 135–6
baits 20–1, 120
  hexaflumuron 242–3
  pheromone 106, 123
  potato 21
  for subterranean larvae *20*
  sugar 117
Baker, T.C. 134
Bakke, A. 106
Bale, J.S. 122
Balogh, J. 62
bamboo traps 175
Banks, C.J. 137
Banksia marginata 82
Barber, H.S. 38, 45
Barbour, D.A. 136
bark sprays 159–60
  index of methods and approaches 167
Barker, M. 147
Barker, M.G. 147
Barlow, N.D. 259
Barndt, D. 45

Baroni-Urbani, C.   222, 226, 228
Barrera, R.   171, 180
barrier traps   45
Bartlett, R.   269
Basidentitermes   223
basin traps   12
Basset, Y.   15–65, 68, 147, 149, **150**,
    155–7
Bauerle, W.L.   149
Baylis, J.P.   29
Bazzaz, F.A.   206
Bean, J.   239
Beasley, V.R.   44
beating trays   69, **76**, 83, 257
Beattie, A.J.   149
bees   128
beetles   4, 46, 65, **76**, 110, 123
  adult, boring by   **79–80**
  ambrosia   **78–80**, **114–15**
  bark   13, **78–9**, 86–8, **114–15**
    attack density distribution of   *102*
    baits for   107
    flight patterns of   109, *109*
    flight trapping of   106, *107*
    interception trapping of   137
    parasitism in   267
    species accumulation curves   99–100,
      100
    survival/emergence data for   104
  carabid   268–9
  carrion   132
  color preference by   128
  dead-wood   66
  distortion by adhesive   129
  dung   132
  eucalyptus longhorn   94
  five-spined engraver   **103**
  ground   37, 137
  jewel   **78**
  ladybird   2, 13, 83
  larch bark   **103**
  longhorn   **78–80**, **114–15**
    egg laying by   90
    larvae of   86, 94
    popular longhorn   **79**
    tropical   101
  marked   210
  olive bark   104
  pine shoot   **79–80**, 86, *87*
  powder-post   **78**, **80**, **114–15**

predatory   11, 60, **75**
roundhead   **79**
rove   37
saproxylic   97
spruce bark   **79**, **103**, 107
surface-active   6
"woodworm"   **78**, 110
Behan-Pelletier, V.   147, 151, 158–9
Belcher, D.W.   20
Belgium   **103**
Belkin, J.N.   170
Bell, J.R.   70
Bellows, T.S.   258, **259**, 261
Belshaw, R.   63
Belshaw, R.D.   149
Benke, A.C.   211–13
benzoic acid, in traps   44
Berlese–Tullgren funnel   26, 229
  Blasdale version   26
Bernklau, E.J.   21
Best, V.   147
Bidlingmayer, W.L.   124–5, 137
Bignell, D.E.   221–2, 225, 229, 236
Binns, M.R.   21, *23*
biotrons   29
Bird, D.F.   206
Bjostad, L.B.   21
Black, H.I.J.   221, 228
Blackburn, T.M.   206
blackfly, citrus   131
Blackwell, A.   134
Blair, J.M.   26
Blanc, P.   149
Blank, R.H.   24
Blanton, C.M.   68, 155
Blasdale, P.   26
Blaustein, L.   186–220
Bloom, P.E.   42, 44, 47
Blossey, B.   16
Blumberg, D.   255
bollworm, pink   118, 136
Bombosch, S.   46, 267
booklice   77, **78**
Borden, J.H.   100
borers   **79–80**, 86
  European corn   129, 265
  lepidopteran   89
  mahogany shoot   89
  pine shoot   89
  teak beehole   110

varicose **79**
wood 109–10
Borneo 230, 233–4, **235**, *236*, **237**
Bostanian, N.J. 45
Bowden, J. 58, 117, 159
Bowie, M.H. 66–7, 257
Bracken, G.K. 17
Braconidae 82, 255, 269
  Aphidiinae 255
Bradshaw, W.E. 171
Bravery, A.F. 238–40
Brazil 232
  cerrado 231
Breene, R.G. 266–7
Brennan, K.E.C. 269
Briggs, J.B. 40, 42, 44
brine (saline solution) 18
  pipe method 18–*19*
  in traps **43**–4
Briones, M.J.I. 29
Britain 121, 125, 240
British Columbia 158
Brodmann, P.A. 264
bromeliads 168, 175, 181
Brower, A.V.Z. 269
Brown, F.S. 137
Brown, M.W. 257
Brown, V.K. 16, 21–2
Brunei **150**
Brust, G.E. 29
budworm, spruce 131, 136
Buffington, M.L. 59, 61–2, 70
Buprestidae **78**–9
Burgess, L. 117
Burke, D. 66
Burmaster, D.E. 212
Burns, D.W. 258
Buxton, R.D. 231
Byerly, K.F. 61–2, 68
Byers, J.A. *108*

cabbageworm 261
California *86*, 110
Calyptera 65, **75**
Cameroon **150**, 221–2, *223*, 225, 228,
  233–4, **235**, **237**
Campion, D.G. 136
Campos, L.A.O. 58
Canada 95, 99–100, 136, 147, *158*

Canaday, C.L. 62–3, 66, 137
canopy *see* forest canopy
Cappuccino, N. 263
Caquet, Th. 191
Carabidae 37, 41, 44, 46–8, 50–1
carbon dioxide 61, 67, 123, 133, 138, 156
  evolution of 21
cardboard traps 238–9
Cardé, R.T. 131, 134, 136
Carpenter, D.R. 176
Carpenter, S.J. 171
Carpenter, S.R. 179–82
Carrias, J.-F. 182
carrion traps 132
Caswell, H. 210
caterpillars 9, 13
  defoliating 11
caves 38
Cedrela 90
Centaurea 24
Central America 89
Cephalotermes *223*
Cerambycidae **78**, 86, 90, 94, 101
Ceratitis capitata 131
Ceratopogonidae 123, 169
Cercopidae 16
Cermak, M. 157
chafer, garden 21
Chalcidoidea **67**, **76**, 270
Chambers, R.J. 267
Chaoborus 187
Chapman, J.W. 120
Chapman, P.A. 45
Chara 196
Charles, E. 147
Charlton, R.E. 136
Cheiropachus quadram 104
chemical knockdown 68, 83, 150–5
  collecting hoops 154, *154*
  collecting trays 153–4
  comparative advantages of 154–5
  fogging 151–2, 269
  index of methods and approaches **166**
  mistblowing 152–5
    Hurricane Major® mistblower *152*–3;
    Stihl® mistblower 153
  pyrethrum **150**
Chen, Y. 258, **266**, 267–8
Cherrill, A.J. 137

Chesson, J.   191
China   **103**
Chironomidae   **67, 76**, 156, **166**
  larvae of   197–8, 206
chloral hydrate, in traps   **43**–4
Choristoneura fumiferana   131, 134, 136
Chrysomelidae   69, **76**
Chrysopidae   257
Chrysops   138
Church, B.M.   117
Churchill, T.B.   269
cicadas   16–17
Cicadidae   16
circle traps   96
Clarke, W.H.   42, 44, 47
Clements, R.O.   19, 25
climatic warming   122
Closs, G.   206
clover   18, 29
Clubionidae   37
Clutter, R.I.   193–4
Coaton, W.G.H.   224
Coccidae   **78**, 84
Coccinellidae   **67, 76**, 257
Coccoidea   90
Coccus pseudomagnoliarum   85
Cochliomyra   132
Cochran, W.G.   207
Cochran-Stafira, D.L.   182
Cohen, J.E.   206
Coleoptera   65–6, **75–6, 78**, 96, **166**
  bark-dwelling   90, 97
  borers   86
  canopy specialists   146
  epigeal   37
  host–plant relationship   101
  larvae of   22, *22*, 30, 88, 101
  night-flying   117
  saproxylic   97
  small   63
  species accumulation curves for   99
  subterranean   16–18, 20, *20*, 24
  trapping of   128, 157–8
  wood-destroying   110
collecting, informed   13–14
collecting hoops   154, *154*
collecting traps (pots)   95–6
Collembola   29, 58, **75–6**, *97*, **166**–7
  canopy specialists   146

densities of   155
digging-in effects   49
extraction of   26–7, *27*, 30, *31*
rhizophagous   17
trapping of   61, 65, 68
Collier, R.N.   130
Collins, N.M.   221–2, 225, 228–30,
    233
color traps   **76**
  spacing of   66–7
column samplers   190
Colunga-Garcia, M.   269
Colwell, R.K.   256
combined traps   129, 157, *158*
Compton, S.G.   157
Conacher, A.J.   221
Conotrachelus nenuphar   96
conservation   3
Constrictotermes   233
Cook, S.P.   265
Coon, B.F.   67, 128
Cooper, R.J.   156
Copeland, R.S.   171
Coppedge, J.R.   132–3
Coptotermes   **79–80**, *223*, 230, 237
  formosanus   232
Cordo, H.A.   16
core samplers   189
corn rootworm, southern   29
Cornelius, M.L.   67
Cossidae   **78**, 110
Cossus   **80**
  cossus   **79**
Costa, J.T.   149
Costa Rica   89, *90*, 172, 258
Costantini, C.   67
Costello, M.J.   257
Cotesia glomerata   261
cotton   61, 68, 266
Coupland, J.B.   125, 133, 138
Coxotermes   *223*
crabs, land   172
cranes   174
crawl traps   96
Crawley, M.J.   204
Crenetermes   *223*
crickets   **78**
Croft, B.A.   136
Croset, H.   210

Crossley, D.A.   26, 146, 148–9, 155
Crossley, J.D.A.   38, 42, 45
Cryptococcus   93
  fagisuga (= *fagi*)   77, 90, 93
Cryptotermes   230
Cubitermes   223
  sankurensis   226
Culicidae   169
Culicoides   118–19, 125, 134
curculio, plum   96
Curculio caryae   29, 96
Curculionidae   24, 86, 88, 96
Curtis, D.J.   43, 49
Cydia pomonella   134
Cynipidae   81
Cytocool®   191

Daane, K.M.   257
Dafni, A.   128
Daktulosphaira vitifoliae   17
damselflies, larvae of   172
Dangerfield, J.M.   226, 231
Dangerfield, P.C.   82
Danum Valley   234, **235**, *236*, **237**
Darling, D.C.   65
Darlington, J.P.E.C.   224, 226, 228
Darwin, C.   2
David, C.T.   136
Davies, J.B.   133
Davies, R.G.   221, 234
Davis, A.E.   65
Dawah, H.A.   258
Dawes-Gromadzki, T.   231
De Barro, P.J.   22, 24, **67**
de Souza, O.F.F.   231
Dean, D.A.   123
Deansfield, R.D.   128
death rate analysis   261
Declining Amphibians Populations Task
    Force (DAPTF)   215
DeFoliart, G.R.   174
Dejean, A.   225
Delia   130
  antiqua   24–5
  radicum   17
Delta traps   134–5
Dempster, J.P.   268
Den Boer, P.J.   38, 47–51
Dendroctonus
  frontalis   105

micans   **79**
valens   109, *109*
Dendy, J.S.   191
Dennehy, T.J.   17
Dennis, P.   40, 44, 51
Desender, K.   38, 44–5, 51
deserts   38
detrended correspondence analysis (DCA)
    97–8
Devy, M.S.   147
Diabrotica undecimpunctata   29
Dial, R.   147–8
Diaz-Aranda, L.M.   257
Dibog, L.   222, *223*, 228
Didham, R.K.   147
diesel fuel   18
Dietrick, E.J.   59
Dietrick vacuum sampler (D-vac)   59
Digweed, S.C.   44, 47–9
Dioryctria cristata   **79**
dippers   188, 195
Diptera   **75–6**, **78**, **166**
  biting   120
  color preference by   128
  densities   155
  large   125
  larvae of   18–19, 169
    rhizophagous   16
    sedentary   68, **76**
  male   123
  night-flying   117
  slow-flying   116
  small   118, 123, 125, 136
  subterranean   23–4
  trapping of   62–3, 65–6, *97*, 118, 128, 157
  in tree holes   174
  and visible light   117
Disney, R.H.L.   65–7, 128
distribution
  aggregated (clumped or contagious)   7,
    21, 45
  random   7
  regular (uniform)   7
Dixon, A.F.G.   4
Doane, C.C.   261
Dodsall, L.M.   24
Dolphin, K.   269
Dondale, C.D.   61
Donovan, S.E.   221
Doty, R.   42, 44

Downey, J.E.   117
dragonflies
    larvae of   172
    libellulid nymphs of   196
drainpipe traps   107
Dransfield, R.D.   226
Dreistadt, S.H.   85–6
drift fences   45–6
drift nets   192
Dromius   51
Dubois, D.   201–2
Duelli, P.   270
dung traps   132
Duplidentitermes   223
Durand, C.M.   258
Dutcher, J.D.   23
dyes, fluorescent   232
Dyna-Fog®   151

earthworms   29
    sampling of   19
Easey, J.F.   224, 228, 231
East, R.   23
Eburnitermes   223
eclosion   17
Ecological Entomology   138
ecological studies, pitfall trapping in   37–57
Edgar, W.D.   **266**, 267
Edwards, P.B.   118
Eggleton, P.   221–53, *223*
Egypt   136
Ekman dredge   190
Elateridae   20, *20*
electrophoretic analysis   267
Elkinton, J.S.   135, 264
Ellington, J.   61–2, 266
Elliott, N.C.   257
Ellwood, M.D.F.   230
elutriation   25
Elvin, M.K.   24
emergence traps   17, 157–9, 192
endoscopes   238
Ephemeroptera   194
epiphytes   159–60
Epstein, M.E.   44, 46, 48
Erbilgin, N.   106
Ericson, D.   38, 42–4, 46, 48
Ernobius mollis   110
Erwin, T.L.   146, 149
EstimateS software   256

ethanol   109, 154, 157
ethics of sampling   215
ethyl acetate   156
ethyl alcohol   238
ethylene glycol   **43–4**
eucalyptus   94, *94*
Eucalyptus marginata   **150**
Euceraphis punctipennis   13
Euchilotermes   223
Eucosma   89
    sonomana   89
Europe   1, 86, 91, 121, 146–7, 153
European Science Foundation Tropical
        Canopy Research Programme
        146–7
Eutermellus   223
Evans, T.A.   231–2
external feeders   83–6
Exterra®   239
extraction methods
    field *see* field extraction methods
    laboratory *see* laboratory extraction
        methods

Fabre, J.H.   2
Faragalla, A.A.   38, 42, 44
Fastigitermes   223
feces *see* frass
Feinsinger, P.   146
Feoktistov, B.F.   44
Fermanian, T.W.   18
Ferrar, P.   231
Ferson, S.   202
Fichter, E.   38, 42, 44
Fidgen, J.C.   80, *81*
field cricket, black   21
field extraction methods   18–23
    baits   20–1
    behavioral   19–20
    chemical   18–19
    hand-sorting   21–3
    index of methods and approaches   **36**
figs, fallen   160
Finch, S.   5, 67, 130
Fincke, O.M.   168–85
Finland   97–8, *98*, 106
Finnamore, A.T.   269
fir, grand   100
Fish, D.   168, 176, 180–1
Fitzgerald, C.J.   238–9

Fleeger, J.W.   189
Fleminger, A.   193–4
flies   4, 133, 169
    apple maggot   131
    biting   64, 67, 133
    blow   132
    fungus   **78**
    horse   138
    Mediterranean fruit   131
    small, distortion by adhesive   129
    sugar beet   128
    tsetse   128, 132–3
    turnip root   130
Floate, K.   256
Floren, A.   68, 147, **150**, 153
Florida   124
flotation techniques   61, 228–9
    compared to Tullgren extraction   *28*
    for egg counts   24
Fluon®   98
fluorescent bulbs, black   61
Foggo, A.   149
foliage bagging   68, **76**
Foraminitermes   *223*
forest   38
    canopy
        access to   147–9; canopy crane   148;
            canopy raft   149; canopy sledge
            (luge)   149; canopy walkways
            148, 174
        aerial and arboreal traps in   157–9
        insect densities in   **150**
        sampling insects from   146–67
        sampling issues   149–50
    cloud   159
    neotropical   170
    plantation   6, 70
    rainforests   67, 70, 100, 159, 228, 232
    temperate   70
    understory
        definition of   58
        insects associated with   58
Forest Research Institute Malaysia   *148*
formaldehyde   215
formalin   18, 127
    in traps   **43**–4
Formicidae   29, 37
Formosa   232
Forrester, G.J.   23
Forschler, B.T.   232

Foster, R.B.   176
Fowler, R.F.   22
Fox, B.J.   269
France   121, 239–41
Frank, J.H.   37, 175, 180–1
frass (feces)   17, 70–**71**, 77, 89–**90**, 101,
        110
French Guiana   101
frogs
    dendrobatid, tadpoles of   169
    territorial   175
Fuchs, A.   238
funnel traps   12, 107, 109, 138
Furculitermes   *223*
fuzzy numbers   202

Gaasch, C.M.   269
Gadagkar, R.   63
Gair, R.   12
Galindo, P.   169, 174–5
gallery patterns   *102*
    species-specific   100–1
galls   80–2
    formers of   **114–15**
    size/population density relationship of
        82
The Gambia   124
Gange, A.C.   16–36, 18–*19*, 21–2, *22*, *27*
Ganio, L.M.   147, 155
gasoline *see* petrol
Gaston, K.J.   146, 206, 258, 269
Gathorne-Hardy, F.J.   221, 224, 234
Gauld, I.D.   258, 269
Gaydecki, P.A.   117
Gee minnow traps   192
Geiger, C.A.   83
Geometridae   95
    larvae of   69
George, K.S.   12
Germany   148
Geurs, M.   25
Giblin-Davis, R.M.   38, 45
Gilbert, F.S.   257
Giller, P.S.   42, 46, 268
Gillespie, D.R.   131
Gillies, M.T.   132, 138
Gist, C.S.   38, 42, 45
Gladwin, J.   104
Glasgow, J.P.   133
Gleditsia triacanthos   69

Glen, D.M. 4
Global Canopy Programme 147
Glossina 132
  morsitans 128
  tachinoides 128
Glyptotermes 230
Goldson, S.L. 17, 24
Gonzalez, R. **103**
Good, J.A. 42, 46
Gora, V. 93
Gordon, R.D. 257
Gould, F. 29
Gould, J.R. **259**, 260–2
Goulet, H. 66
grab samplers 189
  Ekman dredge 190
  Petersen grab 190
Grace, B. 23
Gradwell, G. 96
Graham, H.M. 118
Gramineae 67
grasshoppers **78**
Gratwick, M. 19, 26
Gray, D.R. 84
Gray, H. 62, 137
Great Lakes IPM 96
Greaves, T. 230
Greenslade, P. 41–2, 44–5
Greenslade, P.J.M. 37, 40–2, 44–5, 47–9
Greenstone, M.H. 258, **266**, 267–8
Greg, P.C. 118
grids 46–**7**, 89
Grove, S. 97
grubs
  chafer 17–18, 20, *22*, 30
  white 20, 22
Gryllidae **78**
Guilbert, E. 147, 150
Gurevitch, J. 210
gut detection 257, 267–8
Gutierrez, A.P. 62, 83, 259
gutter traps 45

Haagsma, K. 239
Haarløv, N. 26
habitats
  seasonal changes in 4
  on trees **78**

hackberry, Chinese 85–*6*
Hadrys, H. 181
Hagler, J.R. 258, **266**, 267–8
Hall, D.R. 134
Hall, D.W. 42–4
Hallé, F. 149
Halsall, N.B. 50
Hambler, C. 70
Hammer, M. 26
Hammond, P.M. 38, 42, 44–5, 146, 150, 153
Hand, S.C. 61
Hanski, I. 38, 44–5, 48, 120
Hanula, J.L. 22, 96
Harcourt, D.G. 21, *23*
Harlow, L.L. 209
Harper, A.M. 128
Harper, M.G. 270
Harrington, R. 120–1
Harris, W.V. 230
Harrison, R.D. 29
Hartstack, A.W. 118
Hattis, D. 212
Haufe, W.O. 117
Haugen, L. 175
Haverty, M.I. 231, 239
Havukkala, I. 24
Hawkins, B.A. 256, 258, 269
Hawthorne, D.J. 17
Heath trap 118
Heathcote, G.D. 4, 131
heather 66–7
Heatwole, H. 147
Hébert, C. 95–6
Hedges, L.V. 210
Heimpel, G.E. 268
Heinz, K.M. 265
Heliconia, bracts of 175
Heliothis 118
  zea 129
Hellqvist, C. 87, *87*
Hem, D.G. 125
Hemiptera **78**, *97*, 167
  bark-dwelling 90
  distribution of 80, 91
  population density of 83–4
  rhizophagous larvae of 16
  trapping of 158
Henderson, G. 233
Henderson, I.F. 61

Henderson, P.A.   14, 17, 24–5, 62, 70, 117–19, 124, 139, 159
Hepialidae **78**
Hepialus californicus   17
heptane   25
Herms, D.A.   69–70
Herting, B.   255
Hertz, M.   38, 40, 45
Hess sampler   189
Hester, F.E.   191
Hester–Dendy sampler   191, 196
Heteropsylla cubana   83
Heteroptera   69, **76**, 174, 265
Heydemann, B.   50
Higgins, W.   147
Hilborn, R.   200
Hildrew, A.G.   267
Hill, C.J.   157
Hill, D.   150
Hill, M.G.   **259**, 260
Hinze, B.   233
Hoare, A.   225
Hodek, I.   254, 257
Hodotermes   224
Höft, R.   155–6
Hokama, Y.   188, 192, 195
Hollier, J.A.   149
Holloway, J.D.   158
Holopainen, J.K.   43–5, 47
Holt, J.A.   221, 224, 228, 231
Holzapfel, C.M.   171
Homoptera   **75–6, 166**
    predation of   265
    sessile   260
    trapping of   61, 66, 69
Hondelmann, P.   269
Honêk, A.   38, 40, 42–3, 46–7
Hoogendorn, M.   268
Hopkin, S.P.   17
Hospitalitermes   224, 230, 233
House, G.J.   25
Houston, W.W.K.   45
hoverflies   83
Howard, F.W.   89, **90**
Howell, R.S.   93
Hurlbert, S.H.   150, 174
Husseneder, C.   224
Hutcheson, J.   64
hydrocarbon adhesion   24–5, 30
Hylastes longicollis   109, *109*

Hylobius
    abietis   6, 11
    pales   96
Hylurgops   **103**
Hylurgops palliatus   *102*
Hymenoptera   **75–6, 166**
    aculeate   61
    epigeal   37
    forest-dwelling   58, 81
    gall-forming   82
    night-flying   117
    parasitic   88, 104, 155
    predation by   29
    small   62, 65
    trapping of   63, 66, **67**, *97*, 128, 157
Hypsipyla   89
    grandella   **79**, *90*, **90**
Hyvis®   95

Ichneumonidae   269
incidence counts   11
India   147
Inoue, T.   148
inquilines   225
insects
    activity   **78**
        evidence of   **79–80**
    arboreal habitats   **78**
    bark-boring   101
    canopy, sampling techniques and methods   146–67
    cryptic   155
    densities   *155*
    dimorphic   2
    entomophagous
        assemblages of   254–8
        diversity of   268–70
        index of methods and approaches   **277–8**
    external clues to presence on roots   17–18
    in flight   116–45
    in fruit, seeds, and silk   159–60
    large mobile   155
    migratory   5
    of phytotelmata   168–85
    polymorphic   2
    root-feeding (rhizophagous)   16
    scale *see* scale insects
    sessile   155

Institute of Animal Health, Pirbright   125
Intachat, J.   158–9
interception traps   4, 63, 232, 269
  composite   64, *158*
  index of methods and approaches   **145**
  undetected   137–8
  use in forest canopy   157–9, *158*
  visible   138
  window   65
International Canopy Network (ICAN)   146
inverse prediction   198
invertebrates, epigeal   47
  sampling of   37–8
Ips   **79, 103**, 105, *106*, 107, *108*
  cembrae   *102*, **103**
  grandicollis   96, **103**
  typographus   **103**, 105–6, 107, *108*
Ishii, T.   153
Isoptera   221–53
Italy   239

Jactel, H.   89, 91, *91–2*
Jaffe, K.   44–5
Jakuš, R.   101, *102*, **103**
Janzen, D.H.   269
Japan   147–8, 239
Jarosík, V.   38
Jasienski, M.   206
Jenkins, D.W.   171
Jenkins, T.M.   240–1
Jensen, R.L.   137
Johnson, C.N.   206
Johnson, M.D.   156
Johnson, P.C.   258
Johnson, R.A.   228
Jones, D.T.   221–53
Jones, J.A.   221
Jones, S.A.   **266**, 268
Jones, S.C.   222
Jones, V.P.   131
Joose, E.N.G.   42, 49
Jugositermes   223
Juliano, S.A.   175

Kaila, L.   97–*8*
Kalotermitidae   *223*, 224, 230, 233
Kambhampati, S.   222
Kammen, D.M.   194
Kapteijn, J.M.   49
Katsoyannos, B.L.   131

Kauffman, W.C.   87
Kean, J.M.   259
Keaster, A.J.   20
Keating, K.A.   256
Kegel, B.   48
Keil, S.   134
Keller, M.A.   255
Kelsey, R.   104
Kempson extractor   229
Kendall, D.M.   134
Kendrick, W.B.   26
Kerck, K.   151, 154
kerosene   153
  in traps   44
Kethley, J.   25
ketones   133
Kharboutli, M.S.   38, 46–7
Kieckhefer, R.W.   257
Kirchner, T.B.   207
Kiritani, K.   267
Kirk, W.D.   128
Kirk, W.D.J   66
Kirton, L.G.   239
Kitching, R.L.   17, 146–7, 149–53, 158–60,
      168–9, 175, 181, 270
Kitt, J.T.   255
Kleintjes, P.K.   83
Klironomos, J.N.   26
Knight, J.D.   4
Kolmogorov–Smirnov confidence interval
      211, 212
Koponen, S.   98–9
Kotler, B.P.   191
Kozicki, K.R.   18
Kramer, E.   44
Krebs, C.J.   204, 206–7, 210, 212
Kring, J.B.   131
Kuenen, L.P.S.   136
Kuhn, R.   202
Kulman, H.M.   44, 46, 48
Kuschel, G.   45

Labiotermes   223
laboratory extraction methods   23–8, 159
  behavioral methods   25–8
  dissection of roots   23–4
  flotation methods   24–5
    elutriation   25
    hydrocarbon adhesion   24–5
  index of methods and approaches   **36**

laboratory visualization methods   28–30
  index of methods and approaches   **36**
Labuschagne, L.   17, 21
Lacessititermes   233
ladybirds   83
Lamb, R.J.   62–3
Lamberti, G.A.   190
Långström, B.   87, *87*
Larsen, K.J.   269
Lasius   29
Lauenstein, G.   24
Lawrence, E.   58
Lawrence, K.O.   43, 45
Lawson, S.A.   **103**
Lawton, J.H.   180, 221
leaf miners   11, 13
leaf mines   70
Leather, S.R.   1–15
leatherjackets   18
Lee, K.E.   221–2, 226
Lemieux, J.P.   38, 43–4
Lenz, M.   231
Leong, J.M.   128–9
Leos Martinez, J.   123
Lepage, M.   221
Lepidoptera   63, **75–6**, **78**
  fast-flying   116
  large   118
  larvae of   255
    borers   86, 89, 110
    rhizophagous   16
  male attractants in   134
  mimic species   2
  night-flying   117
    and moonlight   117
  parasitoid assemblages of   258
  shoot boring   88
  trapping of   65, 69, 95, 118, 158–9, 166
  and ultraviolet light   117
  web-spinning   13
Leptomyxotermes   *223*
Lerin, J.   22
Leuthold, R.H.   233
Lewis, C.N.   269
Lewis, V.R.   239
Libellula, larvae of   208, **208**, *209*
life table analysis   258
light traps   4, 116–21, 129, 157–9, 187,
        232
  aquatic   195–6

"black"   117, 119
  comparative catchability   117
  disadvantage of   159
  examples of use   120
  index of methods and approaches   **144**
  influences on catch efficiency and
        effective catching distance   117–18
  mercury vapor   17, 117
  Onderstepoort Veterinary Institute design
        126
  priciples of use   116
  tungsten   17, 117–18
  types of light source   117
Lima, S.L.   197
Limnanthes douglasii rosea   128
Lindgren, B.S.   38, 43–4
Lindgren traps   109
linear transects *see* transects
Linsenmair, K.E.   68, 147, **150**, 153
Linton, Y.M.   125
Lister, A.   268
Lobry de Bruyn, L.A.   221
lobster-pot traps   96, 118
Lolium perenne   18
Long, G.E.   **266**, 267
Longino, J.T.   146, 159
Longipeditermes   233
  longipes   224
Loor, K.A.   174
Lopez, E.R.   **259**, 263–4
Loska, I.   62
Lounibos, L.P.   172, 174–5, 181
Lowman, M.   62
Lowman, M.D.   146–7, 154–5, 174
Lozano, C.   104
Lucilia   132
Luck, R.F.   258
Luff, M.L.   40–1, 44–8, 268
Lunderstädt, J.   93
Lussenhop, J.   26, 29
Lycosidae   37
Lyctidae   **78**
Lyctus   **80**
Lygus   63
  lineolaris   69
Lymantria dispar   70, **71**, 134

Machadotermes   *223*
Macías-Sámano, J.E.   100
Mack, T.P.   38, 46–7

MacKay, R.J.   189
Macleod, A.   60
MacMahon, J.A.   42
Macrotermes   224, 226, 233
  bellicosus   233
  malaccensis   224
Madagascar   221
Madden, K.E.   269
Madge, D.S.   229
Madoffe, S.S.   106
Maelfait, J.P.   38, 44–5, 49, 51
maggot, sugar beet root   22–3
magnesium sulfate   24–5
Maguire, B.   168
Magurran, A.E.   212
mahogany   89–**90**, *90*
Majer, J.D.   49, 68, 147, 150, 155–6, 269
Maki, K.   230
Malaise, R.   64, 138
Malaise traps   4, **75**, *158*, 232, 269
  aerial and arboreal use of   149, 157–9
  basic design for   63–5, 138
  Cornell design   *64*
  Coupland design   133
  Townes model   *64*
Malaysia   228, 234, **235**
Mangel, M.   200
Mankin, R.W.   233
Manly, B.F.J.   210
Margarodidae   91
Marini-Filho, O.J.   146
Maron, J.L.   17, 23
Marshall, D.A.   42, 44
Martikainen, P.   99, 106, *107*
Martin, J.L.   146
Martius, C.   229–30, 232
Masner, L.   66
Masters, G.J.   17
Mastotermes darwiniensis   232
Mathot, G.   226
Matsucoccus feytaudii   91, *91–2*
Matsumoto, T.   221, 225, 228–9, 233
Matthaei, C.D.   192
Matthews, J.R.   64–5
Matthews, R.W.   64–5
May, R.M.   149, 210–11
Mbalmayo Forest Reserve   225, 234, **235**,
  237
McArdle, B.H.   214
McCreadie, J.W.   133

McDougall, G.A.   134
McGeachie, W.J.   117, 128
McGeoch, M.A.   146
McSorley, R.   24, 26
Meads, M.J.   96–7
mealy bugs   16, 77
  wax-covered   24
Medeiros, L.G.S.   230, 232
Melanotus
  communis   20
  depressus   20
  similis   20
  verberans   20
Melbourne, B.A.   38, 45–6, 49
Menalled, F.D.   263
Merrick, M.J.   132
Merritt, R.W.   171, 181, 187, 189
mesocosms, sampling with   191–2
mesohabitats   159–60
Mesostoa kerri   82, *82*
Messing, R.H.   269
methyl bromide   226
methylbutenol   *108*
Meyer, V.W.   226
Meyerdirk, D.E.   131
microarthropods   26–7, 29
Microcerotermes   *223*
  crassus   224
microcrustaceans   181
microhabitats   159–60
microhymenoptera   66, **76**
Microtermes   *223*, 228
midges   118–19, 134, 166
  biting   169
  clouds   156
  slow-flying   123
Mill, A.E.   232
Miller, C.K.   134
Miller, L.J.   269
Miller, L.R.   232
Mills, N.   254–78
Mindarus abietinus   83
minnow traps   187, 195, 208
  Gee model   192
Mitchell, A.   147–8
Mitchell, B.   40, 42, 46
mites   26, 146
  oribatid   158, 166
Miura, T.   233
Moeed, A.   96–7

Moffet, M.   147, 174
molten agar technique   24–5
Mommertz, S.   38, 45, 50
Monserrat, V.J.   257
Moran, V.C.   147, 160
Morin, P.J.   191
morpho-taxa   149
Morrill, W.L.   41–2, 45, 50
Morris, M.   117
Morse, J.G.   **266**, 268
mosquito dipper   188
mosquitoes   67, 117–18, 123–4, 132, 138
   biting   125
   breeding sites   169
   egg rafts   193
   larvae of   170, 173, 181, 195, 198,
      200–1, *202*
   pupae of   200–1, *202*
   and trap shape   125
   in tree holes   175, 177, 180
   woodland   124
moss mats   159–60
   cores from   159–60
      index of methods and approaches
         167
Mosugelo, D.K.   231
moths   110, 131
   distortion by adhesive   129
   European pine shoot   **79**
   ghost   17
   goat   **79–80**
   gypsy   70, **71**, 134, 260–2, 265
   large   136
   larvae of   **78**, 86
      boring by   **79–80**
   migratory   118
   and pheromones   132
   pine beauty   8, 136
      sampling of   12
   pine shoot   **79–80**, 89
   release study of marked specimens   118
   spruce budworm   134
   and ultraviolet light   117
   winter   8–9, 95–6, *96*
   wood   **79–80**, 110
Muirhead-Thomson, R.C.   116, 121
Muller, C.B.   256
Müller, H.   24
Murphy, K.R.   209–10
Murphy, W.L.   130

Murray, P.J.   25
Mycetophilidae   **78**
Myles, T.G.   239
Mymaridae   269
Myors, B.   209–10
Myzus persicae   122

Nadkarni, N.M.   146–7, 159
Naeem, S.   175
Nag, A.   116
Najas   196
Nakamura, K.   261
Naranjo, S.E.   268
Nasco® Whirl-Paks   173
Nasutitermes   223
Nath, P.   116
Natural History Museum, UK   *154*
Nealis, V.G.   88
necrophagy   258, 267
Nelsen, R.B.   207
Neotermes   223, 230
Neuroptera   265
New Jersey trap   118
New, T.R.   269
New Zealand   97, 147
Newton, A.C.   89–*90*
Nicrophorus   132
Niemelä, J.   5, 38, 40–1, 43–8, 269
Nigeria   228
nitric acid   18
nitrogen   182
Noctuidae   118
Noditermes   223
Noirot, C.   224
Norris, R.H.   205, 209–10
North America   88–9, 153
   see also United States
Novak, R.J.   176
Noyes, J.S.   62–6, **67**, 269–70
Nürnberger, B.   210
Nutting, W.L.   222, 231–2
Nyffeler, M.   267
Nyrop, J.P.   29–30

oaks   13, 70, **71**, 81
Obeng-Ofori, D.   38, 40, 42, 50
Obrist, M.K.   270
Obrtel, R.   46, 48
Ocloo, J.K.   231
1-octen-3-ol   134

octenols   133
Odendaal, F.J.   105, *105*
Odonata   156, 166, 169, 173
    predatory   177–8
    territorial   175
Odontotermes   223
Ohiagu, C.E.   228
Ohmart, C.P.   155–6
oilseed rape   66–7
Okwakol, M.J.N.   221
Oliver, I.   149
Oliviera, M.L.   58
Olson, A.C.   264
onion fly   24–5
Operophtera
    bruceata   95
    brumata   95
Ophiotermes   223
Orr, A.G.   169
Orthoptera   **78**, 97
Orthotermes   223
Osenberg, C.W.   209
Otiorhynchus
    ligustici   21, *23*
    sulcatus   17, 21
Otitidae   23
oviposition scars   101
Owen, J.A.   45
Ozanne, C.M.P.   58–76, 146–67, **150**, 269

Paarmann, W.   150–1, 154, 160
Pachypappa   27, *28*
Pachypappella   27, *28*
Packer, L.   65
Pakistan   136
Palmer, I.P.   156
pan traps   5, 257
Panama   148, 168, *169*, 173–4, 177, 180
Panolis flammea   12, 136
Parajulee, M.N.   **67**, 257
parasitism
    apparent   264
    death-rate analysis of   261
    estimating impact of   **259**
    heterogeneous   264
    marginal attack rate   264
    percent parasitism   259–64
    recruitment analysis   261–2
    sampling concerns   263–4
    single-sample estimation of   260–1

species-specific, DNA detection of   264
    stage-specific generational   260–2
    time-specific rate of   262–3
    trap host method   261
parasitoids   66, 104, *106*, 254–78
    assemblages   255–6
    braconid   269
    regional variation   258
Parker, G.G.   147
Parmenter, R.R.   42
Parrella, M.P.   265
Pasoh Forest Reserve   234, **235**
passive traps   5
Pasteels, J.M.   224
Paton, R.   231
PCR analysis   257–8, 268
Pearce, M.J.   222, 231
Pearman, P.B.   191
Peck, R.W.   109, *109*
Peck, S.B.   65
Pectinophora gossypiella   118, 136
Peloquin, J.J.   176
Penev, L.D.   21
Pennsylvania trap   159
Penny, M.M.   49
Pericapritermes   223
Perry, D.R.   174
pest control   3
pest-monitoring programs   38, 44–5
Peters, B.C.   238–9
Petersen grab   190
Petersen, I.   138
Petersen, J.E.   191
Petersen, J.J.   263
petrol (gasoline)   18
Phelps, R.J.   133
Pherocon® trap   135
pheromone traps   12, 17, 107, *108–9*,
    134–6, *135*
    combined   121, 129
    uses of   136
pheromones   45, 67, 106, *108*, 120, 123,
    130
Phloeotribus scarabaeoides   104
Phoracantha   **79**
    semipunctata   94, *94*
Phoridae   **67**, **76**
phosphorus   182
photography   1
phototaxis   13, 63

Phoxotermes   223
Phyllopertha horticola   21
Phyllophaga   22
phylloxera, grape   17
Phymatodes   **80**
Physalaemus pustulosus   174
phytotelmata   168, 170, 174, 180–2
Picado, C.   181
Picea sitchensis   27
Pimm, S.L.   175
pine scale, maritime   91, *91*
α-pinenes   45
β-pinenes   45
pines   *100*, **103**
  plantations   22
  trees   86–7
    fallen shoots of   *87*
Pineus pini   77
Pinus sylvestris   **150**, 242
pipe and bucket system   239, 242
pipe traps   107, *108*
piperonyl butoxide   153
pipette, suction   172
Pirbright trap   125
Pires, C.S.S.   81
Pissodes   **80**
  strobi   88
pitcher plants   168
pitfall trapping   38
  in ecological studies   37–57
pitfall traps   6, 37–57, 98, 128, 269
  baits used in   44–5
  design and application of   39–45, **40, 56**
  funnels in   42, **56**
  killing agents, preservatives, and
      detergents   42–4, **43**, 48, **56**
  materials used in   40–1, **56**
  rims of   42, **56**
  roofs of   41–2, **56**
  shape and size   41, **56**
  spatial arrangement of   **47**
  specialized designs of   45, **56**
  subterranean   45
  types of   **39**
  use of baits in   44–5, **56**
  uses of   38
Pityogenes   **103**
Plantago lanceolata   63
Platypodidae   **78**
Platypus   **80**
Plecoptera   138

Plutella xylostella   120
Poisson distribution   204
Polis, G.A.   214
Polygraphus   **103**
polyisobutylene   130
Pomeroy, D.E.   226
Pontin, A.J.   27
pooter (aspirator)   69
population
  age structure of   3
  density of   3
  dispersion (distribution) of   3
  distribution of   7
  dynamics
    chaotic   123
    prediction of   3
  mortality   3
  natality   3
  trend   3
Porter, E.E.   256
possibility theory   201
Postsubulitermes   223
potassium permanganate   18
Potter, M.F.   238
Powell, W.   267
Prade, H.   201–2
Prairie, Y.T.   206
predation   264–8
  percent predation   265
  prey-specific DNA, detection of   268
  rate of   265–8
predators   66, 254–78
  assemblages   256–8
  density of   265–6
  rate of consumption   266–8, **266**
  regional variation   258
Price, P.W.   81
Proboscitermes   223
Procubitermes   223
Proffitt, J.R.   24
Prohamitermes mirabilis   224
propylene glycol   **43**–4
Protermes   223
protozoans   181
Prueitt, S.C.   89
Prunus padus   4
Pscoptera, bark-dwelling   13
Pseudacanthotermes   223
Pseudococcidae   16
Pseudomicrotermes   223
Psila rosea   130

Psocoptera   65, 68, **76**, **78**, 166–7
Psyllidae   83, 155
    leucaena   83
Pteromalidae   104
Pulvinaria   92–3
    regalis   77, 90–1, *93*
Purcell, M.F.   269
Pyralidae   **78**, 86, 89
pyrethrins   153
pyrethroids   153, 156
pyrethrum   153

quadrats   6, 10–12, *22*, 193, 264
    deep   21
    and suction sampling   61
    for sampling termite mounds   226,
        233–4
    vertical   92
quarantine lists   5
Quercus   **71**
    robur   **150**
Quicke, D.L.J.   269
Quinn, J.F.   256
Quiring, R.   131

radiography   29–30
radioisotopes   231
Raffa, K.F.   38, 45, 106
rainforests   67, 70, 100, 159, 228, 232
Rainio, J.   269
Ramaswamy, S.B.   131
ramp traps   45
Ratcliffe, S.T.   264
Rawlings, P.   119
rearing rooms   104
Rebello, A.M.C.   230, 232
Recher, H.F.   68, 155–6
Recruit®   242
Redak, R.A.   59, 61–2, 70
Reeves, R.M.   258
Reid, M.L.   104
relative abundance   186
Reling, D.   137
Resetarits, W.J.   191
Resh, V.H.   187–8, 190–1, 193–4
Reticulitermes   237
    lucifugus grassei Clement   240, 242
    santonensis   243
Reynolds, B.   148
Rhagoletis pomonella   131
Rhinotermitidae   *223*, 233, 237

Rhoades, M.H.   257
Rhopalosiphon padi   4
Rhyacionia   89
    buoliana   **79**
Ridgway, R.L.   70, **71**, 136
Rieske, L.K.   38, 41, 45
Ring, R.A.   147–8, 156, 159–60
Rinicks, H.B.   67, 128
Ro, T.H.   **266**, 267
Robb, T.   104
Roberts, H.R.   151
Roberts, I.   118
Roberts, R.H.   138
Roberts, R.J.   17, 23
Robinson, G.S.   120
Robinson pattern trap   118, 159
Rodgers, D.   146, 149, 160
Rogers, C.E.   23
Rogers, D.J.   133
Rohlf, F.J.   180, 194, 198, 204, 208, 210,
        212
Roisin, Y.   224
Romero, H.   44–5
root fly, cabbage   17
Rosenberg, D.M.   191, 193–4
Ross, D.W.   89
Rothamsted Experimental Station   121
Rothamsted Insect Survey   4, *119*, 120, *122*
Rothamsted trap   118, *119*, *122*, 124, 159
    data series   120
Rotheray, G.E.   257
rotifers   181
Rowe, W.D.   193
Royama, T.   260–1, 264
rubidium   258
Rudd, W.G.   137
Ruelle, J.E.   225
Russell, D.A.   **259**, 263
Rutherford, J.E.   189–90
Rutledge, P.A.   189
Ryan, R.B   264
rye grass   18, *27*

Saccharicoccus sacchari   24
Safranyik, L.   99, *100*
sage scrub   61
St Ives fluid   18
Sakal, R.R.   180
salamanders   196–7
Salt, D.T.   27–8
Salt, G.   229

Salveter, R.   269
sampling
  access systems   69, 92
  artificial substrate   199
  avoidance of bias in   10
  bagging and clipping   69, 88, 155–7
    index of methods and approaches   166
  of bark, external surface   90–9
    by capturing   95–9
    by counting   91–4
  of bark/sapwood interface   99–109
    by bark removal and log dissection
      101–4
    by emergence trapping   104–6
    by entrance/emergence hole sampling
      104
    by flight trapping   106–9
    by hand-searching   99–100
    by trap-logging   100–1
  by beating   69, 83
  binomial   84
    sequential   22
  by branch clipping   69
  calibration of   194
  canopy insects, techniques and methods
    146–67
  chemical knockdown, index of methods
    and approaches   166
  choice of techniques for   58–9
  climate effect on   64–5
  by column samplers   190
  concepts of   9–12
  by core samplers   189
  dampness (rain/dew) effect on   61, 63,
    65, 118
  design for aquatic insects   186–220
  destructive versus non-destructive   6–7
  devices for aquatic insects   186–220
    area or column samplers   188–90
    mesocosms   191–2
    natural/artificial substrates   190–1
    nets and dippers   187–8
    survey of   187–93
    traps   192
    visual observation and photography
      193
  direct habitat   5
  direct observation   256–7
  efficiency   186
    factors affecting   194–8

  of eggs   24, 92–3, 95–6
  of entomophagous species   258–68
  of epigeal invertebrates   37–8
  equipment for tree holes   171
  errors   193–215
    measurement interactions   193
    random   193, 203–15
    systematic   193–4, 198–203
  ethical considerations concerning   215
  experimental/controlled   2–3
  of external feeders   83–6
  forest canopy issues   149–50
  of galls   80–2
  by grab samplers   189, 199
    Ekman dredge   190
    Petersen grab   190
  by Hess sampler   189
  by Hester–Dendy sampler   191, 196
  index of methods and approaches   **219**
  indirect   257
  informed   13–14
  by Kempson extractor   229
  by kick samples   199
  location effect   64
  by mark–recapture   210
  and measures of unit area   12
  methods for forest understory vegetation
    58–76
    grasses and herbs   59–62
    shrubs   62–8
    tall vegetation, including small trees
      68–70
  passive   59, 62, 151
  patterns of   10
  programs   4
  of roots   16–36
  samples required in   9
  of shoots and twigs   80–6
    external   80–6
    internal   86–90
  by stovepipe (column) sampler   196
  strategy   22, 31, 45–6, **56**
    depletion effect   48–9, **56**
    digging-in effects   49
    duration and temporal pattern   47–8,
      56
    spatial arrangement   46–7, **56**
    and surrounding vegetation structure
      49, **57**
    trap number   46, **56**

stratified   8–9
suction   17, 59–62, **75**, 257
by Surber sampler   189
techniques of   3–4
  for natural tree holes   170–3
of termites   221–53
  approaches to   225–6
  difficulties of   222–6
  index of methods and approaches
    250–3
  methods of   226–33, *227*
  in mounds   226–8
  population density   233–6
  in soil   228–9
  subterranean pests of buildings
    236–40
  transect protocol   234–6
  in trees   230
  using baits   230–1
  using mark–recapture protocols
    231–2
  using traps   232–3
  in wood   229–30
  see also termites
theory and practice of   1–15
tools/techniques   4–5
of trees: shoots, stems, and trunks
    77–115
  detection   77–8
  index of methods and approaches
    114–15
  methods of   78
types of information required in   10
of understory vegetation
  low plants, grasses and herbs   59–62
  medium-height vegetation, including
    shrubs   62–8
  tall vegetation, including small trees
    68–70
units of   9–10
  criteria for   10–12
  estimation in   *214*
use of secondary characteristics in   70,
    76
by vacuum   59–62, 69, **75**, 265
  trees and shrubs   70
by water washing   83
of water-filled tree holes   168–85
  processes of   171–3
wind effect on   118

Sanders, C.J.   134
Sanderson, R.A.   137
Sands, W.A.   221, 224–5, 228, 231
sap-feeders   **114–15**
Saperda populnea   **79**
saprophagy   258, 267
Sarawak   148
savanna   226, 228
scale insects   77, **78**, 83–5, 90
  beech   77, 90, 93
  citriola   85–*6*
  density of cover   93–4
  horse chestnut   77, 90–1, *93*
  maritime pine, vertical distribution of
    91, *91*
  pine, distribution on bark types   91–*2*
  visibility of   91
scarab larvae   29
Scarabaeidae   17–18, 20, 22, *22*, 30
Schaefer, C.H.   188, 198
Schaefer, G.W.   125
Schedorhinotermes   *223*, 224
Scheffrahn, R.H.   239
Scheller, H.V.   41, 44, 46
Schlyter, F.   *108*
Schmitt, J.J.   257
Schoener, T.W.   269
Schowalter, T.D.   68, 146–7, 155–6
Schubert, H.   147
Schuurman, G.   226
Schwartz, S.S.   191, 193, 197
Scolytidae   **78**, 86, 96, 99, 101
Scolytus   **79**
scorpions   172
screw-worms   132–3
Scudder, G.G.E.   64
Scuhravy, V.   44
Seastedt, T.R.   21
Sedcole, J.R.   17
Sentri Tech®   239, 242
Sentricon Colony Elimination System®
    239, 242
Sentry®   242
Service, M.W.   124, 170, 188, 195
Seybold, S.J.   110
Shaw, M.R.   254–5
Sheasby, J.L.   224
Sheppard, A.W.   16
Shlyakhter, A.I.   194
shoot borer, mahogany   **79**

Shufran, K.A.   **266**, 267–8
Siemann, E.   210, 270
Siitonen, J.   99
Silk, P.J.   136
Silva, E.G.   229
Sime, K.R.   269
Simmons, C.L.   38
Simon, U.   58
Simpson, J.A.   **103**
simulation modeling   259
Simulium   125, 133
Singer, M.C.   256
single rope technique (SRT)   147
Sirex   **79–80**
Sitobion avenae   12
Sitona   24
   discoideus   17, 24
   lineatus   17
Sleaford, F.   221
Sleeper, E.L.   38
Slosser, J.E.   **67**, 257
Slovakia   **103**
Smart, L.E.   17
Smith, B.J.   40, 45
Smith, D.T.   133
Smith, R.J.   132
Snodgrass, G.   69
snout beetle, alfalfa   21, *23*
Snow, W.E.   169, 174
soapy water   18
sodium benzoate   66
soils
   sampling   12
   sieving   22, 25
   suspended   159–60
Sokal, R.R.   194, 198, 204, 208, 210, *211*,
      212
Solomon, M.G.   **266**, 267
Sota, T.   173, 175
South Africa   119
South America   89, 147, 221–2
Southward, M.   266
Southwood, T.R.E.   14, 17, 24–5, 62, 70,
      117–19, 124, 139, 146, **150**, 152–3,
      159–60, 199, 203
Spain   104, 125, 239–40
spanworm, Bruce   95
Speight, M.R.   77–115, 136, 146, 153–4
Spence, J.R.   38, 41, 45
Spencer, M.   186–220

Sphaerotermes   *223*
spiders   269
   assemblages   257
   ballooning   123
   environmental sensitivity of   268
   lycosid   267
   predation by   267–8
   wandering   37, 49
   webs of   70
spittlebugs   16
Spodoptera exigua   11
Spragg, W.T.   231
Springate, N.D.   64, 149, 157
spruce   101, *102*, **103**, *107*
   Norway   105, 110
   Sitka   8–9, 27
   white   *81*
Spurr, S.H.   58
Srivastava, D.S.   180
St-Antoine, L.   95–6
Stadler, B.   256
Standridge, N.   147
Staphylinidae   37, 60, **75**
Stary, P.   255
Stein, W.   46
Sterling, P.H.   70
Steven, D.   41
Stewart, A.J.A.   59–61
Stewart, R.J.   188, 198
Stewart, R.M.   18
sticky traps   4, 17, 129–31, 269
   aerial and arboreal use of   157
   baited, effectiveness of   107, *108*
   basic design and use   124–5, 130
   comparative efficiency   131
   Delta pattern   134
   importance of shape and color   131
   index of methods and approaches   **145**
   modifications   130–1
   use on trees   85, 95
Stireman, J.O.   256
Stirling, W.L.   123
Stone, C.   **103**
Stork, N.   149, **150**
Stork, N.E.   83, 147, 150–1, 153
Story, T.P.   128
stovepipe (column) sampler   196
Strickland, A.H.   229
Strickler, J.D.   64–5
Strong, D.R.   17, 23

Su, N.Y.   231–2, 239–40
suction samplers, types of   59–60, 95, 257
suction sampling   17, 59–62, **75**, 95, 270
suction traps   4–5, 117, 120–5, 131
   calibration of catches   123–4
   designs and modifications   121
   factors influencing catches by   124–5
   index of methods and approaches   **144**
   modified designs   123
   Onderstepoort Veterinary Institute design
      126
   standard Rothamsted model   121–3, *122*
   use in indirect sampling   257
   use of baits   125
   wind speed effect on   123–4
Sugihara, G.   212
Sugimoto, A.   221
Sugio, K.   224
Sulawesi   269
Sunderland, K.D.   38, 258, **266**, 267
Surber sampler   189
suspended soils   159–60
   index of methods and approaches   167
Sutherland, W.J.   14, 61
Svensson, B.W.   210
swarms   65
Sweden   87, 105
sweep nets   10, 62–3, **75**, 187, 269–70
   comparative efficiency   69, 196
   compared to vacuum sampling   61
   in indirect sampling   257
   undetected   137
sweeping   149, 154, 265
Swietenia   90
   mahagoni   **90**
Swingfog®   151
Switzerland   148
Swormlure   133
Synacanthotermes   223
Syrphidae   67, **76**, 257
   in sticky traps   130
   in tree holes   172, 174

Tabanidae   65, **75**
Tabanus   138
Tachyporus   60, **75**
tadpoles   191
   in tree holes   173–4
Takematsu, Y.   224
Tallamy, D.W.   138

Tamaki, G.   **266**, 267
Tanglefoot®   95
Tauber, C.A.   257
Tavakilian, G.   100
Tayasu, I.   221
Taylor, H.S.   231
Taylor, J.R.   194, 206
Taylor, L.R.   117, 120–1, 124, 130
Taylor, R.A.J.   137
Teleogryllus commodus   21
Terell-Nield, C.   51
Termes   223
Termigard®   239
Termite Eradication Programme, UK
   241
termites   **79**
   acoustic emissions from   233
   alates   232
   index of methods and approaches
      250–3
   methods for sampling   221–53
   monitoring and baiting program, Devon,
      UK   240–3
   nest units of   224
   population density of   233–6
   processional columns of   233
   recording movements of   233
   sampling by transect protocol   234–5,
      235–6
   as subterranean pests of buildings
      236–40
   swarming   232
   wood boring   **80**
Termitidae   *223*, 234
   Apicotermitinae   *223*
   Macrotermitinae   *223*, 224, 231
   Nasutitermitinae   *223*, 224
   Termitinae   *223*, 224
Termopsidae   224
Tetanops myopaeformis   23, 128
Tetropium   **79**
Thiele, H.-U.   37–8, 50
thigmotaxis   13
Thomas, C.F.G.   42
Thomas, J.D.B.   38
Thompson, D.V.   129
Thoms, E.M.   233
Thoracotermes   223
Thorne, B.L.   232
Thornhill, E.W.   59

Thornhill vacuum sampler   59
Thorp, R.W.   128–9
Thorpe, K.W.   70, **71**
thrips   77
Thysanoptera   257
Tilmon, K.J.   264
Tipulidae   18–19, 24
　larval behavior   19–20
　larval sampling   18–*19*, 26
Tobin, S.C.   147–8
Tomicus   **79**, 87
　piniperda   **76**, **79**, 87, *87*, 88
Topping, C.J.   38, 50
Tortricidae   86, 89
tow nets   124–5, 137–8, 187
Townes, H.   64, 269
Townsend, C.R.   267
Townsend, M.L.   232
transects   188, 264
　linear   46–**7**, 226
　plough   23
　sweeping along   62
　termite protocol   234–5
trap night index (TNI)   159
traps   4–5
　active   5, 66, 150
　　index of methods and approaches   **219**
　aerial and arboreal   157–9
　　index of methods and approaches
　　　166–7
　baited   67, 121, 132–6
　　animals as bait in   133–4
　　designs and modifications   132
　　"graveyard" trials   231
　　index of methods and approaches
　　　145
　　wind effect on   135–6
　bamboo   175
　barrier   45
　basin   12
　beating trays   69, **76**, 83, 257
　Berlese–Tullgren funnel   26, 229
　　Blasdale version   26
　bicolored   65
　cardboard (corrugated)   238–9
　carrion   132
　chemical knockdown   68, 83, 150–5
　　collecting hoops   154, *154*
　　collecting trays   153–4
　　comparative advantages of   154–5
　　fogging   151–2, 269

index of methods and approaches   **166**
　mistblowing   152–5; Hurricane
　　Major® mistblower   *152*–3; Stihl®
　　mistblower   153; pyrethrum   **150**
circle   96
collecting hoops   154, *154*
collecting pots   95–6
color   **76**
　spacing of   66–7
combined   129, 157, *158*
crawl   96
Delta   134–5
dippers   188, 195
　mosquito   188
drainpipe   107
drift fences   45–6
drift nets   192
dung   132
effect of moonlight on   159
emergence   17, 157–9, 192
flight-interception   63
foliage bagging   68, **76**
funnel   12, 96, 99, 105–7, 109, 138, 229
Gee *see* minnow
gutter   45
Heath pattern   118
installation, disturbance caused by   49
interception   4, 63, 232, 269
　composite   64, *158*
　index of methods and approaches
　　145
　undetected   137–8
　use in forest canopy   157–9, *158*
　visible   138
　window   65
light   4, 116–21, 129, 157–9, 187, 232
　aquatic   195–6
　"black"   117, 119
　comparative catchability   117
　design of   118–19
　disadvantage of   159
　examples of use   120
　index of methods and approaches   **144**
　influences on catch efficiency and
　　effective catching distance   117–18
　mercury vapor   17, 117
　Onderstepoort Veterinary Institute
　　design   *126*
　principles of use   116
　tungsten   17, 117–18
　types of light source   117

Lindgren   109
lobster-pot   96, 118
Malaise   4, **75**, *158*, 232, 269
  aerial and arboreal use of   149, 157–9
  basic design for   63–5, 138
  Cornell design   *64*
  Coupland design   133
  Townes model   *64*
minnow   187, 195, 208
  Gee model   192
New Jersey pattern   118
pan   5, 257
passive   5
  index of methods and approaches   **220**
patterns for   46–**7**
Pennsylvania pattern   159
Pherocon®   135
pheromone   12, 17, 107, *108–9*, 134–6, *135*
  combined   121, 129
  uses of   136
pipe   107, *108*
Pirbright pattern   125
pitfall   6, 37–57, 98, 128, 269
  baits used in   44–5
  design and application   39–45, **40**, **56**
  funnels in   42, **56**
  killing agents, preservatives, and
    detergents   42–4, **43**, 48, **56**
  materials used in   40–1, **56**
  rims of   42, **56**
  roofs of   41–2, **56**
  shape and size   41, **56**
  spatial arrangement   **47**
  specialized designs of   45, **56**
  subterranean   45
  types of   **39**
  use of baits in   44–5, **56**
  uses of   38
ramp   45
Robinson pattern   118, 159
Rothamsted pattern   118, *119*, *122*, 124, 159
  data series   120
sticky   4, 17, 129–31, 269
  aerial and arboreal use of   157
  baited, effectiveness of   107, *108*
  basic design and use   124–5, 130
  comparative efficiency   131
  Delta pattern   134
  importance of shape and color   131

index of methods and approaches   **145**
mini-sticky   85
  modifications   130–1
  use on trees   85, 95
stovepipe (column)   196
suction   4–5, 117, 120–5, 131
  calibration of catches   123–4
  designs and modifications   121
  factors influencing catches by   124–5
  index of methods and approaches   **144**
  modified designs   123
  Onderstepoort Veterinary Institute
    design   *126*
  standard Rothamsted pattern   121–3, *122*
  use in indirect sampling   257
  use of baits   125
  wind speed effect on   123–4
sweep nets   10, 62–3, **75**, 187, 269–70
  comparative efficiency   69, 196
  compared to vacuum sampling   61
  in indirect sampling   257
  undetected   137
time-sorting   45
tow nets   124–5, 137–8, 187
trawl   120
trunk   96–7, 107
  window   96–7, *98*
water (pan)   4–5, 66, **67**, 125–9, *127*, 131
  basic design and use   127–8
  importance of pan color   128
  index of methods and approaches   **144**
  modifications   128
  types of use   128–9; in indirect
    sampling   257
window   4, 65–6, **76**, 106, *107–8*, 137, 157
yellow water   4, 257, 269
trawl traps   120
tree holes   *169–70*
  abiotic variables in   **179**
  analogues of   175–6
  artificial   174–80, *176–8*
    important considerations for   179–80
    sampling techniques for   178–9
  classification of   168–9
  comparative data, problems of
    interpretation   181
  index of methods and approaches   185
  natural   168–74
    experimental considerations   173–4

tree holes, natural (*cont'd*)
    sampling techniques for   170–3
    species diversity with height   174
    water-filled
        sampling of   168–85
        statistical methods for   180
trees
    index of methods and approaches
        114–15
    major stem habitats on   **78**
    sampling from shoots, stems, and trunks
        77–115
Treloar, A.   62, 137
Tretzel, E.   50
Trexler, J.C.   187, 191–2, 196
Trexler, J.D.   192
Trialeurodes vaporarium   131
Trichoplusia ni   118
Trichoptera   138
    night-flying   117
    and ultraviolet light   117
Trichosirocalus troglodytes   63
Trifolium repens   18
Trinervitermes   228
Trinidad   133
tropics   146
trunk traps   96–7
    window traps   96–7, *98*
Trypodendron   **79–80**
Tuberculitermes   223
Tuberculoides annulatus   131
Tuck, K.R.   120
Tullgren extraction   27
    compared to flotation   28
Tullgren funnel   26–8, 61, 159
Tunset, K.   137
Tupy, J.L.   110
Turchin, P.   105, *105*
Turner, A.M.   187, 191–2, 196
Twombly, S.   210

Uetz, G.W.   37–8, 42–3, 46–7
ultrasound   110
Unguitermes   223
United States   83, 96, 109, 118, 132, 136,
        146–7, 232, 239
    Arizona   81
    Missouri   20
    north-western   148
    Oregon   89, 109, *109*

Unzicker, J.D.   37–8, 42–3, 46–7
Usher, M.B.   66–7, 231

vacuum sampler   *60*
vacuum sampling   59–62, 69, **75**, 265
    of trees and shrubs   70
Vale, G.A.   133
van den Berg, H.   **266**, 267
van den Berghe, E.   44
Van Driesche, R.G.   255, 259, **259**, 260–3
van Huizen, T.H.P.   137
van Lenteren, J.C.   259
van Roermund, J.W.   259
Vance, G.M.   192
Vandenberg, N.   257
Varga, L.   168
variance–mean ratios   7–8
Varis, A.L.   43–5
Varley, G.   96
Veliidae   174
Vent-Axia® fan   123
Verkerk, R.H.J.   221–53
Verrucositermes   223
Vessby, K.   132
video technology for *in situ* observation   29
Villani, M.G.   29–30
Villemant, C.   269
visualization methods *see* laboratory
        visualization methods
Vitale, G.   169
Vlijm, L.   42
Vogt, W.G.   132
von Ende, C.N.   182

Waage, B.E.   43
Waage, J.K.   264
Wagner, T.   147
Wainhouse, D.   86, 93, 136, 146
Walker, E.D.   64–5, 171, 179–82
Walsh, G.B.   44
Walter, D.E.   24–6, 146, 151
Wang, B.   263
Ward, R.H.   20
Ward, S.A.   11
Waring, P.   117
Washburn, J.O.   179
Washino, R.K.   188, 192, 195
wasps   104
    chalcid   258
    distortion by adhesive   129

fig   157
gall   **79**, 81–2
wood   **79–80**, **114–15**
Watanabe, H.   147
water chemistry   171–2
water (pan) traps   4–5, 66, **67**, 125–9, *127*, 131
  basic design and use   127–8
  importance of pan color   128
  index of methods and approaches   **144**
  modifications   128
  types of use   128–9
    in indirect sampling   257
Watt, A.D.   1–15, 221
Way, M.J.   94, *94*
Wearing, C.H.   83
Webb, R.E.   95, 131
weevils   63, **80**, 86, 88, 96
  black vine   17
  large pine   6, 11
  pea and bean   17
  pecan   96
  smaller pecan   29
  white pine   88
Weissling, T.J.   233
Welsh, A.H.   207
Weseloh, R.M.   265
Weslien, J.   105–*6*
West Africa   221–2, *223*
wet-sieving   18, 28, 30
Wheater, C.P.   268
White, E.G.   17
White, T.C.R.   69
whitefly   265
  greenhouse   131
Whitfield, G.H.   23
Whitfield, J.B.   269
Whitmore, R.C.   156
Whittaker, T.M.   61
Wigglesworth, V.B.   63
Wilcox, C.   191
Willett, T.R.   268
Williams, C.B.   118
Williams, J.B.   269
Williams, L.   264
Williams, P.   21
Willoughby, B.E.   23
willow   13
Wilson, E.O.   221
Wilson, L.F.   22

Wilson, L.T.   62
Winchester, N.N.   64, 146–8, 156, 158–60
wind-speed effect   124–5
window traps   4, 65–6, **76**, 106, *107–8*, 137, 157
windthrow   65
Winkler extraction   159
wireworms   20, *20*
  larval bait for   21
Wise, I.L.   62–3
Wittman, P.K.   146
Woiwood, I.P.   120–1, 159
Wood, S.N.   210
Wood, T.G.   221–2, 225–6, 228–9
Woodcock, B.A.   37–57
"woodworm"   **78**, **80**, 110
Wratten, S.D.   50
Wright, A.F.   59–61
Wright, R.J.   29
Wylie, F.R.   77, 89, 94
Wyss, E.   257
Wytham Wood study   96

X-ray analysis   29, 110
X-ray computed tomography   29
Xylechinus   101
  pilosus   *102*
Xyleutes   **80**
  ceramica   110
Xyloterus   **103**
  lineatus   *102*

Yamamura, K.   258
Yamashita, Z.   153
Yanoviak, S.P.   168–85
Yasuda, K.   45
Yeargan, K.V.   24
Yee, W.C.   133
yellow water traps   4, 257, 269
yield–effort curves   214
Young, M.   116–45
Young, M.R.   117, 128, 137

Zaidi, R.H.   268
Zar, J.H.   194, 210
Zhang, Q.H.   102, *102*, **103**
Zhou, X.   123
Zhu, Y.C.   264
Zolubas, P.   *108*